International Conference on Intelligent Control, Measurement and Signal Processing (ICMSP 2019)

IOP Conference Series: Materials Science and Engineering
Volume 799

Xi'an, China
27 - 29 December 2019

ISBN: 978-1-7138-3507-3
ISSN: 1757-8981

Printed from e-media with permission by:

Curran Associates, Inc.
57 Morehouse Lane
Red Hook, NY 12571

Some format issues inherent in the e-media version may also appear in this print version.

Institute of Physics
Dirac House, Temple Back
Bristol BS1 6BE UK

Phone: 44 1 17 929 7481
Fax: 44 1 17 920 0979

techtracking@iop.org

Additional copies of this publication are available from:

Curran Associates, Inc.
57 Morehouse Lane
Red Hook, NY 12571 USA
Phone: 845-758-0400
Fax: 845-758-2633
Email: curran@proceedings.com
Web: www.proceedings.com

TABLE OF CONTENTS

1. ADVANCED MEASUREMENT TECHNOLOGY AND APPLICATION

IOP Conf. Series: Materials Science and Engineering **799** (2020) 011001 doi:10.1088/1757-899X/799/1/011001

Preface

The 2019 International Conference on Intelligent Control, Measurement and Signal Processing (ICMSP 2019), jointly organized by Xi'an Shiyou University and China Petroleum Logging Co., Ltd, was grandly held during December 27-29, 2019, in Xi'an, China. ICMSP 2019, held in two days, was a high level and high standard academic conference. Under the full supports of the hosts, every session of the conference was successfully completed by the joint efforts of all representatives attending the conference and all staff.

We are honored to invite Dr. Jie Li from University of Tsukuba, Japan, Prof. ZhenMao Chen from Xi'an Jiaotong University, Prof. Zahriladha Zakaria from Universiti Teknikal, Malaysia, Prof. Alex Noel Josephraj from College of Engineering, Shantou University, and Dr. Tuan Guo from Jinan University as the keynote speakers to professionalize this conference. The spotlights of the conference lie on the great progress in the fields of Measurement and Control Technology and Intelligent Instruments, Photoelectric Detection and Measurement Technology, Intelligent Control and Automation, Signal and Information Processing, Artificial Intelligence and Robots and other interdisciplinary areas. This conference aims to encourage more young academic researchers to share their new ideas, thoughts and research results, enhance academic research and exploration, and promote the transmission of academic results.

At last, ICMSP 2019 would like to thank all guests and speakers for sharing their wonderful talks and all participants coming from distant places without complaints for attending this conference in Xi'an. We also want to thank all sponsors for offering the materials and funds to this conference. Thanks to your great supports, we're sure International Conference on Intelligent Control, Measurement and Signal Processing can be held better in the future!

Prof. Jie Wu
College of Electronic Engineering, Xi'an Shiyou University
The Chairman of the Organizing Committee

IOP Conf. Series: Materials Science and Engineering **799** (2020) 011001 doi:10.1088/1757-899X/799/1/011001

Committee member

President of the General Assembly
Prof. Tiantai Li, President of Xi'an Shiyou University

Chairman of Academic Committee
Prof. Jie Li, Fellow of The Engineering Academy of Japan

Vice chairmans of Academic Committee
Prof. Zhen Mao Chen, Xi'an Jiaotong University (Non-destructive Testing)Prof. Hui Xin Zhou, Xi'an University of Electronic Science and Technology (Photoelectric Detection)
Prof. Maode Yan, School of electronics and control engineering, Chang'an University
Tian Zhi Tang Chief Engineer of CNOOC Logging Co., Ltd.

Members of Academic Committee
Prof. Ruirong Dang, School of Electronic Engineering, Xi'an Shiyou University
Researcher Tuan Guo, Institute of Photonics Technology/Jinan University
Prof. Haiwei Fu, School of Science, Xi'an Shiyou University
Prof. Jing Zhou, School of Electronic Engineering/Xi'an Shiyou University
Prof. Fei Li, School of Electronic Engineering/Xi'an Shiyou University
Prof. Fanhua Shang, Xidian University (Artificial Intelligence)
Prof. Hong Li, College of Electronic Information Engineering /Xi'an Technological University
Prof. Hanshan Li, College of Electronic Information Engineering /Xi'an Technological University
Prof. Binjian Wang, College of Physics and Optoelectronic Engineering /Xidian University

Chairman of the Organizing Committee
Prof. Jie Wu, College of Electronic Engineering, Xi'an Shiyou University

Vice chairman of the Organizing Committee
A. Prof. Guowang Gao, College of Electronic Engineering, Xi'an Shiyou University
A. Prof. Long Jiao, Department of Science and Technology/Xi'an Shiyou University
Prof. Yanhong She, School of Science /Xi'an Shiyou University

Members of the Organizing Committee
Yingna Guo, College of Electronic Engineering, Xi'an Shiyou University
Keman Liu, College of Electronic Engineering, Xi'an Shiyou University
Jing Li, College of Electronic Engineering, Xi'an Shiyou University
Zhiping Ren, College of Electronic Engineering, Xi'an Shiyou University
Danwu Wu, College of Electronic Engineering, Xi'an Shiyou University
Yinchuan Wu, College of Electronic Engineering, Xi'an Shiyou University
Jingtian Xu, College of Electronic Engineering, Xi'an Shiyou University
Bing Zhu, College of Electronic Engineering, Xi'an Shiyou University
Qi Zhu, College of Electronic Engineering, Xi'an Shiyou University

Peer review statement

All papers published in this volume of *IOP Conference Series: Materials Science and Engineering* have been peer reviewed through processes administered by the proceedings Editors. Reviews were conducted by expert referees to the professional and scientific standards expected of a proceedings journal published by IOP Publishing.

IOP Conf. Series: Materials Science and Engineering **799** (2020) 012001 doi:10.1088/1757-899X/799/1/012001

Research on the Signal Characteristics of Internal Gas Leakage in Safety Valve

Ma Haodong[1], He Rui[2], Chen Dianjing[3] and Ma Haodong [1*]

[1]Beijing Aerospace Propulsion Institute, Beijing, 100076, China

[2]Beijing Aerospace Propulsion Institute, Beijing, 100076, China

[3]Beijing Aerospace Propulsion Institute, Beijing, 100076, China

[*]Ma Haodong's e-mail: mahaodong1994@163.com

Abstract. Based on the acoustic emission theory, this paper explores the signal characteristics of internal gas leakage in safety valve. In this paper, a simulation test-bed and an acoustic emission signal detection system for internal leakage in safety valve are designed and built. Through theoretical analysis and experimental research, the influence of factors such as inlet pressure of safety valve, size of leakage hole of sealing surface are explored, and the quantitative relationship between the gas volume leakage rate and the average signal level ASL of the characteristic parameters of acoustic emission signal is established.

1. Introduction

The safety valve is an automatic valve, which uses the pressure of the medium itself to discharge the specified amount of fluid to prevent the pressure of pressure bearing devices and equipment such as boilers, pressure vessels or pressure pipes from exceeding the predetermined safety value then causing overpressure and damage, so as to ensure the normal operation of the equipment and the personnel safety [1]. However, when the safety valve is affected by medium corrosion, erosion aging, improper operation and other factors, the sealing surface is easy to be damaged, resulting in internal leakage, causing medium loss and waste, even serious safety accidents. Traditional detection methods need experienced technicians to listen to the sound to judge or remove the safety valve from the pipeline for off-line detection, which is time-consuming and laborious, and cannot grasp the internal leakage of the safety valve in time.

Acoustic emission detection technology is a dynamic non-destructive detection method, with the advantages of convenient detection, no need to stop production and low cost [2-5]. It is of great significance to use this technology to monitor the internal leakage of the safety valve so as to repair and replace the damaged safety valve in time, prevent medium waste and reduce the probability of safety accidents.

2. Basic theory of acoustic signal detection for internal leakage in safety valve

2.1. Acoustic emission theory

Acoustic emission refers to a physical phenomenon [6] in which an object or material is subjected to deformation or external force, producing a transient stress wave due to the rapid release of elastic energy, as shown in Figure 1. When acoustic emission occurs in materials, each acoustic emission signal emitted by the acoustic source contains the information of the internal structure or defect

IOP Conf. Series: Materials Science and Engineering **799** (2020) 012001 doi:10.1088/1757-899X/799/1/012001

property and state change of materials. Therefore, sensitive instruments can be used to receive and process the acoustic emission signal. Through the analysis and research of the characteristic parameters of the acoustic emission source, the position, state change degree and changing trend of the internal defects of materials or structures can be inferred. The acoustic emission signal of internal leakage in safety valve carries the information of leakage point. The acoustic emission signal is picked up by the acoustic emission sensor, and the leakage degree of the safety valve can be judged by analyzing and processing the signal.

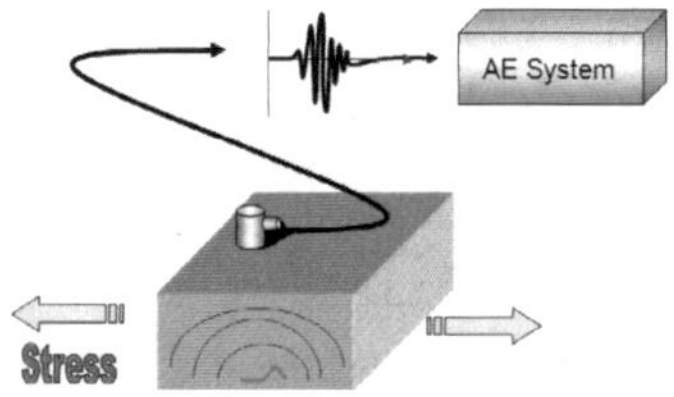

Figure 1. Acoustic emission

2.2. Time-frequency characteristics of acoustic emission signal of internal leakage in safety valve

The acoustic emission signal of internal leakage in safety valve belongs to the continuous acoustic emission signal. In order to extract the information representing the characteristics of internal leakage in safety valve from the acoustic emission signal in time domain, it is necessary to use the average value of signal characteristics rather than the instantaneous value. The average signal level ASL represents the average value of signal level in the sampling time, which can be used as the criteria for determining the internal leakage in safety valve [7]. The average signal level ASL is expressed as follows:

$$ASL = 20\log\left(\frac{RMS}{10^{-6}}\right) - pre \tag{1}$$

Where: ASL is the average signal level of acoustic emission signal (dB), RMS is the effective value voltage (mV), and pre is the amplification factor of preamplifier.

The effective value voltage RMS is the mean square root value of the signal in the sampling time. For the discrete acoustic emission signal, there are N samples, which are x [0], x [1], x [2], ..., x [n-1]. The mean square root value equation can be expressed as:

$$RMS = \left(\frac{1}{N}\sum_{n=0}^{N-1} x\left[n\right]^2\right)^{0.5} \tag{2}$$

The leakage rate of safety valve can be estimated by the acoustic emission signal of internal leakage, and the acoustic power is needed as the intermediate variable. For the safety valve, the acoustic power of the internal leakage signal is approximately equal to the average energy of the leakage signal, and the relationship is simply defined as follows [8]:

$$P \propto RMS^2 \tag{3}$$

Where: P is the acoustic power, W.

The logarithm of acoustic power is linear with the logarithm of leakage rate [9], namely:

$$\log RMS = b\log Q + c \tag{4}$$

Where: Q is the leakage rate, ml/min; b and c are coefficients respectively, and their values are related to safety valve type, leakage type, leakage hole size, inlet pressure, type of discharge medium, size of safety valve body and other factors.

According to equations (1) and (4):

$$ASL = b\log Q + c \tag{5}$$

The characteristic frequency, peak frequency and other characteristic values of acoustic emission signal of internal leakage in frequency domain can also provide the characteristics of internal leakage

IOP Conf. Series: Materials Science and Engineering **799** (2020) 012001 doi:10.1088/1757-899X/799/1/012001

in safety valve. These characteristic values can be obtained by conducting Fourier transform (FFT) on the acoustic emission signal of internal leakage in time domain.

3. Test device and detection system

The detection system for internal leakage in safety valve [10] consists of internal leakage simulation test-bed of safety valve and signal acoustic emission detection system for internal leakage in safety valve.

The internal leakage simulation test-bed is composed of nitrogen gas source, pressure regulating valve, pressure gauge, flowmeter and internal leakage simulation prototype of safety valve, as shown in Figure 2.

Figure 2. The internal leakage simulation test-bed

Spring loaded safety valve of HTO series of 3K4 specification developed by Beijing Aerospace Power Research Institute is selected as the internal leakage simulation prototype of safety valve, with nominal inlet pressure of 600 pounds. In order to simulate the different leakage state of safety valve, the safety valve clack is processed. There are four valve clacks used in the test: three of them have been slotted, with the size of 0.15mm * 0.5mm, 0.3mm * 0.5mm and double holes of 0.15mm * 0.5mm * 2, respectively, to simulate contact surface damage and particle accumulation; the other one has been manually polished on the sealing surface by sandpaper to simulate sealing surface scratch. Because the leakage rate of gas in safety valve is very small, the volume leakage rate of gas is measured by soap-film flowmeter.

The acoustic emission detection system for internal leakage signal of safety valve adopts the acoustic emission system produced by American Physical Acoustics Company, including sensor, preamplifier, acoustic emission acquisition card and supporting acoustic emission software. The center frequency of acoustic emission sensor is 30kHz, and the gain of preamplifier is 40dB.

4. Analysis of test results

For the internal leakage detection test of safety valve, first, fix the internal leakage simulation prototype of safety valve on the internal leakage simulation test-bed, and connect it to the internal leakage detection system of safety valve. Open the air source, adjust the pressure at the inlet of the prototype and control the gas flow. When the leakage is stable, measure the flow through the soap-film flowmeter, and record the acoustic emission signal of internal leakage in the safety valve through the acoustic emission system. Change the inlet pressure of safety valve (0.01MPa, 0.02MPa, …, 0.1MPa) to measure and test different leakage states. Finally, through the acoustic emission signal detection system for the internal leakage in the safety valve, the signal is collected, amplified and filtered to extract the required signal eigenvalues. Replace the leakage holes of different types and sizes of safety valves and repeat the above test process.

4.1. Time domain and frequency domain characteristics of acoustic emission signal of internal leakage in safety valve

The leakage holes of different types and sizes of safety valves are selected to carry out the internal leakage simulation test under different pressure. Collect the acoustic emission signals of internal leakage in time domain and frequency domain, and compare and analyze the signals. By analyzing the

IOP Conf. Series: Materials Science and Engineering **799** (2020) 012001 doi:10.1088/1757-899X/799/1/012001

acoustic emission signals of internal leakage in time domain and frequency domain, we can preliminarily judge whether the safety valve is in internal leakage state. Due to different types of leakage holes, the inlet pressure required for the test will change accordingly. Taking several cases in figures 3-7 as examples, the time domain waveform and frequency spectrum of the acoustic emission signal of internal leakage are selected to make a comparative analysis of the acoustic emission signal of internal leakage in safety valve.

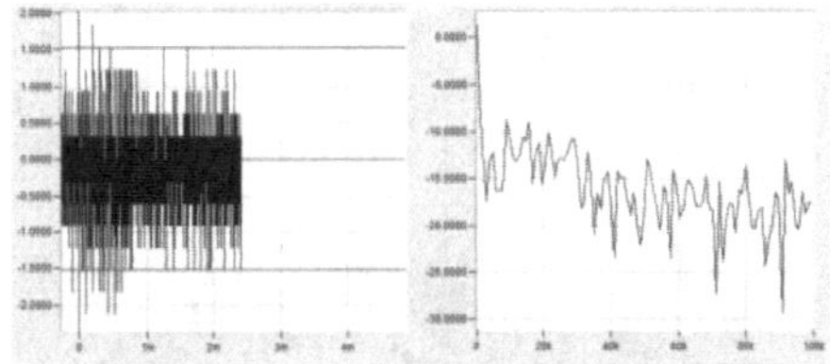

Figure 3. The size of 0.15mm * 0.5mm square leakage hole with the inlet pressure 0 MPa

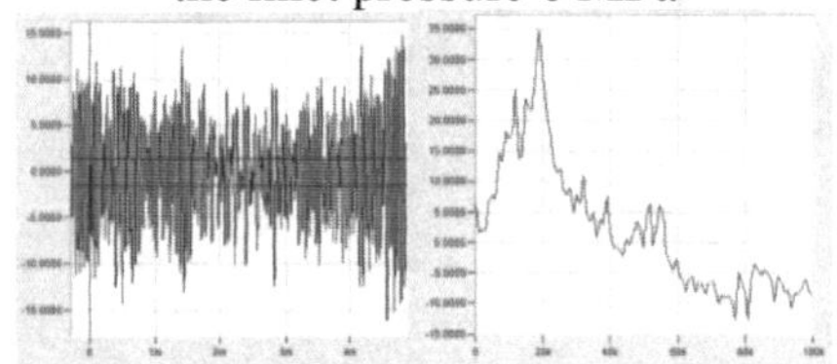

Figure 4. The size of 0.15mm * 0.5mm square leakage hole with the inlet pressure 0.03 MPa

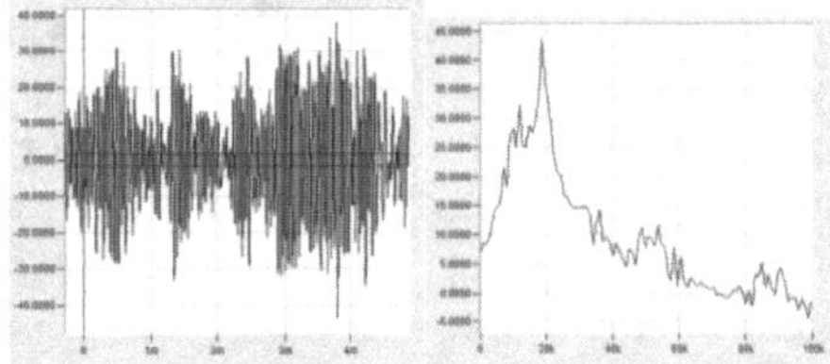

Figure 5. The size of 0.15mm * 0.5mm square leakage hole with the inlet pressure 0.05 MPa

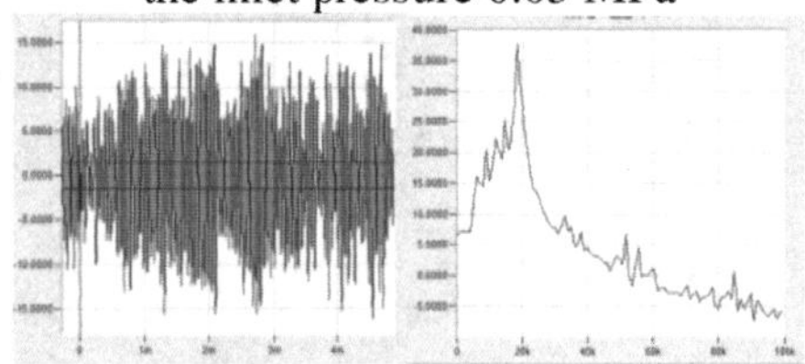

Figure 6. The size of 0.3mm * 0.5mm square leakage hole with the inlet pressure 0.03 MPa

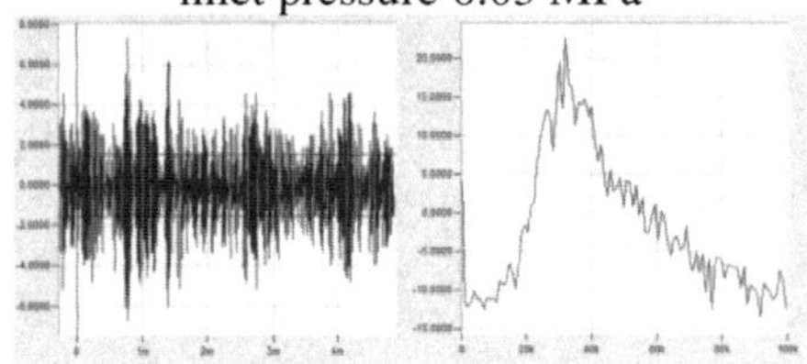

Figure 7. The polished hole with the inlet pressure 4.5 MPa

Based on the analysis of the acoustic emission signals of internal leakage of the above types of leakage holes under different inlet pressures, the following conclusions are drawn:

(1) The peak frequency range of internal leakage in safety valve is 20-30kHz, which is related to the type of leakage holes. For square slotted leakage holes, the peak frequency is near 20kHz; for manually polished leakage holes, the peak frequency is near 30kHz.

(2) For the same type of leakage holes, the peak frequency of internal leakage signal in the safety valve does not change with the size of the hole diameter and the inlet pressure. But for the same leakage hole, the amplitude of peak frequency increases with the increase of inlet pressure.

(3) By observing whether the acoustic emission signal has peak value in frequency domain, it can preliminarily judge whether the safety valve has internal leakage.

4.2. Relationship between acoustic emission signal and leakage rate in safety valve

If the leakage rate of safety valve can be calculated according to the characteristic value of acoustic emission signal of internal leakage in the safety valve, the degree of internal leakage or damage of sealing surface of safety valve can be judged more directly. According to equation (5), the relationship between the ASL value of acoustic emission signal of safety valve and the logarithm of leakage rate is linear. Therefore, the leakage rate of safety valve can be estimated by measuring the ASL value of the acoustic emission signal of internal leakage in safety valve. Taking 0.15mm * 0.5mm square leakage hole as an example, by collecting the ASL value of acoustic emission signal under different leakage rates, and using origin to make a linear fitting between the two, the curve is fitted, as shown in Figure 8:

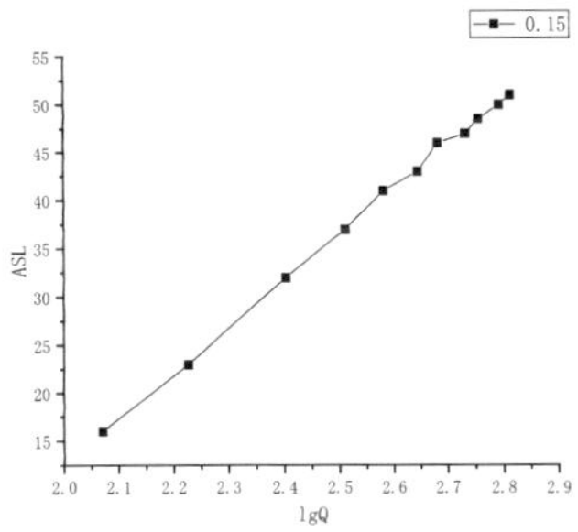

Figure 8. The relationship of leakage rate and ASL of 0.15mm*0.5mm and 0.3mm*0.5mm square leakage hole

Take the leakage rate and ASL value into equation (5), then obtain the coefficients b = 47.5734 and c = -82.4719. For square holes, the relationship between leakage rate and ASL is $ASL = 47.5734\log Q - 82.4719$. The leakage rate can be calculated according to $Q = e^{\frac{ASL+82.4719}{47.5734}}$. Therefore, for the internal leakage of safety valve, the leakage rate and average signal level can be fitted according to $ASL = b\log Q + c$, and the leakage rate of safety valve can be estimated by measuring the average signal level ASL of the acoustic emission signal of internal leakage.

5. Conclusion

Through theoretical analysis and experimental research, this paper explores the influence of factors such as the inlet pressure of safety valve, the size of leakage holes, the type and the number of leakage holes on the acoustic emission signal, and establishes the quantitative relationship between the gas volume leakage rate and the average signal level ASL of the characteristic parameters of acoustic emission signal in the process of internal leakage in safety valve, and draws the following conclusions:

(1) It is the first time to apply acoustic emission technology to the internal leakage detection in safety valve. The acoustic emission technology can effectively detect the internal leakage state in safety valve, which provides a new method for the internal leakage detection of safety valve.

(2) The peak frequency range of internal leakage in safety valve is 20-40kHz. The specific value is related to the type of internal leakage. The influence of hole size and inlet pressure is not significant. The peak frequency of contact surface damage and particle accumulation is about 20kHz, while the peak frequency of contact surface scratch is about 30kHz.

(3) For multi-hole leakage, the ASL value of acoustic emission signal is related to the leakage rate of single leakage hole, but it is independent of the total leakage rate, and there is no superposition of acoustic emission signals of each leakage hole.

(4) For internal leakage in safety valve, given the average signal level ASL value of internal leakage in safety valve, the leakage rate of safety valve can be estimated by equation.

References

[1] Yukun, Z. Dianjing, C. Ying, Y. (2016) Safety Valve Technology. Mechanical Industry Press, Beijing

[2] Xuefeng, F. Hua, L. Zhimin, X. (2007) On Line Detection Method of Valve Leakage Based on Acoustic Emission Technology. Chemical machinery, 34(1):52—54,58.

[3] Rongsheng, G. (1998) Development Status of Acoustic Emission Technology. Nondestructive testing, 6: 151— 158.

[4] Ruifeng, Y. Tiehua, M. (2006) Research and Application Progress of Acoustic Emission Technology. Journal of North University of China, 27(5) : 456 — 460.

[5] Qianxia, G. Luping, L. Hongde, R. (2011) Experimental Study on Quantitative Relationship between Valve Leakage Fault State and Acoustic Emission Signal Characteristics. Journal of Thermal power engineering, 26(5):5 82-587.

[6] Zhenlin, L. Zhiqiang, B. (2013) Inner Leakage Detection of Ball Valve in Natural Gas Pipeline Based on Acoustic Emission, 32(6)

[7] Delian, K. Jianfeng, Y. (2010) Application of Acoustic Emission Technology in on-line Monitoring of Valve Leakage. Beijing University of Chemical Technology

[8] Kaewwaewnoi W, Prateepasen A, Kaewtrakulpong P. Investigation of the Relationship between Internal Fluid Leakage through a Aalve and the Acoustic Emission Generated from the Leakage[J]. Measurement, 2010, 43(2): 274-282.

[9] Guang, D. Xinying, W. (2002) Acoustic testing method for inner leakage of pressure valve. Journal of Daqing Petroleum Institute. 25(3):95-98.

[10] Ying, Z. Guang, D. (2005) Estimation of Internal Leakage Rate of Gas Valve by Acoustic Detection Method. Journal of Daqing Petroleum Institute. 29(4):79-82.

IOP Conf. Series: Materials Science and Engineering **799** (2020) 012002 doi:10.1088/1757-899X/799/1/012002

Analysis of Importance of Engineering Survey in Construction Engineering Quality Management

Liu Liguo[1], Zhang Caixia[2]

[1]Shandong Normal University, 250358

[2]Jinan Vocational College, 250103

Abstract: Engineering survey is an important technology applied in the quality management of construction engineering. By conducting engineering survey, it can provide necessary data support and measurement methods for the early exploration, mid-term construction, and later acceptance quality management of construction engineering. In the process of construction quality management, how to apply the engineering measurement technology reasonably becomes an urgent problem to be solved. This article will give an overview of engineering survey, and describe the important role of it, and give some opinions and suggestions.

1. Introduction

Engineering survey mainly refers to the exploration during the construction, such as the measurement of the foundation, the measurement of the terrain fluctuations and the geological characteristics, the measurement of the points during the construction phase, the measurement of the installation position of the equipment as well as the measurement and acceptance of the various data during the acceptance phase. It is precise that engineering survey can provide comprehensive and multi-level exploration for the quality management of construction engineering, so it plays its due role of construction engineering quality management. The following will be an overall overview of the engineering survey, and explain the construction engineering quality management and how to get better application of the introduction of several ways or methods to contribute to the construction engineering quality management.

2. An overview of engineering survey

Engineering survey refers to the general term for the surveying and mapping of various data in engineering construction. It includes both engineering survey and parameter measurement during the design process as well as various surveys performed during the construction and quality management stages. Measure and test the engineering height, pressure capacity and engineering quality between joints. In the process of engineering measurement, the properties can be divided into engineering control survey and topographic survey and design stage, construction survey and equipment installation survey in the construction stage, and completion survey in the management stages according to the work order and nature of the survey. Among them, the engineering control and topographic survey and design phase refer to the measurement of the geological soil type, terrain fluctuations, division of the construction area, environmental construction conditions, and the impact of the engineering on surrounding residents in the area where the construction engineering is located. With terrain measurement, the factors can be determined, such as the area of the construction engineering, the soil and terrain conditions of the construction, and the degree of influence of the surrounding residents. Appropriate foundation types, construction equipment, and methods for the engineering can be chosen. It provides necessary data support for the construction, including the scope and the bearing capacity of the foundation and the degree of environmental impact, so that the design scheme meets the construction conditions. In terms of construction and equipment installation measurement during the construction phase, it mainly refers to the daily progress of the construction engineering, such as the measurement of the engineering height, the width of the building, the parameters of the construction materials, and the carrying capacity, especially for buildings, the detailed positioning measurement and elevation measurement of structures, the measurement of the axis of high-rise buildings, the installation and positioning measurement of structures and accessories, and the measurement of deformation of important buildings and structures. It requires technicians to use measurement equipment to carry out various parameters of the construction process to ensure that the quality to meet the construction requirements. The measurement after the completion mainly refers to the deformation of the subsequent buildings and the drawing of the as-built. The technicians need to draw the as-built drawings according to the form and various measurement parameters after it is completed. The follow-up quality management is to strengthen the measurement of building deformation, improve the efficiency of construction engineering quality management[1].

In addition, engineering survey also has the characteristics of higher measurement accuracy requirements and closer connection with the construction progress. When conducting engineering surveys, it is necessary to check and verify the measuring instruments, use the calibrated instruments simultaneously, and follow the principles of whole first and then local. For the engineering survey, the main axis of the engineering needs to be measured first, and the object next to the construction engineering is used as a reference point to determine the location. If it is necessary to measure the correlation line inside and control the measurement accuracy, multiple measurements will take the average value to perform the measurement. When measuring the deformation of a construction project, it is also necessary to strictly control the measurement accuracy to provide more accurate data for engineering changes. As far as the relationship between engineering measurement and construction progress is concerned, there are many uncertain factors, so the engineering measurement needs to be continued. During the construction process, earth excavation and construction material placement will affect the environment of the construction site, causing errors in various data. Therefore, it is necessary to continuously perform engineering measurements according to the construction progress to control the errors.

3. The important role of engineering survey in quality management of construction

(1) Providing data guarantee for construction engineering quality management

The first important role of engineering survey is to provide data guarantee for construction engineering quality management. The preliminary survey of construction engineering, the selection of construction material specifications, the arrangement of construction progress, the position and height of load-bearing piles, the bearing capacity of the building. And the construction conditions of soil all require data from engineering surveys. For example, in the selection of construction materials, technicians can understand the performance parameters, such as length, width, height, hardness, and bearing capacity of the material through engineering measurements, and compare their performance parameters with the requirements of the design plan to keep data at all stage. Especially for the quality management of building engineering positioning and decoration, engineering surveys are needed to provide reference points or areas such as the main axis for building positioning. According to the requirements of the engineering design plan, data exploration is performed on the foundation construction stage and regional positioning parameters. In this process, it is necessary to pay attention to the use of appropriate measuring instruments, control the measurement error of each data and try to improve the measurement accuracy. In the process of exterior decoration, technicians need to use engineering measurement technology to repair the deformation and quality problems that occurred in the early stage of construction for the height control of exterior walls and construction line measurement. Engineering measurement is carried out in the heavy sag, etc., and the indoor and outdoor elevation control lines are controlled to improve the users' satisfaction[2].

(2) Providing monitoring and management methods for the whole construction engineering

As shown in Figure 1, it is a schematic diagram of construction engineering quality management. Another important role of engineering survey is to provide monitoring and management methods for the entire building construction. Quality management needs to be based on accurate reference and standard data. Comparing the measurement data with the data specified in the design plan, and check whether the error of the actual construction measurement and the ideal data is within the allowable range. If the error is outside the allowable range, the management personnel need to repair and remedy the problem accordingly to improve the quality as much as possible. In addition, the engineering survey at the suitable time period can also reflect the situation, and provide monitoring and management methods.

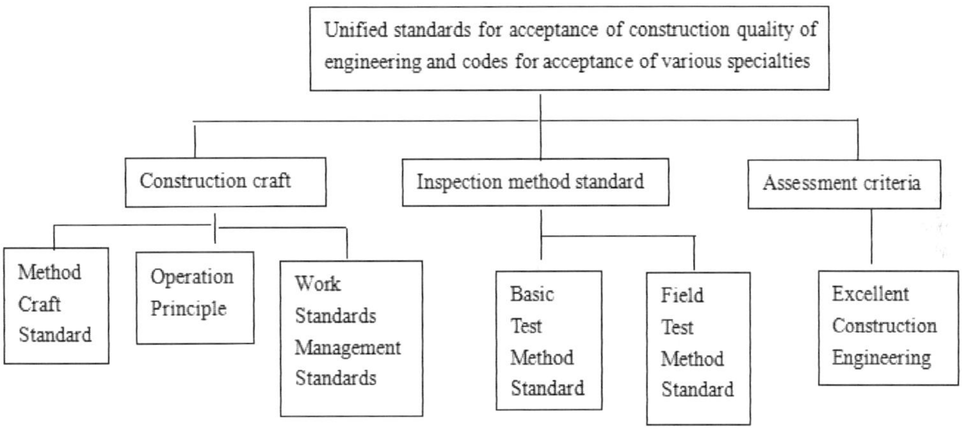

Figure 1 Schematic diagram of construction engineering quality management

4. How to make engineering survey play its due role in quality management

(1) Application of engineering survey in the positioning and stage of foundation construction

In order to play its due role in the quality management for engineering survey, the first step is to apply engineering survey during the construction positioning and foundation construction stages. Before the construction, technicians need to survey various parameters of the project. The focus is on the location of the engineering, the surrounding construction conditions, and the construction of the infrastructure. In the process of engineering survey, technicians need to use horizon instead of reference and use them to measure the height and verticality. For the determination of the location, it is necessary to use the landmark buildings or objects in the surrounding area as reference points, and use software to make a three-dimensional model of the engineering, and mark the construction in the model height, location, etc. to build a basic construction model for construction and management personnel.

For example, technicians need to apply engineering survey during the construction positioning and foundation construction stages, construct reference lines such as the building's main axis and height vertical line in accordance with the operating specifications of the measuring instruments used, and lay out ground points in the construction area. The location and number of points need to meet the requirements of foundation construction, and the load-bearing facilities should be carried out with the points, so that the engineering survey can play its due role in the positioning and foundation construction stages, and improve the construction quality[3].

(2) Strengthening the application of engineering survey in the construction phase of the main body of the construction engineering

Another measure that needs to be taken is to strengthen the application of engineering survey in the main construction phase of construction engineering. The main construction refers to the upper part of

IOP Conf. Series: Materials Science and Engineering **799** (2020) 012002 doi:10.1088/1757-899X/799/1/012002

the foundation, which is used to receive, bear and transmit all the upper loads. It can effectively ensure the integrity, stability and safety of the superstructure and carry the main structure of the construction. In the main construction phase, it is necessary to use steel, sand, cement, masonry materials, mortar, and necessary dopants. Engineering survey can depict the main structure of the construction to obtain the load capacity and form of the building data, and technicians can use the measurement data to inspect and screen the material mix ratio, hardness parameters, and material carrying capacity of the main body during the construction phase, and select materials that meet the load requirements of the construction engineering to better carry out engineering construction.

For example, technicians can use measuring instruments to make a virtual description of the main structure, and use computers to calculate and analyze the material types and quantities of each part, and estimate the material losses in the main construction to determine the approximate quantities of construction materials for the main structure. Through the comprehensive description and analysis of it, the technical personnel can obtain accurate engineering measurement data to select more suitable construction materials and proportion.

(3) Adopting engineering measurement technology to carry out quality acceptance

In addition to the application of engineering survey, the construction engineering positioning and foundation construction stage and the strengthening of its application in the main construction stage well as the adoption of engineering survey technology for the follow-up quality acceptance are also important measures to promote its application. The follow-up quality acceptance includes management of construction accuracy and engineering deformation. After the completion of construction, technicians need to measure various parameters of the project, such as control height, wall redraping and spindle line to carry out quality acceptance.

For example, technicians can measure parameters such as the height, the sag of the external wall surface, and the deformation of the building. After obtaining accurate measurement data, compare it with the construction standards and design drawings to control good measurement accuracy and timely feedback on problems in the acceptance so that it can be properly resolved[4].

5. Conclusion

Engineering survey can play a role in the positioning, foundation construction, main construction and follow-up quality acceptance. However, technicians need to choose appropriate measuring instruments and methods to measure various parameters to provide necessary data support for quality management and improve the efficiency of it.

Reference

[1] Chen Jingjing. 2017. Analysis of the Importance of Engineering Measurement in Construction Quality Management [J]. Science and Technology Innovation Guide, (34)

[2] Sun Liye. 2017. On the Importance of Engineering Survey in Construction Quality Management [J]. World Nonferrous Metals, (4): 203-204

[3] Zhang Youlin, Cao Zhen. 2019. Engineering Survey and Error Control Analysis in Construction [J]. Agricultural Staff, (2Z): 228-228

[4] Chen Jianxing. 2018. Analysis and Discussion on the Application of Digital Mapping Technology in Geological Engineering Survey [C] / / Academic Conference on Construction Technology and Management

An Optimization Method for Measurement Matrix Based on Double Decomposition

Zhaoyang Mao[1], Lan Li[1]*

[1]School of Science, University of Xi'an Shiyou, Xi'an 710065, China

Zhaoyang Mao; email: zhymao_xsy@163.com.

*Corresponding Author: Lan Li; email: lanli98@126.com.

Abstract: This paper introduces a novel method of measurement matrix in compressed sensing. In order to overcome the difficulties associated with coherence of measurement matrix, we propose double optimization methods by eigenvalue decomposition and singular value decomposition under mild conditions. An efficient algorithm (SVD-EIG) is used to recover sparse inputs from the optimized measurement matrix, based on the adaptation of the optimized matrix by eigenvalue decomposition. Lastly, compared with the other methods as the same sampling rate, we demonstrate through simulations that SVD-EIG algorithm can improve accuracy and probability of the reconstruction.

1. Introduction

Compression sensing(CS)[1-5] is an emerging paradigm in signal/image processing, which senses a sparse signal by taking a set of incomplete measurement.

Measurement matrix is very important to recover CS signal/image. Choosing measurement matrix is mainly divided into two categories. The first is random measurement matrix[3], such as random Gaussian matrix, Bernoulli random matrix and so on. The second is deterministic measurement matrix, such as partial Hadamard matrix. Recent work has explored coherence between perception matrix and representation matrix. These methods construct Gram matrix by product of perception matrix and representation matrix, and research the coherence of Gram. The off-diagonal entries in Gram are the inner products. In[6] off-diagonal entries are minimized by thresholding, but this way always spends more time and produces larger values in the matrix. An optimization method for measurement matrix based on eigenvalue decomposition[7] and singular value decomposition [8], study the relationship between coherence and eigenvalues of Gram matrix, and experiment results show that these methods rely on initial value.

This paper suggests an efficient algorithm (SVD-EIG) for recovering spare input from the optimized measurement matrix, based on an adaptation of the optimized matrix by eigenvalue decomposition. This method reduces the coherence by eigenvalue decomposition and singular value decomposition, therefor it improves the accuracy and probability of the reconstruction.

This paper is organized as follows. In next section we briefly recall the basic concept of compressed sensing. Section 3 shows our main result about optimal measurement matrix by combining singular values and eigenvalues. Section 4 presnets SVD-EIG algorithm based on Orthogonal Matching Pursuit (OMP)[9]. Section 5 simply shows the experimental results of the algorithm, compare with several classical algorithm. In section (6), it is concluded that our novel method based on singular values and eigenvalue is very effective in terms of iteration time and computational complexity.

2. Preliminary

2.1. CS

Consider the real signal $\mathbf{x} \in \mathbb{R}^n$, $\|\mathbf{x}\|_0$ stands for the number of nonzero entries of vector $\mathbf{x}$. If $\|\mathbf{x}\|_0 = k < n$, then $\mathbf{x}$ is k-sparse. When the vector $\mathbf{x}$ is not sparse, we can always find a sparse representation matrix(sparsity basis) $\mathbf{\Psi}$ such that

$$\mathbf{x} = \mathbf{\Psi s} \tag{1}$$

where $\mathbf{s}$ is a sparse vector, which contains nonzero coefficients. The linear system is as follows

$$\mathbf{y} = \mathbf{\Phi x} = \mathbf{\Phi \Psi s} \tag{2}$$

where the matrix $\mathbf{\Phi} \in \mathbb{R}^{m \times n}$ ($m \ll n$) and $\mathbf{y}$ is a set of m linear measurements. Measurement matrix $\mathbf{A}$ denotes by the $\mathbf{\Phi\Psi}$, then

$$\mathbf{y} = \mathbf{As} \tag{3}$$

To recover such a signal $\mathbf{s}$, a natural a choise is seek a solution of the l_0 minimization problem

$$\min_{\mathbf{s}} \|\mathbf{s}\|_0 \quad s.t \quad \mathbf{y} = \mathbf{As} \tag{4}$$

But equation (4) is NP hard problem, so we want to find a way to approximate Equation (4). A general method is

$$\min_{\mathbf{s}} \|\mathbf{s}\|_1 \quad s.t \quad \mathbf{y} = \mathbf{As} \tag{5}$$

Equation (4) is equal to Equation (5) when $\mathbf{A}$ satisfy the RIP (Restricted Isometry Property) condition.

Definition 1. If the matrix $\mathbf{A}$ satisfies the RIP condition of k-order, If there exists a smallest constant $\delta_k \in (0,1)$, for all k-sparse vector $\mathbf{s}$

$$(1-\delta_k)\|\mathbf{s}\|_2^2 \le \|\mathbf{As}\|_2^2 \le (1+\delta_k)\|\mathbf{s}\|_2^2$$

The matrix $\mathbf{A}$ satisfies the RIP condition of k-sparse.

□

In practice it is difficult to verify that a matrix $\mathbf{A}$ satisfies the RIP or calculate the corresponding RIP constant δ_k. In this respect, results based on coherence are appealing, which is very useful for sparsity. The coherence defined by

$$\mu(\mathbf{A}) = \max_{i,j} \frac{|\langle \mathbf{A}_i, \mathbf{A}_j \rangle|}{\|\mathbf{A}_i\| \cdot \|\mathbf{A}_j\|} \tag{6}$$

Here, we denote the i th column of $\mathbf{A}$ and Euclidean norm of $\mathbf{A}$ by $\mathbf{A}_j$ and $\|\mathbf{A}_j\|$, respectively.

To guarantee a unique solution to Equation (3), we have the condition based on the coherence of $\mathbf{A}$

$$\|\mathbf{x}\|_0 \le \frac{1}{2}(1+\frac{1}{\mu}) \tag{7}$$

with $\|\mathbf{x}\|_0$ being the number of nonzero entries in $\mathbf{x}$.

2.2. Gram Matrix

The different way to obtain the coherence is from the Gram matrix $\mathbf{G} = \mathbf{A}^T\mathbf{A}$, using the arbitrary dictionaries for normalizing each of its columns. The off-diagonal entries on $\mathbf{A}$ are the inner product

IOP Conf. Series: Materials Science and Engineering **799** (2020) 012003 doi:10.1088/1757-899X/799/1/012003

from Equation (6). The mutual coherence is the off-diagonal entry g_{ij} with the largest magnitude, i, j represent the rows and columns of matrix, respectively. Assume $\mu(\mathbf{A}) = \max_{i,j \atop i \neq j} |g_{ij}|$ *and* $g_{ij} = \mathbf{A}_i^T \mathbf{A}_j$ There exist some problems:

- It's impossible that measurement matrix is global incoherent when μ is small.
- It's impossible that measurement matrix is global coherent when μ is large.

So we will consider global coherence of $\mathbf{A}$ in next section.

Lemma 1. If signal $\mathbf{x}$ is k-sparse and satisfied

$$\|\mathbf{A}\|_0 < \frac{1}{2}\left(1 + \frac{1}{\mu(\mathbf{A})}\right)$$

then there is unique signal $\mathbf{x}$ from $\mathbf{y}$ and $\mathbf{y} = \mathbf{A}\mathbf{x}$. Where $\|\mathbf{A}\|_0$ represents the zero norm of a matrix.

3. Choosing Measurement Matrix

3.1. Eigenvalue decomposition

In this section, we describe mutual coherence, which reflects average performance. Inspired by [7], we consider global coherence of matrix $\mathbf{A}$ from eigenvalue. The result of reconstruction is associated with coherence of $\mathbf{\Phi}$ and $\mathbf{\Psi}$. In other words, the smaller global coherence of measurement matrix $\mathbf{A}$, the better effect of reconstruction algorithm.

Theorem 1. Suppose $\mathbf{\Phi} \in \mathbb{R}^{m \times n}$, $rank(\mathbf{\Phi}) = m$, and sparse representation matrix $\mathbf{\Psi} \in \mathbb{R}^{n \times n}$, Let $\mathbf{A} = \mathbf{\Phi}\mathbf{\Psi}$ and Gram matrix $\mathbf{G} = \mathbf{A}^T \mathbf{A}$. If The $\mathbf{G}$ is semidefinite matrix and $\lambda_s (s = 1, \ldots, m)$ are eigenvalues, then

$$\sum_{s=1}^{m} \lambda_s = n \tag{8}$$

$$\sum_{s=1}^{m} (\lambda_s)^2 = \sum_{i,j=1}^{n} (<\mathbf{A}_i, \mathbf{A}_j>)^2 \tag{9}$$

where $\mathbf{A}_i$ and $\mathbf{A}_j$ represent the i columns and j columns of normalized $\mathbf{A}$, respectively.

The purpose is to minimize the squares sum of the off-diagonal entries of $\mathbf{G}$. According to Equation (8) and Equation (9) we obtain an optimized problem

$$min \sum_{i \neq j} (g_{ij})^2 = \sum_{s=1}^{m} (\lambda_s)^2 - \sum_{i=1}^{n} (g_{ij})^2 \quad s.t \quad \sum_{s=1}^{m} \lambda_k = n \tag{10}$$

When $\lambda_s = \frac{n}{m}$, the solution to problem Equation (10) is

$$\sum_{i \neq j} (g_{ij})^2 = \sum_{i=1}^{m} (\frac{n}{m})^2 - n \tag{11}$$

The object function of Equation (10) existS minimize value.

3.2. Singular value and Condition number

In the previous subsection, we give a method of optimized measurement matrix. But we must have the initial matrix. The general method is to produce Gaussian random matrices or Bernoulli matrix, depends on the initial measurement matrix. Next, we will show that construct the measurement matrix $\mathbf{A}$.

To analyse the condition number we obtain the singular value decomposition (SVD) of measurement matrix $\mathbf{A}$:

$$\mathbf{A} = \mathbf{U\Sigma V}^T \tag{12}$$

where $\mathbf{U} \in \mathbb{R}^{m\times m}$ and $\mathbf{V} \in \mathbb{R}^{n\times n}$ is orthogonal matrix. The singular value matrix $\mathbf{\Sigma}$ is defined as:

$$\mathbf{\Sigma} = \begin{pmatrix} \sigma_i & 0 \\ 0 & 0 \end{pmatrix}, \quad i \in \{1,\ldots,m\} \tag{13}$$

where $\sigma_i = diag\{\sigma_1, \sigma_2, \ldots, \sigma_m\}$ denotes the diagonal matrix.

Suppose $\mathbf{A} \in \mathbb{R}^{m\times n}$, $\sigma_1,\ldots,\sigma_m$ are singular value of $\mathbf{A}$. The condition number $cond(\mathbf{A}) = \sigma_{max}/\sigma_{min}$. σ_{max} and σ_{min} represent the maximum singular value and minimum singular value, respectively.

$$\| \mathbf{A}\|_2 = \sigma_{max} \quad \| \mathbf{A}\|_2 = \sigma_{min} \tag{14}$$

If the condition number $cond(\mathbf{A})$ is smaller, then the coherence of $\mathbf{A}$ becomes smaller.

4. SVD-EIG Algorithm

Firstly, the random Gaussian matrix are optimized by SVD as initial matrix. Secondly we use the eigenvalue decomposition to improve the performance of matrix.

```
Algorithm 1 SVD-EIG
```

Input: Observations $\mathbf{y}$, Object image $\mathbf{x} \in \mathbb{R}^{n\times n}$, Measurement matrix $\mathbf{\Phi} \in \mathbb{R}^{m\times n}$, Sparse representation matrix $\mathbf{\Psi} \in \mathbb{R}^{n\times n}$, Error ò.

Output: Recovered image $\mathbf{x}$

Initialize: $\mathbf{\Phi} \leftarrow$ random Gaussian matrix, $\mathbf{\Psi} \leftarrow$ DCT matrix, $\mathbf{A} = \mathbf{\Phi\Psi}$.

Svd_Optimize $\mathbf{A} = \mathbf{U\Sigma V}^T, \mathbf{\Sigma} \leftarrow \mathbf{\Sigma} = min(\sigma), \mathbf{A} \leftarrow \mathbf{U\hat{\Sigma}V}^T$

Function SVD-EIG($\mathbf{\Phi}, \mathbf{\Psi}$, ò)

  While halting criterion false ò **do**

    $\mathbf{G} = \mathbf{A}^T\mathbf{A}$ and Eigenvalue decomposition $\mathbf{G} = \mathbf{VHV}^T$

    $\mathbf{H} \leftarrow diag(\mathbf{H}) = \sqrt{\dfrac{n}{m}}$

    $\hat{\hat{D}} = \mathbf{VHV}^T$

    $\grave{o}_0 = \sum off - diag(\mathbf{D})$

    If $|$蝌$- \sum (n/m)^2 - n| <$ **then**

      $\mathbf{\Phi} = \mathbf{D\Psi}^{-1}$

      Return $\mathbf{\Phi} \leftarrow \mathbf{\Phi}$

    EndIf

  EndWhile

EndFunction

While $i = 1, \ldots,$ **do**

  $\mathbf{x} = OMP(\mathbf{y}, \mathbf{A}, k)$

EndWhile

The experimental platform is matlab. In algorithm1, $\hat{\mathrm{A}}$ is Estimate value, $\tilde{\mathrm{A}}$ is normalized by each of its columns, $off - diag(\cdot)$ is Off-diagonal element of $(\cdot)$. Since our focus is on optimizing the measurement matrix, we select a simple row as the code of OMP Refer to Section 3 for details of calculations.

IOP Conf. Series: Materials Science and Engineering **799** (2020) 012003 doi:10.1088/1757-899X/799/1/012003

5. Implementation

Now we demonstrate the performance of SVD-EIG via simulations. The object image is chosen 512×512 pixel Lena. Initial matrix $\mathbf{\Phi}$ is random Gaussian matrix, and sparse representation matrix $\mathbf{\Psi}$ is discrete cosine transform (DCT) matrix. We select the OMP as reconstruction algorithm, because the main purpose of this paper is to design measurement matrix.

We examine the performance of random Gaussian, eigenvalue decomposition and SVD-EIG in recovering an image $\mathbf{x}$ of size 512×512.

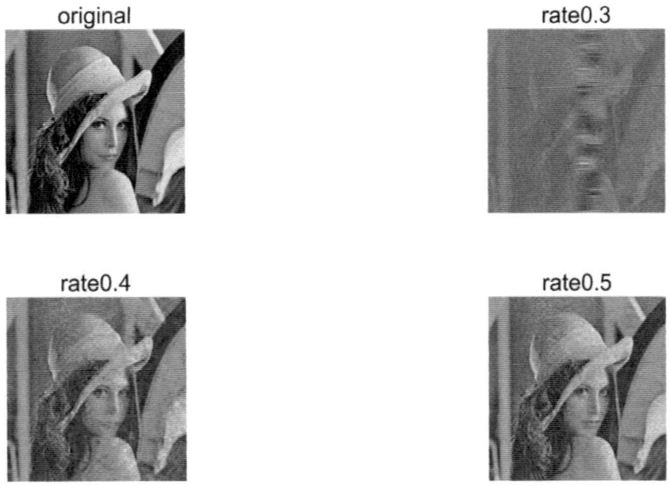

Figure 1. Random Gaussian matrix.

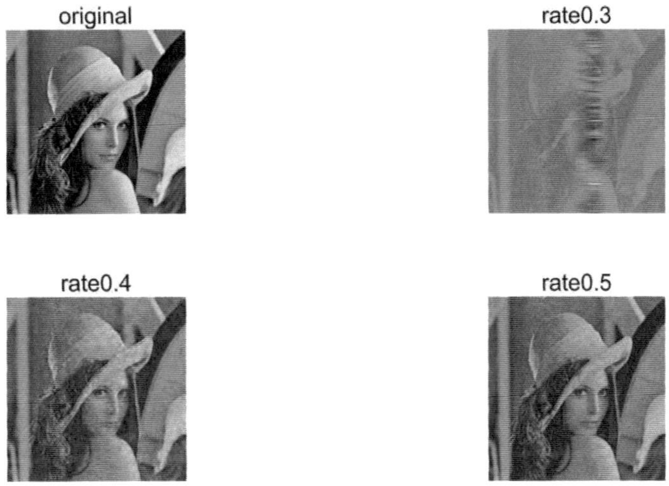

Figure 2. Eigenvalue decomposition.

IOP Conf. Series: Materials Science and Engineering **799** (2020) 012003 doi:10.1088/1757-899X/799/1/012003

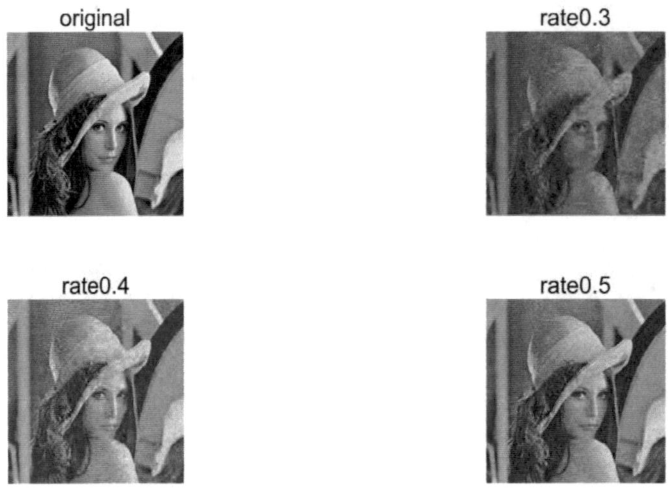

Figure 3. SVD-EIG.

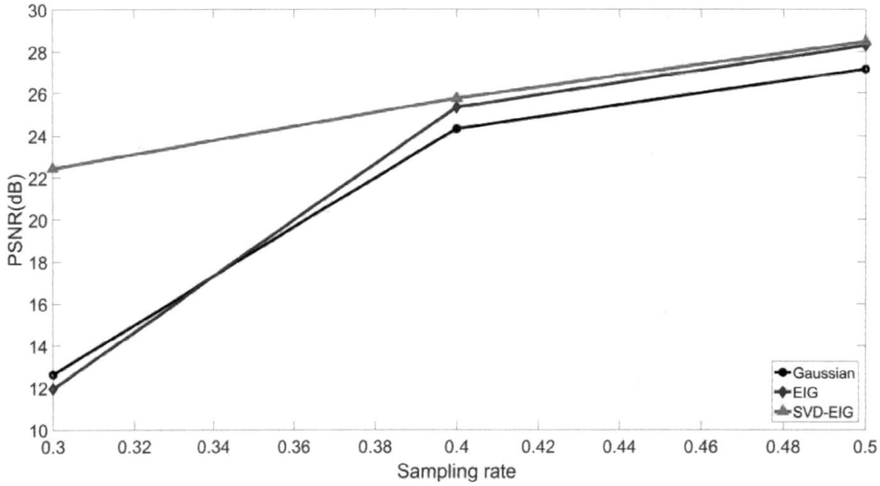

Figure 4. Comparison of PSNR (X-axis is the sampling rate, Y-axis is the PSNR value)

In the simulation in fig.1 and fig.2, we show the recovery result under the random Gaussian matrix, The measurement matrix is optimized by eigenvalue decomposition at different sampling rates(0.3,0.4,0.5), respectively. In the fig.3, we keep the same conditions except for the measurement matrix and we can clearly observed that the recovery are better compared with the other algorithms. Moreover, we can also see from fig.4 that the PSNR value of SVD-EIG algorithm is very high.

6. Conclusion

In this paper, we suggest to improve the performance of image optimization method by singular value decomposition and eigenvalue decomposition. For same rate of measurements, the method of SVD-EIG can lead to better performance than using the random Gaussian and eigenvalue. We also demonstrate that SVD-EIG is able to enhance the quality of recovery as long as Gram matrix is sufficient incoherent. In addition, the experimental result also explains that the double optimization algorithm has less iterations and low computational complexity. There are some drawbacks in this algorithm. Running our code will spent more time compered with other algorithms.

Acknowledgements

This work was financially supported by Natural Science Basic Research Plan in Shaanxi Province of China (No. 2017JM1046) and China Scholarship Council.

IOP Conf. Series: Materials Science and Engineering **799** (2020) 012003 doi:10.1088/1757-899X/799/1/012003

References

[1] Candès, E.J., Romberg, J., Tao, T. (2006) Robust uncertainty principles: exact signal reconstruction from highly incomplete frequency information. J. IEEE Transactions on Information Theory., 52: 489-509.

[2] Romberg, J. (2008) Imaging via compressive sampling. J. IEEE Signal Processing Magazine., 25: 14-20.

[3] Candès, E.J. (2006) Compressive sampling. J. Marta Sanz Solé., 17: 1433-1452.

[4] Candès, E.J., Wakin, M.B. (2008) An introduction to compressive sampling. J. IEEE Signal Processing Magazing., 25: 21-30.

[5] Herman, M.A., Strohmer, T. (2009) High-resolution radar via compressed sensing. J. IEEE Transactions on Signal Processing., 57: 2275-2284.

[6] Elad, M. (2007) Optimized Projections for Compressed Sensing. J. IEEE Transactions on Signal Processing., 55: 5695-5702.

[7] Zhao. R.Z., Qin, Z., Hu, S.H. (2012) An optimization method for measurement matrix based on eigenvalue decomposition. J. Signal Processing., 28: 653-658.

[8] Li, Z., Cui, S. (2018) Observation matrix optimization algorithm in compressive sensing based on singular value decomposition. J. Journal of Computer Applications., 38: 568-572.

[9] Yagyensh, C.P., Ramin, R., Perinkulam, S.K. (1993) Orthogonal matching pursuit: recursive function approximation with applications to wavelet decomposition. J. Proceedings of 27th Asilomar Conference on Signals, Systems and Computers., 1: 40-44.

IOP Conf. Series: Materials Science and Engineering **799** (2020) 012004 doi:10.1088/1757-899X/799/1/012004

Development of Thin Film Thermocouple Test System for Transient Temperature Measurement

Zhu Xi*, An Wanqing, Liu Chang, Yang Xiaoyuan, Su Xinming, Zhou Yan, Li Zhenwei, Liu Zeyuan

Beijing Institute of Spacecraft Environment Engineering, Beijing, 100094, China

*Corresponding author's e-mail: 525891876@qq.com

Abstract. In order to achieve accurate temperature measurement in a fast temperature change environment, a set of thin film thermocouple test system for transient temperature measurement is designed and developed. The system consists of cold junction compensation and data collector. The temperature of the film thermocouple can be obtained in real time through the upper acquisition software. The hardware composition and software design of the thin film thermocouple test system are given. The data acquisition terminal and coordinator program are designed to control the temperature signal of the MAX31855 to collect the sensor, realizing the real-time transmission of temperature data between the sensor and the computer. In order to verify the actual application effect of the system, the matching, stability and reliability of the temperature acquisition system and monitoring software, the FLUKE-9144 dry calibration furnace was selected as the heat source for the temperature test. The test results show that the system has fast acquisition speed, high measurement accuracy and good temperature stability.

1. Introduction

With the continuous advancement of equipment technology in China, it is increasingly necessary to obtain accurate physical parameters of products or environments to correct simulation models or to clarify design boundaries. At present, how to measure instantaneous temperature changes in an environment above 1000 °C is still a difficult problem in the development of equipment. For example, the instantaneous measurement of the barrel temperature of the projectile launch, the temperature measurement of the inner wall of the cylinder of the internal combustion engine, and the temperature measurement of the thermal protection structure of the high-speed aircraft require that the high temperature sensor be small in size, without damaging the test piece or affecting the environment to be tested. For explosive temperature measurement, rocket engine tail flame temperature measurement, combustion chamber gas temperature measurement, etc., the heat is rapidly released, and the temperature of the measured target is instantly increased by several hundred degrees (usually in the order of milliseconds or even microseconds), which requires The temperature sensor has excellent dynamic response.

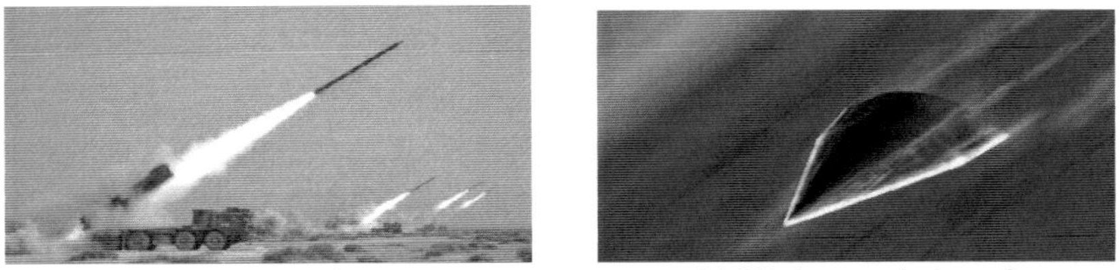

(a) Shell launch (b) High-speed aircraft return

Figure 1. transient temperature test occasion

In order to adapt to the high temperature environment, the traditional contact type high temperature sensor usually uses the armor method to protect the sensor itself, which not only increases the volume of the sensor, but also changes the heat capacity of the sensor, and cannot accurately detect the instantaneous temperature change. The optical-based non-contact temperature measurement method is not only inferior in accuracy, but also cannot be applied to temperature measurement inside a closed structure. A thin film thermocouple refers to a temperature sensor that uses the Seebeck effect to measure temperature and has a hot junction thickness of only micrometers. The high temperature sensor based on thin film thermocouple technology has the characteristics of small volume, small heat capacity, short thermal response time, high precision and high sensitivity, and it has small interference and damage to the measured target and test environment, and is suitable for narrow space and difficult installation. When the temperature changes instantaneously. This paper introduces the composition of the film thermocouple test system, and develops the film thermocouple upper acquisition software. Finally, the system is tested. The test results show that the system has fast acquisition speed, high measurement accuracy and good temperature stability.

2. Test system components

2.1. Cold junction compensation design

The thermocouple output thermoelectric potential reflects the operating temperature relative to the reference terminal, while the purpose of the temperature measurement is to measure the operating temperature at 0 °C. Therefore, only when the reference end of the thermocouple is placed in the ice water mixture, the temperature of the reference end can be kept at 0 °C regardless of the ambient temperature. At this time, the thermoelectric potential of the thermocouple corresponds to the temperature of the standard index table. It is the temperature of the measuring point (ie the working end). In practical engineering applications, it is difficult to ensure that the reference end of the thermocouple is always at 0 °C, which is the reason and significance of the thermocouple compensation for the cold junction during the temperature measurement process.

In this paper, the MAX31855 chip is used to collect the cold junction temperature. The MAX31855 is a thermocouple to digital output converter from Maxim, USA. The chip has a 14-bit analog-to-digital converter (ADC). It has cold junction compensation and correction for external thermocouples. Maxim offers a range of models for different types of thermocouples such as B, K, and N. Each model is optimized and tuned for the specific thermocouple type. The film thermocouple sensor of this system belongs to type B thermocouple, and the model is MAX31855KASA+ chip. In this system, its function is to perform cold junction compensation and analog-to-digital conversion of the film thermocouple in the sensing temperature designed by the project. Figure 2 shows the pin configuration of the MAX31855, and Figure 3 shows the internal structure of the MAX31855.

IOP Conf. Series: Materials Science and Engineering **799** (2020) 012004 doi:10.1088/1757-899X/799/1/012004

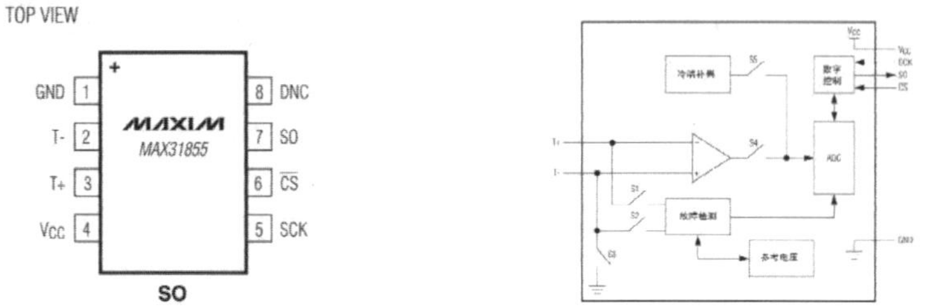

(a) Pin configuration diagram (b) Internal structure diagram

Figure 2. MAX31855 structure

The MAX31855 includes a signal conditioning hardware circuit that adjusts the weak thermocouple signal to match the input to the analog-to-digital converter ADC input channel. Before the electrical signal of the thermocouple is converted to its corresponding temperature value, the cold junction compensation of the thermocouple is required. In order to ensure that the temperature value obtained after the conversion is the actual working terminal temperature. For a standard K-type thermocouple, the Seebeck coefficient is approximately 41 μV/°C, and the thermocouple output characteristics are approximated according to equation (1).

$$V_{OUT} = (41.276\mu V/°C) \times (T_R - T_{AMB}) \quad (1)$$

In the formula, V_{OUT} is the thermocouple bipolar output voltage; T_R is the thermocouple working terminal temperature; T_{AMB} is the thermocouple reference terminal temperature.

The MAX31855 performs cold junction compensation and correction of external thermocouples by sensing the temperature at the reference junction. The die temperature inside the chip is first detected by the chip internal temperature sensor. The output voltage of the external thermocouple is then measured by an analog-to-digital converter ADC and converted to the equivalent temperature of the thermocouple before compensation. By the superposition of this value and the internal die temperature of the chip, the operating temperature of the external thermocouple based on 0 °C is calculated. It can be seen from the above that when applying the MAX31855 in engineering, the reference end of the external thermocouple should be placed as close as possible to the MAX31855 chip, so that the temperature is as close as possible to the die temperature of the chip to reduce the reference temperature measurement error.

2.2. *Data collector design*

The data acquisition terminal device function backplane is mainly composed of CC2530 pin interface circuit, indicator circuit, MAX31855 thermocouple to digital output converter, reset circuit and power supply circuit. Figure 3 is a schematic diagram of the function of the data acquisition terminal function of the thin film thermocouple sensor temperature test system designed for this project. The indicator circuit is used to indicate the working state of the thermocouple signal acquisition terminal. The function of the MAX31855 thermocouple to digital output converter is to convert the mV-class weak signal of the thermocouple into the thermocouple temperature signal before compensation, and collect the thermocouple cold junction temperature signal and fault diagnosis of the thin film thermocouple. The power supply circuit uses the HT7533 high-precision voltage regulator chip to convert the 3.7V power supply of the lithium battery into the 3.3V power supply voltage required by the system, and at the same time filter out the power supply interference signal. The physical data diagram of the data acquisition function of the film thermocouple sensor temperature test system is shown in Figure 4.

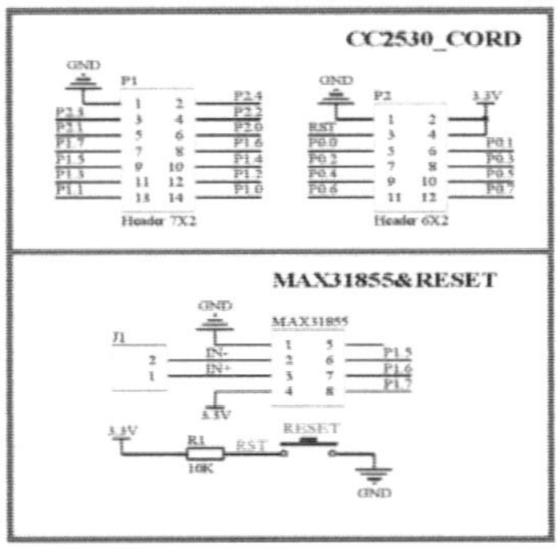

Figure 3. Schematic diagram of the data acquisition device

Figure 4. Data acquisition terminal function device

3. PC software design

3.1. Software overall structure design

The software is modularized for the functions and usage requirements of the system. The designed modules mainly include: user information module, data monitoring module, data processing module and database part. The software structure diagram is shown in Figure 5.

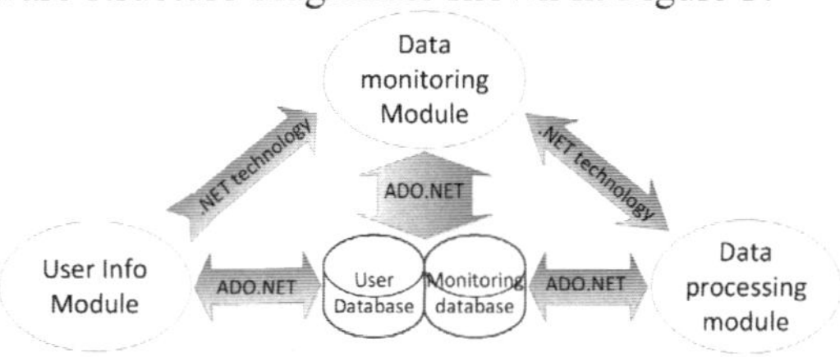

Figure 5. Sensor temperature test system software structure diagram

The PC software uses the Visual Studio development tools to apply the WPF user interface framework application developed in C# language. The PC software adopts a modular design and has a good human-computer interaction interface. The software implements the functions of receiving, storing and deleting the data transmitted by the coordinator, and displays the measured temperature data in real time in both digital and graphical forms. For non-standard thermocouples, the software can manually set the Seebeck coefficient, and can set the upper and lower limits of the cold end and hot end. When the temperature exceeds the upper limit or lower than the lower limit, the software alarm light flashes and the alarm sound is played. The crystal report control is used to realize the export of multiple formats of sensors to measure temperature data.

3.2. Data Monitoring Module Design

The data monitoring module is the core module of the application software system, and provides the main functions realized by the system. The data monitoring interface is shown in Figure 6. It realizes functions such as data acquisition and processing, data display, database access, temperature alarm, fault alarm, Seebeck coefficient setting, and serial port automatic retrieval.

1) Data acquisition is to read the data of the serial port buffer through the serial port of the system. After the data is separated by a certain algorithm, the actual cold end and hot end temperature data sets are obtained.

2) The data display is to draw the data curve in real time by embedding the ZedGraph control in the Win Form application in the WPF framework application. At the same time, the temperature data of the current sensor is displayed in real time through the software text box control.

3) Database access is to save the data read by the serial port and additional information to the monitoring database.

4) The Seebeck coefficient setting is to set the Seebeck coefficient of the self-sensing sensor thermocouple through the text box input control. After the software reads the Seebeck coefficient set by the user, the data collected by the serial buffer is calculated. The corresponding sensor is measured to measure the temperature.

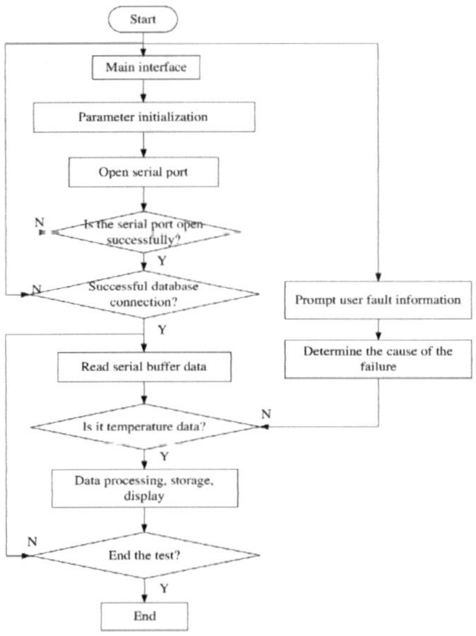

Figure 6. Data monitoring module software flow chart

3.3. *Data processing module design*

The data processing module provides users with functions such as data query, data deletion, and report generation. The data query function uses the conditional query method to perform a secondary query. The data processing module software flow chart is shown in Figure 7. The data query interface is shown in Figure 8. The software uses the Crystal Report Control to implement a variety of format file export sensor measurement temperature data.

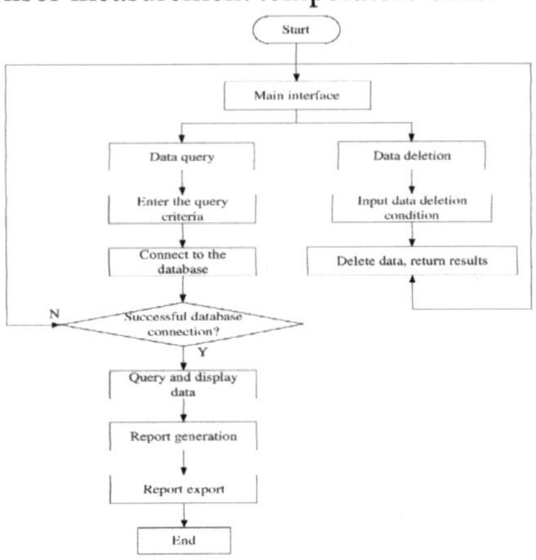

Figure 7. Software flow chart of data processing module

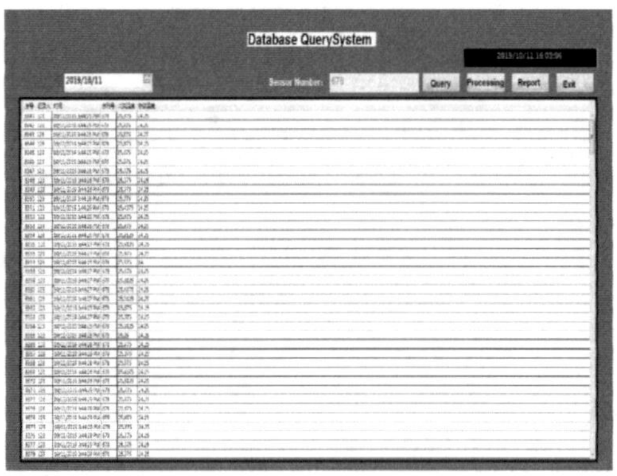

Figure 8. Data query interface

4. Test results and analysis

In order to verify the practical application effect of the developed film thermocouple temperature sensor, the matching, stability and reliability of the temperature acquisition system and monitoring software, the FLUKE-9144 dry calibration furnace was selected as the heat source for the temperature test, and the calibration furnace was set up. The temperatures were 100 °C, 200 °C, 300 °C and 400 °C, respectively. Through the data processing of the monitoring software, the temperature measurement curve of the temperature test system is shown in Fig. 9. From the four sets of temperature measurement curves, it can be seen that the thin film thermocouple sensor can quickly respond to temperature changes, and the signal is sent to the monitoring software for processing and display in time through the temperature measurement system. The temperature measurement system has good temperature stability and temperature measurement. The error does not exceed ±0.5 °C.

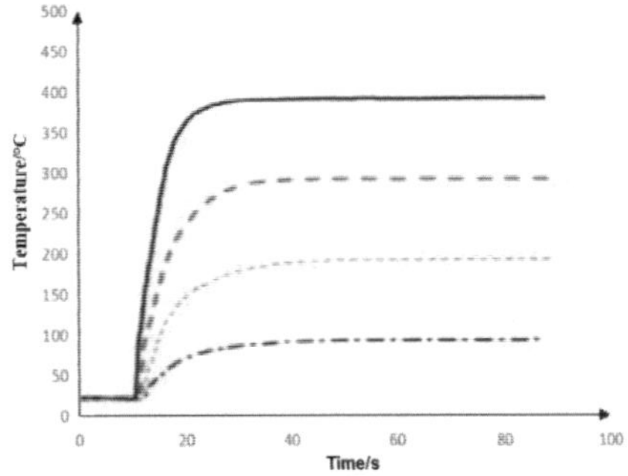

Figure 9. Temperature measurement system test curve

5. Conclusions

This paper introduces a thin film thermocouple test system for transient temperature measurement, which is studied from the aspects of system composition and upper acquisition software. The cold junction compensation, data acquisition terminal and coordinator program of the acquisition system are designed to realize the real-time transmission of temperature data between the sensor and the computer. The system PC software based on C/S architecture has been developed, which has the advantage of fast response speed. The system has the advantages of fast acquisition speed, high measurement accuracy and good temperature stability. It can meet the application of gun barrel temperature measurement, cylinder inner wall temperature measurement and high-speed aircraft thermal protection structure temperature measurement.

References

[1] Yu-Rong G E, Dong-Hua W U. Application of digital temperature sensor to detecting temperature of motor train set[J]. Diesel Locomotives, 2005.
[2] Mohanty A R, Fatima S. Shaft misalignment detection by thermal imaging of support bearings[J]. IFAC-PapersOnLine, 2015, 48(21):554-559.
[3] QIAN J, CHEN J, ZHANG C, et al. Design of a 0.8V low power CMOS temperature sensor for RFID-based train axle temperature measurement[C]. IEEE International Conference on Solid-State and Integrated Circuit Technology. IEEE, 2010:1404-1406.
[4] Martiny N, Rheinfeld A, Geder J, et al. Development of an All Kapton-Based Thin-Film Thermocouple Matrix for In Situ, Temperature Measurement in a Lithium Ion Pouch Cell[J]. Sensors Journal IEEE, 2014, 14(10):3377-3384.
[5] Forney, L.J. and G.C. Fralick (1995), "Multiwire thermocouples in reversing flow," Rev. Sci. Instrum. 66 (10) 5050-5054.
[6] Wang W L, Zhang H, Du X J, et al. PCB-integrated thin film thermocouples for transient temperature measurement[J]. Electronics Letters, 2016, 52(13):1140-1141
[7] Choi H, Li X. Experimental Investigations of Laser Micromachining of Nickel Using Thin Film Micro Thermocouples[J]. Journal of Manufacturing Science & Engineering, 2008, 130(130):021002.

IOP Conf. Series: Materials Science and Engineering **799** (2020) 012005 doi:10.1088/1757-899X/799/1/012005

Overview of the Development of Rotary Steerable Systems

Fei Li*, **Xueying Ma**

School of Electronic Engineering, Xi'an Shiyou University

*Corresponding author's e-mail: lif@xsyu.edu.cn

Abstract: Rotary steerable system remains the development focus of current drilling technology. It is essential for China's unconventional Oil and Gas development. This paper outlines the research and development history, classification, development status of the rotary steerable system, and summarizes the advantages and disadvantages of different systems. Moreover, the prospects of the intelligent RSS technology is explored, and the outlook for the future development of the steerable drilling technology is presented.

1. Introduction

With the continuous decline of conventional oil and gas production, exploration and development have gradually turned to unconventional, offshore and deepwater operations and became increasingly difficult; as a result, demands for improved technologies and equipment remain high. Unconventional oil and gas exploration often uses complicated trajectory wells such as Extended Reach Drilling (ERD), multi-lateral horizontal, and sidetracking wells. More specifically, shale gas drilling often utilizes horizontal drilling, rotary steerable systems (RSS), and other technologies.

Rotary steerable systems are widely used in drilling applications. Three different mechanisms are currently used by rotary systems to control the direction in which the bit drills: push-the-bit [1]–[3], point-the-bit [4], and hybrid steering tools [5], [6]. The point-the-bit RSS tools have better navigation control in the situations where the hydraulic limitation is present because of high mud weight in deep gas drilling environment or in soft formations where other technologies have difficulties in achieving planned trajectories.

This paper aims to present an overview of current RSS technologies and provide insights into the future RSS development in China.

2. Current Status of RSS

RSS tools can be classified into Push-the-bit and Point-the-bit according to the action of the steering mechanism, and the hybrid mechanism uses both push and point technologies. Moreover, these two steering technologies can be further divided into mechanical and hydraulic according to the offset mechanism. As Figure 1 shows, the RSS technologies can be categorized with different characteristic features such as non-rotating and strapdown, dynamic and static, and internal push and external push force.

IOP Conf. Series: Materials Science and Engineering **799** (2020) 012005 doi:10.1088/1757-899X/799/1/012005

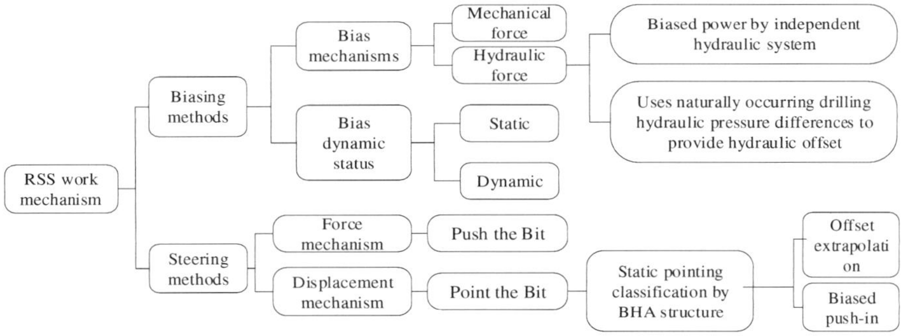

Figure 1. Classification of RSS tools

At present, foreign oilfield service companies such as Schlumberger, Baker Hughes, Halliburton, and Weatherford have their RSS tools, which have been widely used in oilfield services. Schlumberger's RSS PowerDrive system dominates over 1/3 of the market. Other representative tools include Baker Hughes' Auto Trak X-treme system and Halliburton's GeoPilot system. Table 1 provides the details of the classification of the RSS technologies.

Table 1. Classification of the RSS technologies

Company	RSS	Push-the-bit	Point-the-bit
Schlumberger	Archer	✓	✓
	Xceed		✓
	PD X6	✓	
	ICE	✓	
Haliburton	Geopilot		✓
Baker Hughes	Autotrac	✓	
Weatherford	Revolution		✓
	Magnus	✓	
NOV	Vector	✓	
Terravici	Terravici		✓
Gyrodata	Well Guide		✓
APS	SureSteer	✓	
Sanvean	Scott	✓	

Some Chinese research institutions and oilfield service companies also started independent research and development and made some technological breakthroughs in the same period. However, compared with the level of foreign service companies, there is still a significant gap. So far, there is no Chinese RSS tool, which has reached the commercial service level at a worldwide level yet. The research work on RSS in China started in the 1990s, and many research institutions and universities have been involved in the research and development of this technology. With the support of the national "863" project, the research institutes of the China Petroleum & Chemical Corporation Shengli Petroleum Administration Drilling Institute, China National Offshore Oil Corporation, and other scientific research institutions and related colleges and universities have conducted systematic research on RSS systems. Some breakthroughs have been made in the technology. Prototypes were developed, and field tests were conducted; however, the efforts are still required to improve the reliability of the tools for commercial services.

3. Comparison of RSS tools

At present, foreign oilfield service companies such as Schlumberger, Baker Hughes, Halliburton, and Weatherford, have their RSS technologies. A comparison of these RSS technologies is given in Table 2.

IOP Conf. Series: Materials Science and Engineering **799** (2020) 012005 doi:10.1088/1757-899X/799/1/012005

Table 2. Comparison of advantages and disadvantages of different RSS technologies

Type of RSS	Name of RSS	Advantages	Disadvantages
Static push the Bit	Auto Trak RCLS	(1) For high-angle directional wells, ultra-large displacement directional wells, and highly challenging horizontal wells, drilling costs are relatively low. (2) Less harmful vibration and harmful heat generated during drilling, and longer tool life. (3) Real-time direction control accuracy is high and reliable.	(1) In soft formations or wells with enlarged holes, lack of adequate support, the ribs are prone to stick; (2) Due to the significant friction, challenging to transfer weight on bit. (3) The miniaturization is impossible, the structure is complicated, and DLS is relative low.
Dynamic push the Bit	PowerDrive X6	(1) As the component rotates, the friction is reduced, and the problem of WOB transfer is improved; (2) The structure is relatively simple; (3) Good directional performance and high drilling efficiency.	(1) Wears on drill bits and its bearing are severe, resulting in a short working life of the drill tool; (2) When the drilling hydraulic pressure difference is small, it will cause the problem of insufficient thrust. (3) During the drilling process, the ribs will generate harmful heat due to dynamic friction while maintaining the extrapolation.
Static bias extrapola tion pointing Static biased push-in pointing	Well-Guide RSS Revolution Geo-Pilot Power Drive-Direct	(1) The bending degree of the rotating mandrel can be adjusted to obtain the expected build rate; (2) Since it is not necessary to push the ribs against the wellbore to change the direction, its guiding effect is not affected by the formation environment; (3) More regular holes can be drilled, reducing spirals and protrusions.	(1) Since the rotating mandrel is subject to high strength alternating stress, the mandrel is prone to fatigue damage; (2) Because of its many rotating parts and high requirements for component resistance, the manufacturing cost and drilling cost of drilling tools are relatively high.
Dynamic point the bit	Still experimental, not yet in production	(1) The number of downhole motors can be reduced to a certain extent, and a more significant weight on the bit can be used; (2) Since there is no rest point on the wellbore wall, the wear rate of the biasing mechanism can be reduced; (3) Low friction and torque, high mechanical drilling speed and the lateral load on the drill and its bearings are small.	(1) The drilling cycle is long, the structure is complicated, and the production cost of drilling tools is high; (2) Because the dynamic biasing mechanism is to work continuously, the service life of the drilling tool is high.

Table 3 presents the comparison of different 4.75 inch RSS tools with the high DLS capacity. 4.75 inch RSS tools are of smallest OD but highest DLS in drilling application. Among all the tools, Archer from Schlumberger can deliver 18 deg/100 ft DLS, which is the highest. The eXact tool from Baker Hughes is tailored for high-temperatures and high-pressure applications. The sensor distances from bits are between 5.9 and 11.25 ft. Most of these tools can rotate up to 250 rpm. Figure 2 shows the performance of these tools in a pentagon view.

IOP Conf. Series: Materials Science and Engineering **799** (2020) 012005 doi:10.1088/1757-899X/799/1/012005

Table 3. Comparison of different 4.75 inch RSS with the high DLS capacity

Company	RSS Series	Max DLS (°/100ft)	Maximum temp (degC))	Max pressure (psi)	Sensor distance (ft)	Maximu m RPM (rpm)
Baker Hughes	4.75" AutoTrak eXact	12	165	30,000	5.9	250
Gyrodata E	Well-Guide 4-100	12	150	20,000	11.25	250
Halliburton	Geo-Pilot XL RSS 5200Series	14	150	20,000	10	250
Schlumberger	PowerDrive Archer 475	18	150	20,000	8.2	250
Weatherford	Revolution 475 RSS	10	150	25,000	9	200

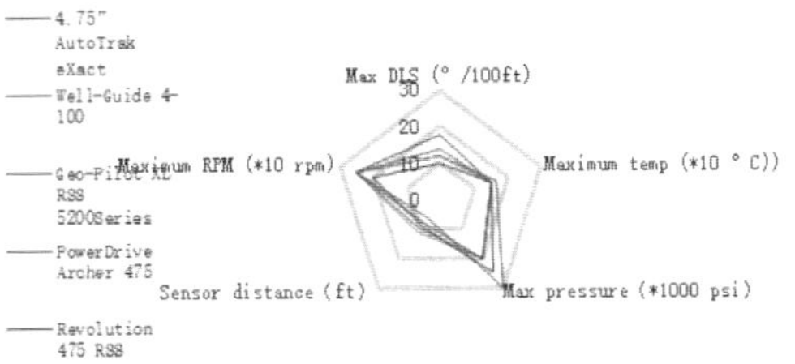

Figure 2. Comparison of performance indicators of 4.75 inch RSS tools

At present, RSS tools are essential to oil and gas drilling activities; however, only a few companies have RSS technologies. Therefore, in order to maintain and expand the current international market leader position, major oilfield service companies continue improving efficiency and market shares. The market share of several major oil service companies in the directional drilling market is shown in Figure 3. Schlumberger takes around 1/3 of the market, and the 2nd place is Haliburton, with 17% market share.

Schlumberger owns push-the-bit, point-the-bit, and hybrid RSS technologies. Figure 4 provides a comparison of the technical specification of different RSS families from Schlumberger.

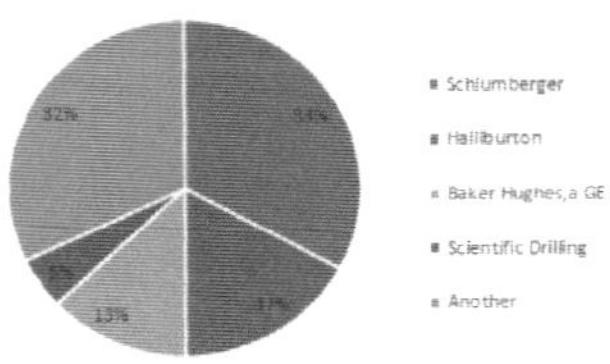

Figure 3. Comparison of the market shares of rotary steering tools by oil service companies

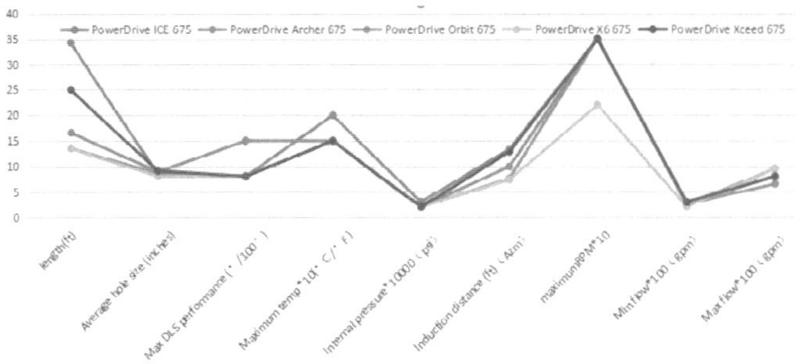

Figure 4. Comparison of the technical specification of different 4.75 inch RSS from Schlumberger

4. Prospects of RSS Technology

The prospects of RSS developments are forecasted as below.

4.1. Near-bit measurement technology.

In the current oil extraction process, the measurement of bit parameters directly affects the optimization of the drilling process. Therefore, near-bit parameter measurement technology is an essential direction for the development of new and automated drilling technologies. This technology installs a sensing device inside the drill bit and transmits the geological data in the formation of the drill bit to a control unit, which might reside in RSS tools. The combination of the near-bit measurement and RSS technologies can achieve the purpose of optimizing the drilling path.

4.2. Smart drill pipe technology.

In the actual oil extraction process, in order to ensure the safety and stability of the drilling process, relevant personnel must monitor the real-time data during drilling, which requires a higher data transmission rate. Therefore, in order to ensure the overall rate and stability of data transmission, to provide strong support for improving drilling data analysis capabilities, to meet drilling data analysis, and to ensure productive drilling data, the development and application of intelligent drill pipe technology are extremely important. With higher data transmission, more data of status and surveys from RSS tools can be seen on the surface; surface crews will be able to make an appropriate decision based on data analysis.

4.3. High Dogleg Severity RSS

High DLS RSS tools are of significance for unconventional oil and gas drilling. With higher DLS, RSS tools can shorten curve sections and have a higher exposure path of the reservoir. Foreign oilfield service companies have owned the technology; China still needs to invest in the development of high DLS RSS tools.

5. Conclusion

In summary, China has made considerable development and progress in RSS technology. However, there is still a significant gap compared with foreign technologies after years of research and development efforts by researchers in China. Chinese oil and gas companies heavily depend on imported foreign products.

This paper reviewed the development status of current RSS tools and future trends. The review of the RSS can improve the understanding of the technologies and promotes the future innovation of researchers and engineers.

References

[1] J. D. Barr, J. M. Clegg, and M. K. Russell, "Steerable Rotary Drilling With an Experimental System," in *SPE/IADC Drilling Conference*, Amsterdam, Netherlands, 1995, doi:

10.2118/29382-MS.
[2] G. C. Downton and D. Carrington, "Rotary Steerable Drilling System for the 6-in Hole," in *SPE/IADC Drilling Conference*, Amsterdam, Netherlands, 2003, doi: 10.2118/79922-MS.
[3] R. Meli *et al.*, "Integrated BHA System Drills Curve/Lateral in One Run at Record ROP Saving Seven Days Rig Time," in *IADC/SPE Drilling Conference and Exhibition*, Fort Worth, Texas, USA, 2014, doi: 10.2118/167920-MS.
[4] S. Schaaf, C. R. Mallary, and D. Pafitis, "Point-the-Bit Rotary Steerable System: Theory and Field Results," in *SPE Annual Technical Conference and Exhibition*, Dallas, Texas, 2000, doi: 10.2118/63247-MS.
[5] S. K. Andersen, K. M. Winterstø, M. Eidem, and I. Robson, "Development and Implementation of a Slimhole Very-High-Dogleg Through-Tubing Rotary-Steerable Drilling System," in *SPE Annual Technical Conference and Exhibition*, New Orleans, Louisiana, 2009, doi: 10.2118/124674-MS.
[6] H. H. Bryan, J. Cox, D. Blackwell, F. W. Slayden, and S. Naganathan, "High-Dogleg Rotary-Steerable Systems: A Step Change in Drilling Process," in *SPE Annual Technical Conference and Exhibition*, New Orleans, Louisiana, 2009, doi: 10.2118/124498-MS.

IOP Conf. Series: Materials Science and Engineering **799** (2020) 012006 doi:10.1088/1757-899X/799/1/012006

A novel coplanar system with a minimal anisotropic effect

Miaoyu Zhang [1,2*], Jie Wu [1,2] and Liu Yanping [1,2]

[1]School of Electronic Engineering, Xian Shiyou University, Xian, Shaanxi, 710065, China

[2]Key Laboratory of Shaanxi Province for Gas-Oil Logging Technology, Xian Shiyou University, Xian, Shaanxi, 710065, China

[*]Corresponding author's e-mail: myzhang@xsyu.edu.cn

Abstract. To obtain the response of the novel coplanar system, we use an analytic method to solve horizontal component of electromagnetic field in the anisotropic stratigraphic model by introducing the magnetic Hertz vector and the magnetic Hertz scalar from Maxwell equations. This paper examines anisotropic response characteristics between novel coplanar system and traditional coplanar system. The negative response area of the proposed coplanar system is decreased. Compared this with traditional coplanar system, the anisotropic effects have the same trend; the affected amplitude of novel coplanar system is far less than that of the traditional coplanar system. The results of experiments demonstrate the superiority of the proposed coplanar system.

1. Introduction

With the development of fine oil and gas exploration, the obviously electrical anisotropic reservoirs, such as conglomerate formation, thin layer, caving and fracturing types, have become an important exploration direction. At the same time, it is urgent resource to be developed for enhancing or stabilizing well productivity at late period of high water cut stage. The conventional measurement techniques and interpretation theories are based on isotropic assumptions of formation conductivity. It is easy to miss such reservoirs so that much energy is wasted. The tri-axial array induction logging is an effective method to measure the conductivity of sandstone and mudstone and can obtain the horizontal and vertical resistivity. However, the response of coplanar coil system has entirely different characteristics from that of coaxial coil system; the response of coplanar coil system is often more complex and more sensitive to the change in drilling liquid conductivity, invasion zones conductivity, formation conductivity, and formation anisotropy. These parameters may even cause sign flip in the response under many conditions, thereby causing negative response at the bed boundary and seriously nonlinear. Zhang et al. studied the multi-component induction logging's response in dipping well anisotropic model, which shows that the cross-components are influenced by the dipping anisotropy[1]. Wei et al. and Ertan Peksen used different ways to obtain analytical expressions in cylindrically stratified anisotropic media respectively, which shows the same response of coplanar system may be produced by entirely different formation conductivities in isotropic or anisotropic formations[2][3]. These will become much complicated to interpret or process the measured data. The present paper relates to a new coplanar subarray configuration, which consists of two transmitters and one receiver coil[4]. To obtain the response of the novel coplanar system, we use an analytic method to solve horizontal component of electromagnetic field in the anisotropic stratigraphic model by introducing the magnetic Hertz vector and the magnetic Hertz scalar from Maxwell equations. Meanwhile, we identify and analyse the advantages of the novel coplanar system in its ability to reduce an anisotropic effect.

2. Anisotropy of the novel Coplanar EMI systems

The novel coplanar coil system adopts a configuration shown in Figure.1, which consists of two transmitters and one receiver coil. The overall coplanar system also consists of 8 subarrays, but all of subarrays share one receiver coil [5]. Table 1 lists parameters for each subarray. Consider a borehole electromagnetic induction system equipped with transverse coil located on the origin, the transverse coil can be equivalent to a magnetic dipole source $\boldsymbol{M}$ because its radius is much smaller than its distance from the receiver coil when an alternating current passes through it during induction logging [6]. Figure.2 shows a transverse magnetic dipole on the origin with conductivity of the horizontal and vertical denoted by σ_h and σ_v respectively.

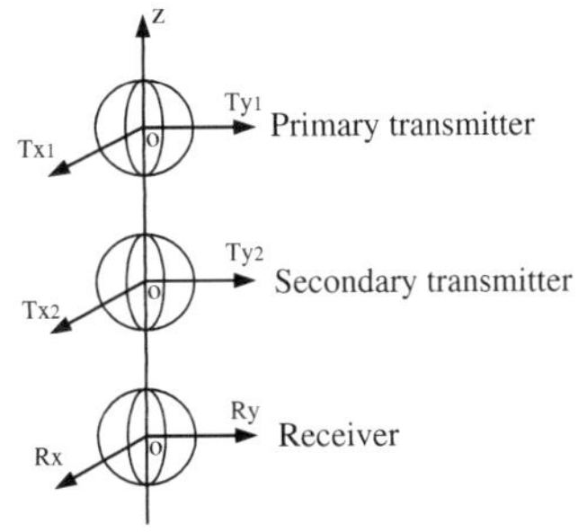

Figure 1. Configurations of a novel coplanar array.

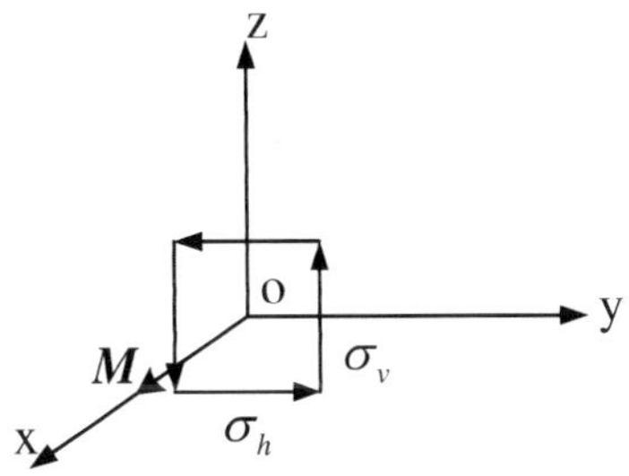

Figure 2. Horizontal magnetic dipole on the borehole axis.

Table 1. Parameters of subarray in the novel coplanar system

Subarray	Primary transmitter		Secondary transmitter	
	L/m	Turns	L/m	Turns
1	0.1500	50	0.0750	50
2	0.2250	62	0.1125	62
3	0.3000	74	0.1500	74
4	0.3750	40	0.1875	40
5	0.5250	34	0.2625	34
6	0.6750	80	0.3375	80
7	0.9750	124	0.4875	124
8	1.8000	124	0.9000	124

The electrical anisotropy means the different conductivity of the formation in different directions, including horizontal resistivity in the horizontal direction (its reciprocal is the horizontal conductivity σ_h) and vertical resistivity (its reciprocal is the horizontal conductivity σ_v) in the vertical direction[7]. It is proved in the published classical literature that the magnetic moment consisting of a Hertz vector of a dipole emitter has the form of $\boldsymbol{M} = \left(M_x, M_y, M_z\right)$[8]. Its magnetic strength is a time-harmonic field generated by a transmitting coil with a magnetic moment $\boldsymbol{M}$. Neglecting the influence of other interfering factors and supposing the formation is homogeneous, infinite and anisotropic, the conductivity tensor can be expressed as[9]

$$\sigma = \begin{bmatrix} \sigma_v & 0 & 0 \\ 0 & \sigma_v & 0 \\ 0 & 0 & \sigma_h \end{bmatrix} \quad (1)$$

The magnetic moment of the horizontal Hertz vectors are[10]

IOP Conf. Series: Materials Science and Engineering **799** (2020) 012006 doi:10.1088/1757-899X/799/1/012006

$$M_x = M \quad , \quad M_y = M_z = 0 \tag{2}$$

The Hertz vector of horizontal magnetic dipole can be expressed as[3]

$$\begin{aligned} \Pi_x &= \frac{M_x}{4\pi}\frac{e^{\mathrm{i}k_v s}}{\lambda s} \\ \Pi_y &= 0 \\ \Pi_z &= \frac{M_x}{4\pi}\frac{x}{\rho^2}\left(\lambda z\frac{\mathrm{e}^{\mathrm{i}k_v s}}{s} - z\frac{\mathrm{e}^{\mathrm{i}k_h r}}{r}\right) \end{aligned} \tag{3}$$

The Hertz scalar of horizontal magnetic dipole is[11]

$$\Phi_x = \frac{M_x}{4\pi}\frac{\mathrm{i}k_h x}{\rho^2}\left[\mathrm{e}^{\mathrm{i}k_v s} - \mathrm{e}^{\mathrm{i}k_h r} + \frac{\rho^2}{r^2}\left(1 - \frac{1}{\mathrm{i}k_h r}\right)\mathrm{e}^{\mathrm{i}k_h r}\right] \tag{4}$$

where $\rho = \sqrt{x^2 + y^2}$, $s = \sqrt{\rho^2 + \lambda^2 z^2}$, $\lambda^2 = \frac{\sigma_h}{\sigma_v}$, $r = \sqrt{\rho^2 + z^2}$, $k_h^2 = i\omega\mu\sigma_h$, $k_v^2 = i\omega\mu\sigma_v$, $\mu = 4\pi \times 10^{-7}$ H/m.

From Maxwell' equation, we know[12]

$$\boldsymbol{H} = i\omega\mu_0\sigma_h \boldsymbol{\Pi}_{\mathbf{m}} + \nabla\Phi \tag{5}$$

Substitute equation (3) into equation (5), the equation (6) is obtained

$$\begin{bmatrix} H_{xx} \\ H_{xy} \\ H_{xz} \end{bmatrix} = \begin{bmatrix} \mathrm{i}\omega\mu_0\sigma_h\Pi_{xx} \\ \mathrm{i}\omega\mu_0\sigma_h\Pi_{xy} \\ \mathrm{i}\omega\mu_0\sigma_h\Pi_{xz} \end{bmatrix} + \begin{bmatrix} \frac{\partial\Phi_x}{\partial x} \\ \frac{\partial\Phi_x}{\partial y} \\ \frac{\partial\Phi_x}{\partial z} \end{bmatrix} \tag{6}$$

Through proper simplification, the following equations are obtained

$$\begin{aligned} H_{\mathrm{xx}} &= \frac{\mathrm{e}^{\mathrm{i}k_v s}}{4\pi}\left[\frac{\mathrm{i}\omega\mu_0\sigma_h}{\lambda s} + \frac{\mathrm{i}k_h s - k_h k_v x^2}{s\rho^2} - \frac{2\mathrm{i}k_h x^2}{\rho^4}\right] \\ &- \frac{\mathrm{e}^{\mathrm{i}k_h r}}{4\pi}\left[\frac{\mathrm{i}k_h r - k_h^2 x^2}{r\rho^2} - \frac{2\mathrm{i}k_h x^2}{\rho^4} - \frac{\mathrm{i}k_h}{r^2} + \frac{\left(k_h^2 x^2 + 1\right)}{r^3} + \frac{3\mathrm{i}k_h x^2}{r^4} - \frac{3x^2}{r^5}\right] \end{aligned} \tag{7}$$

$$H_{\mathrm{xy}} = \frac{\mathrm{e}^{\mathrm{i}k_v s} xy}{4\pi\rho^2}\left[\frac{-k_h k_v}{s} - \frac{2\mathrm{i}k_h}{\rho^2}\right] - \frac{\mathrm{e}^{\mathrm{i}k_h r} xy}{4\pi}\left[\frac{-k_h^2}{r\rho^2} - \frac{2\mathrm{i}k_h}{\rho^4} + \frac{k_h^2}{r^3} + \frac{3\mathrm{i}k_h}{r^4} - \frac{3}{r^5}\right] \tag{8}$$

$$H_{\mathrm{xz}} = -\frac{xz\mathrm{e}^{\mathrm{i}k_h r}}{4\pi r^3}\left[k_h^2 + \frac{3\mathrm{i}k_h}{r} - \frac{3}{r^2}\right] \tag{9}$$

The induced electromotive force in the receiver is obtained by applying Faraday's electromagnetic induction law.

$$V = i\omega\mu(N_R A_R H_{Tx1R} + N_R A_R H_{Tx2R}) \tag{10}$$

After calibration, we get the formation apparent conductivity [13].

$$\sigma_a = -\frac{V - V_m}{K} \tag{11}$$

where V_m is the direct coupled electromotive force, K is instrument constant.

3. Results and discussion

The response characteristics are investigated when the horizontal resistivity is $4.0\mathrm{S/m}$ ($\sigma_h = \dfrac{1}{R_h} = 0.25\mathrm{S/m}$)and the anisotropic coefficients are 1.0,1.2,1.5,2,2.5,3.0,3.5,4.0,4.5,5.0 respectively. Figure.3 and Figure.4 are the vertical conductivity influence for the two coplanar systems when horizontal conductivity is fixed and the instrument inclination angle are 0°, 90° respectively.

The apparent conductivity of each subarray increases with the increase of the vertical conductivity for the two coplanar systems when formation horizontal conductivity is fixed and the instrument inclination angle is 0°, especially the short array is apparent. The influence of anisotropy coefficient on the novel coplanar coil system is the same as that of the traditional coplanar coil system. However, the affected amplitude is slightly smaller than that of traditional coplanar coil systems. The influence of anisotropy on both coplanar systems is weakened when formation horizontal conductivity is fixed and the instrument inclination angle is 90°. The influence of anisotropy decreases when the conductivity is high. It's not affected by anisotropy when $\sigma_v = 0.1\mathrm{S/m}$ for the traditional coplanar system and $\sigma_v = 0.2\mathrm{S/m}$ for the novel coplanar system. The longest subarray 8 of the traditional coplanar system almost unaffected by the anisotropy coefficient because of instrument rod and shield coil.

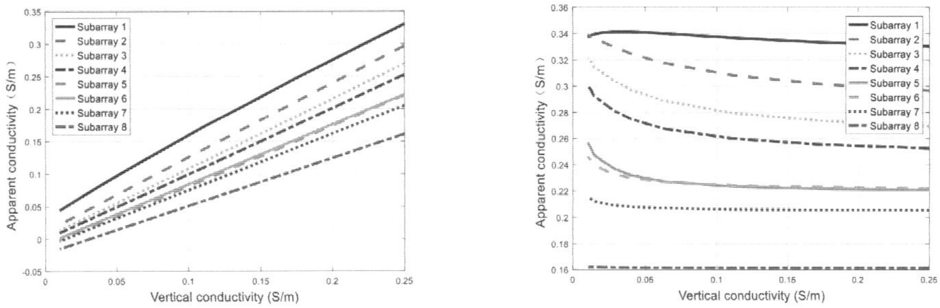

(a) Instrument inclination is 0° (b) Instrument inclination is 90°

Figure 3. Vertical conductivity influence for the traditional coplanar system when formation horizontal conductivity is fixed.

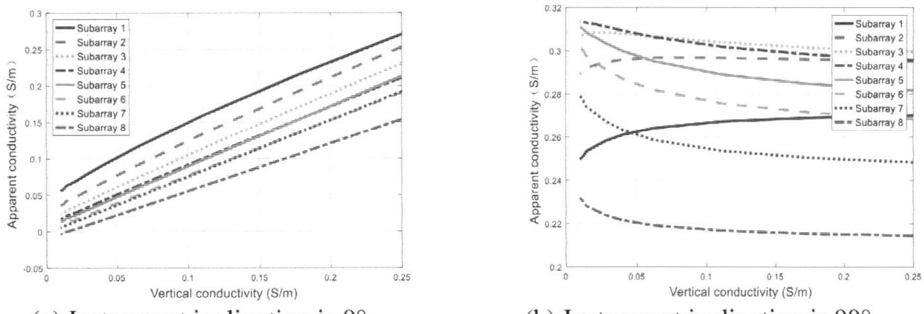

(a) Instrument inclination is 0°. (b) Instrument inclination is 90°.

Figure 4. Vertical conductivity influence for the novel coplanar system when formation horizontal conductivity is fixed.

The response characteristics are investigated when the vertical resistivity is $4.0\mathrm{S/m}$ ($\sigma_v = \dfrac{1}{R_v} = 0.25\mathrm{S/m}$) and the anisotropic coefficients are 1.0,1.2,1.5,2,2.5,3.0,3.5,4.0,4.5,5.0 respectively. Figure.5 and Figure.6 are the horizontal conductivity influence for the two coplanar systems when vertical conductivity is fixed and the instrument inclination angle are 0°, 90° respectively.

Figure.5 and Figure.6 show comparison of apparent conductivity for the two coplanar systems. Note that the coplanar systems are still affected by the horizontal conductivity. It is found that there are eddy currents along the vertical direction and a new horizontal direction by decomposition of coplanar transmitting coil vortex. Only the apparent conductivity of subarray 1 increases with the increase of horizontal conductivity and the remaining subarrays are reduced nonlinearly for the traditional coplanar system. However, the long arrays (subarray 5,6,7 and 8) of the novel coplanar system show a trend toward reduction. Comparing with the traditional coplanar system, the apparent conductivity of the novel coplanar system is higher than that of the traditional coplanar system. The influence of horizontal conductivity is less than that of traditional coplanar coil systems for long arrays.

The apparent conductivity increases nonlinearly with the increase of the horizontal conductivity. It also shows a nonlinear decreasing trend affected by skin effect for long arrays, even there exist negative values for the subarray 8 of the traditional coplanar system. However, the apparent conductivity of each subarray increases for the novel coplanar coil system, there is no decreasing and the anisotropy is linear.

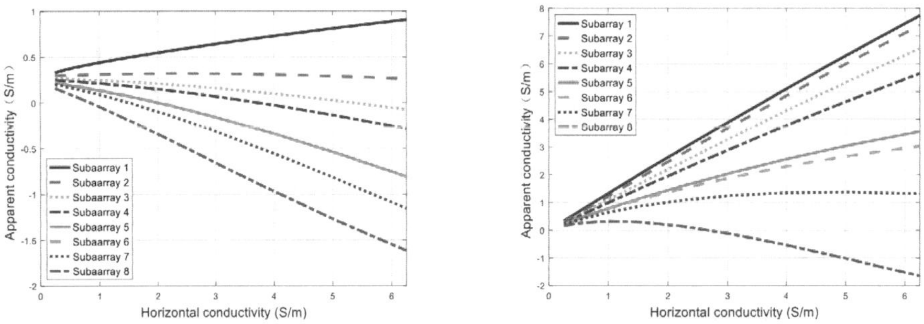

(a) Instrument inclination is 0°. (b) Instrument inclination is 90°.

Figure.5 Horizontal conductivity influence for the traditional coplanar system when formation vertical conductivity is fixed.

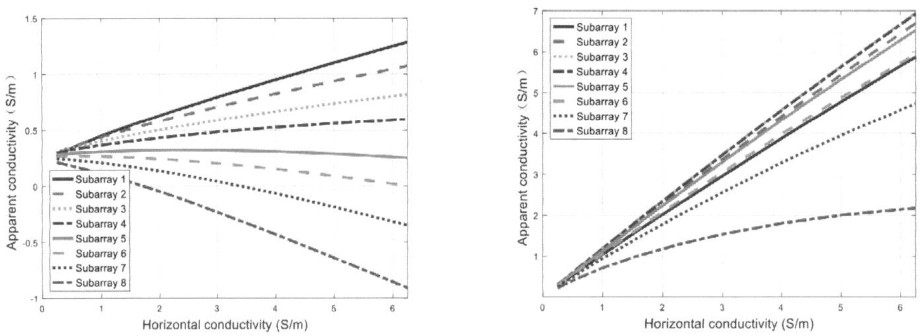

(a) Instrument inclination is 0°. (b) Instrument inclination is 90°.

Figure.6 Horizontal conductivity influence for the novel coplanar system when formation vertical conductivity is fixed.

4. Conclusions

The response characteristics between the novel coplanar system and the traditional coplanar system generally show the same trend, which are affected by both horizontal conductivity and vertical conductivity. The different influence of subarrays also relates to their spacing. Compared with the traditional coplanar system, the effect of the novel coplanar system decreases with anisotropy in vertical well; and there is no negative response in the horizontal well. Simulation experiments demonstrate the superiority of the novel coplanar system.

Acknowledgments
This article is supported by the National Natural Science Foundation of China under Grant No.41704106 and Shaanxi Natural science foundation research under Grant No.2018JM4014.

References
[1] Zhang, Y., Shang, Q.L., Lin, L., et al. (2012) Finite volume simulation of MCIL's response in dipping well anisotropic formation model. Journal of Jinlin University (Earth Science Edition), 42(2): 417-423.
[2] Wei, B.J., Wang, CH. Y., Dang. F., et al. (2016) Simulating response of mult-component induction logging tools by dyadic Green's functions in cylindrically stratified anisotropic media. Journal of China University of Petroleum, 40(2): 59-69.
[3] Peksen, E. (2004) Methods for interpretation of tensor induction well logging in layered anisotropic formations. The University of Utah.
[4] Zhang, M.Y, Guo, B.L. and Wu J. (2018) An improved structure of coplanar coil system and its logging response. Journal of applied geophysics,158: 29-40.
[5] Zhang, M.Y., Guo, B.L. and Wu, J. (2018) Response characteristics of new coplanar coil system in thin beds. Oil geophysical prospecting, 53(3): 578-586.
[6] Wang, H.N., Tao, H.G., Yao J.J.,et.al. (2012) Efficient and reliable simulation of multicomponent induction logging response in horizontally stratified inhomogeneous TI formations by numerical mode matching method. IEEE Transactions on Geoscience and Remote Sensing, 50(9): 3383-3395.
[7] Somlor, S., Hartanto, R.S., Schmitz, A.,et.al. (2015) A novel tri-axial capacitive-type skin sensor. Advances Robotics, 29(21):1375-1391.
[8] Wang, H.N., So, P., Yang SH.W., et.al. (2008) Numerical modeling of multicomponent induction well-logging tools in the cylindrically stratified anisotropic media. IEEE Transactions on Geoscience and Remote Sensing, 46(4): 1134-1147.
[9] Khoshbakht, F.,Rasaie,M.R., Shekarifard, A. (2016) Investigating induction log response in the presence of natural fractures. Journal of Petroleum Science and Engineering, 145:357-369.
[10] Kaufman, A. and Dashevshy Y.A.(2003) Principles of induction logging. Elsevier Publishing, New York.
[11] Beiranvand R. (2014) Magnetic field uniformity of the practical tri-axial Helmholtz coils systems. Review of Scientific Instruments, 85(5):055115.
[12] Li, M.Y., Yue, X.G., Hong, D.CH., et.al. (2015) Simulation and analysis of the symmetrical measurements of a tri-axial induction tool. IEEE Geoscience and Remote Sensing Letters, 12(1):122-124.
[13] Hong, D.CH., Xiao J.Q., Zhong,G.Y., et.al. (2014) Characteristics of the sum of cross-components of triaxial induction logging tool in layered anisotropic formation. IEEE Transactions on Geoscience and Remote Sensing, 52(6):3107-3115.

IOP Conf. Series: Materials Science and Engineering **799** (2020) 012007 doi:10.1088/1757-899X/799/1/012007

Content-based Re-ranking Method for Recommendation

Xin Huang[1,2], Jiameng Bai[1,2], Jia Peng[1] and Min Li[1]

[1]State Key Laboratory of Information Security, Institute of Information Engineering, CAS, Beijing, 100093, China

[2]School of Cyber Security, University of Chinese Academy of Sciences, Beijing, 100049, China

[*]Corresponding author's e-mail: pengjia@iie.ac.cn

Abstract. The recommendation system is one of the effective tools to solve information overload. Most of the current deep learning recommendation algorithms only focus on the accuracy of the recommendation results by learning users' preferences. However, the diversity of the recommendation results is neglected, resulting in homogeneous recommendation results, which reduces users' satisfaction. Diversity can not only effectively avoid over-fitting, but also comprehensively consider multiple dimensions to improve the quality of recommendation results. Therefore, we propose a content-based re-ranking method (CBR) for recommendation systems. The proposed method can make full use of the data set, and effectively supplement and re-rank the recommended results produced by some deep learning algorithms based on the edge information such as tags to be recommended from different perspectives, thus can effectively improve the diversity while preserving the accuracy of recommendation results. Experimental results demonstrate the significant improvements of the proposed re-ranking method.

1. Introduction

With the development of the Internet, people are in an era of information explosion. The recommendation system is one of the most effective tools for solving information overload. It is committed to learning the potential interests of each user by establishing a binary relationship between users and items.

Recommendation algorithms based on deep learning realize the abstract expression of data and the generation of high-level semantic concepts through a multi-layer feature extraction mechanism. They achieve better results than traditional recommendation algorithms in terms of accuracy and recall rate. Jun Wang[1] proposed a two-layer model IRGAN consisting of the generative model and discriminant model, and it performs best. However, like some other recommendation algorithms, the model only uses the scoring matrix for feature modeling, which wastes semantic-rich edge information such as tags, resulting in a low diversity of recommendations. Some algorithms use the tag information to make recommendations. Xu[2] uses DSSM to learn tags and map users and items to the semantic space. Han Xiao[3] and others proposed a two-layer feature generation model SAR, mapping users and items to the same semantic space, and then performing semantic similarity calculation. These two models are slightly worse in terms of accuracy, but they prove that tags can reflect the content of items.

In 2013, Castagos[4] conducted a survey to provide users with a diverse list of recommendations, analyzing users' satisfaction. The study found that the diversity of recommendations significantly improves users' satisfaction with the recommendation system. Deep learning algorithms produce accurate recommendation results, but they usually neglect the diversity of recommendations, thus have the following deficiencies[5-7]. Firstly, they usually use only part of the data information, resulting in low data utilization. Secondly, their recommendation results are more homogeneous, affecting user satisfaction.

To solve these problems, we propose a content-based re-ranking method (CBR). CBR divides the dataset into scoring information and tag information and adds probabilistic models to content-based recommendations to effectively use the tag information. It can make full use of the dataset and make recommendations from multiple perspectives while preserving the high accuracy of given deep learning recommendation results, thus produce more diverse results. Besides, the cold start problem can be solved because there is no requirement for user ratings for content-based recommendations.

In summary, the main contributions of this paper are as follows:

- **Diversity**. We add probabilistic models to content-based recommendations to increase the likelihood of recommending different tags. This method ensures the accuracy of the given results and further improves diversity.
- **Universality**. Our proposed method is based on the existing results, so it can be applied to a variety of deep learning algorithms, which has a certain universality.
- **Usability**. We combine and re-rank deep learning results and content-based results, and set parameter adjustments to make it a realistic recommendation system that is suitable for various situations.

2. Related work

In order to increase the diversity of recommendations, the current main research can be divided into the following three based on the different implementation methods: adding semantic information, using multiple recommendation algorithms, and re-rank the result.

- **Adding semantic information**. Rao Junyang[8] integrated semantic information into the content-based recommendation system, thereby enhancing the relevance between users and items, and calculating the correlation between users and projects to improve the quality.
- **Using multiple recommendation algorithms**. Cao Yi[9] combines content-based filtering and collaborative filtering, and can adopt different recommendation strategies at different stages of **the** recommendation system. Li Zhongjun[10] proposed a new model, the prediction matrix of content-based recommendation is the input of collaborative filtering, thus predicting the user's score on the new project.
- **Re-rank the result**. Adomavicius[11] firstly used collaborative filtering to get the results, secondly filtered the results according to certain strategies. Finally, the results were re-ranked according to the popularity and average score of the items to balance the accuracy and overall diversity. Chen Tianyi[12] used collaborative filtering and content filtering to remove the similar recommendation results to improve the diversity.

At present, a large amount of research has focused on one of the above aspects, and relatively few studies have incorporated it. This paper is dedicated to integrating above methods, while ensuring the accuracy of the recommendation results, and increasing the diversity of recommendations.

3. Content-based re-ranking method

The method consists of three parts, the main structure flow chart is shown in Figure 1. The data set used in this model is movielens. Movielens is split into three parts: movie tags, user rating data, user tags. Different methods are recommended for different data:

- For rating data, the relationship between the user and the movie is modeled by using deep learning models such as IRGAN to obtain the candidate set *RD* for its' powerful feature extraction capabilities.
- For movie tags, firstly pre-processing the tag, and then the cosine similarity is used to calculate the feature vector similarity. In this section, the candidate movie is sampled according to the

weight of the feature and a probabilistic model, so the diversity of the recommendation can be effectively improved.

- For user tags, the potential association between users' intuitive attributes such as age, occupation, and other attributes can be found. Thus, the recommendation according to the users' preference with similar tag characteristics can effectively improve the diversity of the recommendation and bring certain surprise to the user, and can also solve the users' cold start problem.

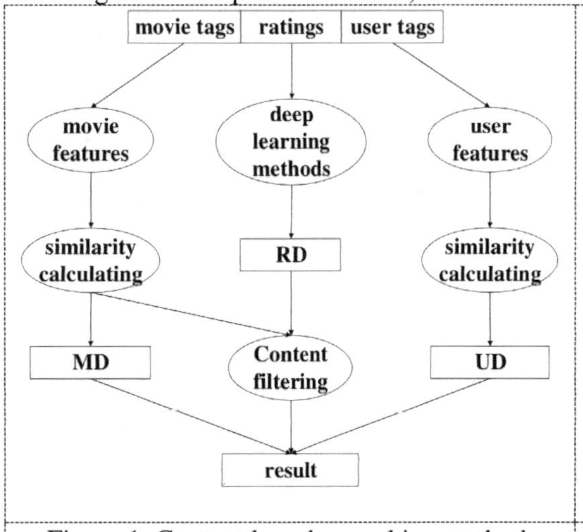

Figure 1. Content-based re-ranking method.

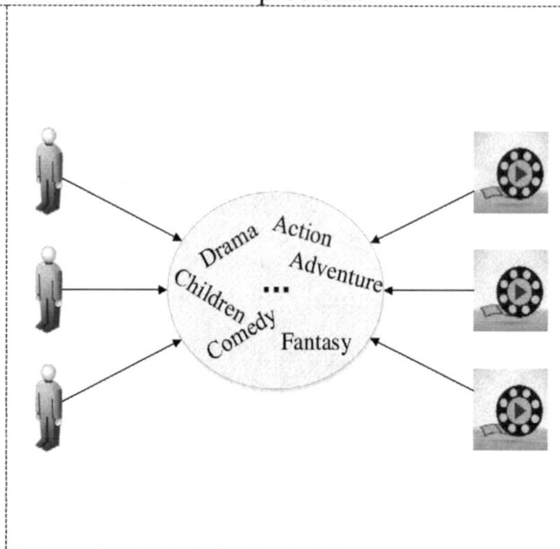

Figure 2. User-movie feature mapping.

3.1. content-based recommendation

The method proposed in this paper combines the content of the movie and the user attribute to recommend. We extract the features of movies and users, then calculates the similarity.

3.1.1. content-based recommendation for movie. It is generally believed that the tags of a movie represent the type of the movie and can reflect the actual content of the movie. Therefore, content-based recommendation for movies needs to be based on the tags of the movie.

Feature construction. In this method, n-gram is used to construct feature words. Unlike the n-gram model in traditional documents, movie tags in movielens are independently scattered and unordered, so the construction features are described by a combination of multiple tags. For example, the tags of "Say Anything (1989)" is "Comedy | Drama | Romance", whose bi-gram are (Comedy Drama), (Comedy Romance) and (Drama Romance). In addition, since there are some movies with only one tag, in this model, if an n-gram is used to construct the feature, it actually contains 1-gram, 2-gram, ..., and n-gram together to form a feature vector and then calculate the weights. This article uses *tf-idf* to calculate the weight of features:

$$tf - idf = tf_{w,d} * idf_w = \frac{n}{N} * log\frac{D}{d_w} \tag{1}$$

Where w represents a term, d represents a document, n represents the number of occurrences of the term w in the document d, and N represents the total number of terms of the document d. D represents the number of documents in the document set, and d_w represents the number of documents in the document set containing the term w.

Feature similarity calculation. Both the user and the movie can be mapped to the feature space, represented by a feature vector. A movie can be directly represented as a feature vector by the weight of the feature. The user will have behavior on the movie, and the tags of the movie that the user has scored highly are used as the feature tag of the user itself, and can also be mapped to the feature space, as shown in Figure 2.

The user can use the feature vector to represent the matrix $U(m\times k)$, a total of m rows and k columns, m represents the number of users, k represents the number of features constructed. Where U_{ij} is expressed as the weight of feature j for user u_i . The movie can be represented as a matrix $M(n\times k)$, n represents the number of movies, k represents the number of features constructed. After obtaining the feature vectors of the user and the movie, the relevance between the user and the movie are calculated using the cosine similarity. As shown in Figure 3, it can be intuitively seen that user u_1 is significantly more relevant to m_1 than movie m_2.

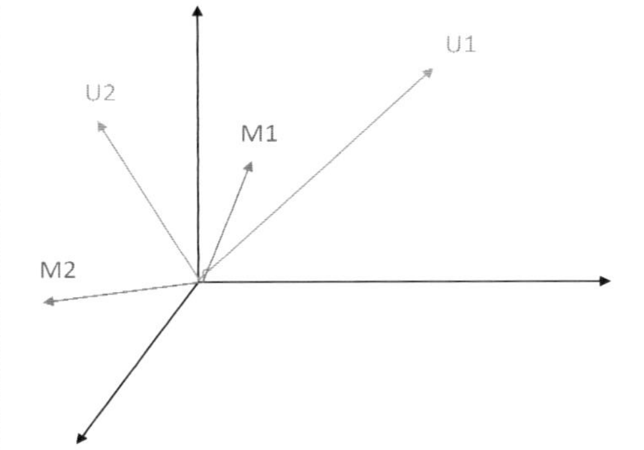

Figure 3. Feature space cosine similarity example.

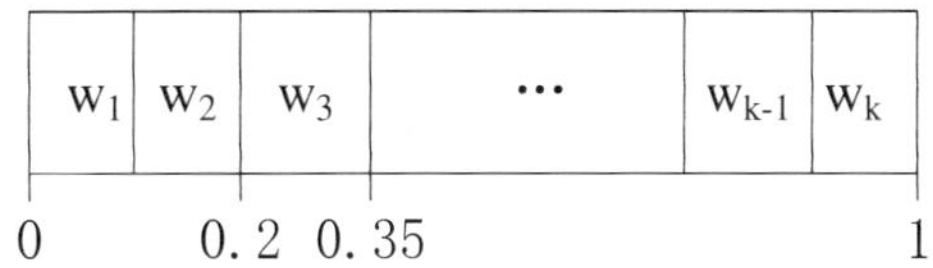

Figure 4. Random selection feature.

For any user u_i and any movie m_j, the correlation degree weight integral rel can be calculated as:

$$rel = log\ cos\ (\boldsymbol{U}_{i*}, M_{j*}) \tag{2}$$

The log is used to prevent the product from being too small when calculating the cosine similarity.

Recommend algorithm. In this method, the main part of the recommendation result set is derived by the deep learning algorithm. The content-based recommendation is mainly to make some supplementary recommendations to *RD* to enhance its diversity. Given *RD*, and all the features of *RD* are regarded as a document, and the times of occurrence of the features is counted. The more the feature appears, the better this feature can represent the user's preference. A lower frequency indicating that the feature does not reflect the user's real interest. In order to improve the diversity of recommendations, it is necessary to punish the features with more occurrences, so that it is not easy to be selected in the process of film content-based recommendation, and vice versa, it should reward the features with fewer occurrences, making it easier to be recommended.

Construct the feature vector $\boldsymbol{r}$ with k columns correspond to k-dimensional features, and each column value corresponds to the frequency at which the feature appears. Then the user feature weight vector $\boldsymbol{w}'$:

$$w_j' = \frac{U_{ij}}{log\ (max\ (e\ ,\ r_i))} \tag{3}$$

Assuming $sum = \sum_{i=1}^{k} w_i'$, normalization can be obtained $\boldsymbol{w} = \frac{w'}{sum}$. Changing the value of $\boldsymbol{w}$ to make the value of $\boldsymbol{w}_i$ is equal to the value of the original $\boldsymbol{w}_i$ plus the value of $\boldsymbol{w}_{i-1}$. So that it maps to a range of 0-1. The calculation process like a Fibonacci sequence:

$$for\ i = 1\ to\ k-1: \quad w_i = w_i + w_{i-1} \tag{4}$$

Given a random number ε_1, wen can select a feature according to $\boldsymbol{w}$. As shown in the Figure 4, if $\varepsilon_1 = 0.3$, then we choose the feature f corresponding to w_3 to recommend. After selecting the feature, we need to select the movie that is relevant to the feature for recommendation. The recommendation process is similar to the feature selection process described above. For feature f, the correlation vector with the movie is $\boldsymbol{M}_{*f}$. Add the probability value of each movie and normalize it, then obtain the weight vector $\boldsymbol{t}$ of the feature-movie. Take a random number between 0 and 1 to get the corresponding movie in $\boldsymbol{t}$ for recommendation. Assuming the random number is ε_2, the recommended procedure is shown in Figure 5.

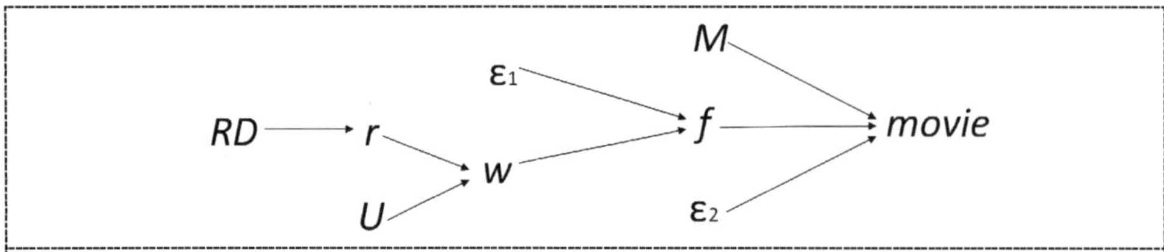

Figure 5. Content-based movie recommendation process

The final recommended movie set *MD* is generated by repeating several times.

3.1.2. *content-based recommendation for user.* Content-bases recommendation for user is to solve the problem of cold start of the user. Besides, it can make full use of the user's tag information to carry out personalized recommendation. As shown in Figure 6, this process is similar to movie content-based recommendation. But there still has two differences.

- User content-based recommendation is based on the user's tag to build features, such as gender, age and other attributes.
- The model obtains the user-tag matrix *U*, U_i resents the feature vector of the user u_i, and can directly calculate the user similar to the target user by the cosine similarity. Then recommending the movie similar to the user, and obtaining the user-recommend result set *UD*.

3.2. Combing and re-ranking

After obtaining *RD*, *MD* and *UD*, as Figure 7 shows, the results need to be combined by a certain strategy to form a final recommendation result set ***result***.

For a recommendation system, the most important thing is to recommend the items that users like, and secondly to recommend as many categories as possible to enhance the user experience. Therefore, *RD* with high accuracy should be placed in the top position in the original order, with a proportion of α. After that, *MD* is placed, and the proportion is β. Finally, the *UD* is placed, and the proportion is γ, and $\alpha + \beta + \gamma = 1$.

For users with more rating records, their preferences are relatively easier to model, so the weight of α should be appropriately increased to obtain better performance. If the user scores less, it is not easy to obtain their preferences, and the weight of α should be appropriately reduced to solve the cold start problem. When β=1, the model is transformed into a full content-based recommendation model. When α=1, the model is transformed into a complete deep learning recommendation model.

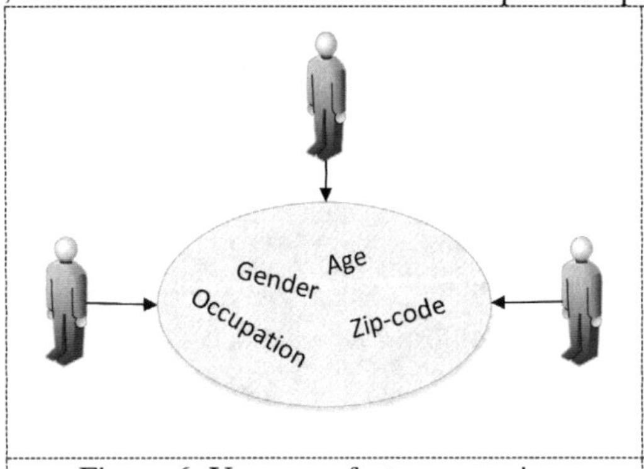

Figure 6. User-user feature mapping.

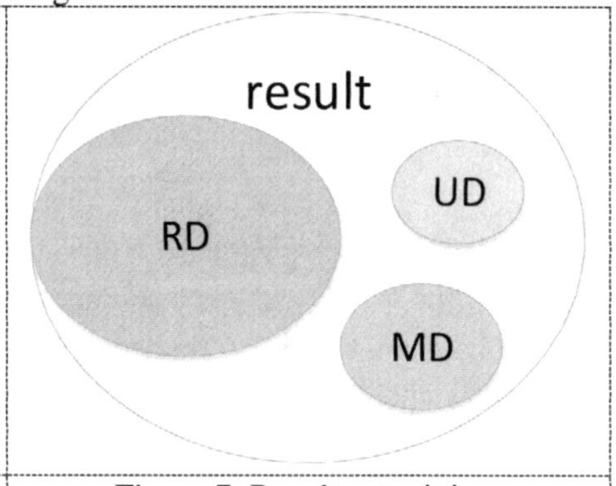

Figure 7. Result set mixing.

4. Experimental result

4.1. Experimental Settings

The experimental environment for this article is configured for Windows 10, Python 3.6.

The data set used movielens 100k containing 943 users' 100,000 ratings for 1,682 movies. In addition, it also includes the title of the movie, as well as the tags of the movie, some attributes of the user, and so on. The evaluation methods used in this paper include the accuracy rate P, the normalized discounted cumulative gain NDCG. At the same time, the number of different categories included in the recommendation results is recorded as "genres", indicating the diversity of recommendations. In this experiment, based on the experimental results, the values of α β and γ are set to be 60%, 20% and 20%. As for n-gram to generate features, the recall rate for content-based recommendations is shown in Table 1.

Table 1. Recall rate for n-gram multiplied by 100.

	1-gram	2-gram	3-gram	4-gram
recall@5	**14.37**	12.26	10.94	10.56
recall@10	**23.53**	19.85	18.12	17.60

Based on the results, 1-gram is used to construct the feature in the actual modeling process.

4.2. Experiments

In this paper, deep learning methods IRGAN and BPR[13] are selected for baselines. Both models use only the ratings data of the movielens data set. CBR method was used to supplement and re-rank the result sets produced by IRGAN and BPR. Accordingly, we name the two relevant model CBR-IRGAN and CBR-BPR. The results of the experiments are as follows.

Table 2. Evaluation results of IRGAN, CBR-IRGAN, BPR, CBR-BPR multiplied by 100

	P@5	P@10	NDCG@5	NDCG@10	genres@5	genres@5
IRGAN	37.17	31.39	40.03	37.18	2	3
CBR-IRGAN	35.74	30.94	38.46	37.03	**4**	**7**
BPR	30.03	26.47	32.21	31.89	3	4
CBR-BPR	29.97	26.64	31.89	31.78	**4**	**7**

Comparing IRGAN and BPR, it can be seen that P and NDCG of IRGAN and BPR decrease obviously with the increase of the recommended number, indicating that the correlation is weakened with a lower ranking. There is still a certain gap between the recommended accuracy of BPR and IRGAN. However, the average number of Genres in BPR is more than that of IRGAN, indicating that IRGAN is more specific and accurate in modeling user preferences, but the recommended types are relatively homogeneous and not diverse.

Comparing CBR-IRGAN and CBR-BPR, the overall performance of the CBR method is closely related to the performance of the deep learning algorithm, so the diversity and accuracy of the recommendation are still inseparable.

After adding the CBR method to the IRGAN and BPR models, it can be seen from the experimental results that the accuracy of the model has a little degree of influence, but the impact gradually decreases as the number of recommendations increases. However, it can be seen that the CBR method has a very significant improvement in the number of types of recommended movies compared to the original model. In the case where the recommended accuracy is not much different, the increase in diversity will greatly enhance user satisfaction. Therefore, the CBR method has broad application prospects.

5. Conclusion and future work

Most of the recommended methods currently available are blindly considering the accuracy of the recommendation, while ignoring the diversity of recommendations. This paper proposes a CBR method that can re-rank the existing results and add content-based personalized recommendation results to enhance the diversity of recommendations. The method can fully utilize the information of the data set, and learn the user's interest from various dimensions, and it can also solve the problem of recommendation over-fitting effectively which has a positive significance.

However, the concept of diversity is still vague at present. Many studies on diversity are subjective, and the actual user demand for diversity is not easy to quantify. And the research on the diversity of recommendations does not form a systematic theory.

CBR can solve certain diversity problems at present, but it still needs further investigation on how to adjust the weighting parameters automatically of different recommendation result sets. At the same time, the recommended accuracy is still insufficient compared with the original recommendation method when the recommended number is small. We leave more theoretical investigation on these problems for future work.

Acknowledgments

This work is partially supported by National Key Research and Development Program of China (No.2019QY1704).

References

[1] Jun Wang, Lantao Yu, Weinan Zhang, Yu Gong, Yinghui Xu, Benyou Wang, Peng Zhang, and Dell Zhang. (2017) IRGAN: A Minimax Game for Unifying Generative and Discriminative Information Retrieval Models. SIGIR. Tokyo. pp. 515-524.

[2] Xu Z, Chen C, Lukasiewicz T, et al. (2016) Tag-aware personalized recommendation using a deep-semantic similarity model with negative sampling. CIKM. Indianapolis. pp. 1921-1924.

[3] Han Xiao, Minlie Huang, Xiaoyan Zhu. (2017) SAR: A Semantic Analysis Approach for Recommendation. ICJAI. Melbourne.

[4] Castagnos, S., Brun, A. and Boyer, A. (2013) When Diversity Is Needed...But Not Expected!. In: Advances in Information Mining and Management. Lisbon. pp. 44-50.

[5] Ho, Y.-C., Chiang, Y.-T. and Hsu, Y.-J. (2014) Who Likes It More? Mining Worth-Recommending Items from Long Tails by Modeling Relative Preference. In: ACM International Conference on Web Search and Data Mining. New York. pp. 253-262.

[6] Lee, K. (2015) Escaping Your Comfort Zone: A Graph-Based Recommender System for Finding Novel Recommendations among Relevant Items. Expert Systems with Applications, 42(10): 4851-4858.

[7] di Noia, T., Ostuni, V., Rosati, J., Tomeo, P. and di Sciascio, E. (2014) An Analysis of Users' Propensity toward Diversity in Recommendations. In: 8th ACM Conference on Recommender Systems. Foster City. pp. 285-288.

[8] Rao Junyang, Jia Aixia, Feng Yansong. (2014) Ontology-based News Personalized Recommendation. Acta Scientiarum Naturalium Universitatis Pekinensis, 50(1): 2-7.

[9] Cao Yi. (2007) Research on a Hybrid Recommendation Model Based on Collaborative Filtering and Content Filtering. Changsha.

[10] Li Zhongjun, Zhou Qihai, Shuai Qinghong. (2009) Recommender System Model Based on Isomorphic Integrated to Content-based and Collaborative Filtering. Computer Science, 36(12): 142-145.

[11] Adomavicius, G. and Kwon, Y. (2012) Improving Aggregate Recommendation Diversity Using Ranking-Based Techniques. In: IEEE Transactions on Knowledge and Data Engineering. 896-911.

[12] Chen Tianhao, Shuai Jianmei, Zhu Ming. (2014) A Film Recommendation Method Based on Collaborative Filtering. Computer Engineering, 40(1): 55-58.

[13] Steffen Rendle, Christoph Freudenthaler, Zeno Gantner, and Lars Schmidt-Thieme. (2009) BPR: Bayesian personalized ranking from implicit feedback. In UAI. pp. 452-461.

IOP Conf. Series: Materials Science and Engineering **799** (2020) 012008 doi:10.1088/1757-899X/799/1/012008

Real-time Monitoring System of Leakage Current for Electric Submersible Pump

Nan Song[1], Min Wang[2*], Bo Dang[3], Shenghu Liu[4], Zhiping Ren[5]

[1]College of Electronic Engineering, Xi'an Shiyou University, Xi'an, Shaanxi,710065, China

[2]College of Electronic Engineering, Xi'an Shiyou University, Xi'an, Shaanxi,710065, China

[3]College of Electronic Engineering, Xi'an Shiyou University, Xi'an, Shaanxi,710065, China

[4]College of Electronic Engineering, Xi'an Shiyou University, Xi'an, Shaanxi,710065, China

[5]College of Electronic Engineering, Xi'an Shiyou University, Xi'an, Shaanxi,710065, China

[*]Corresponding author's e-mail: 182030183@stumail.xsyu.edu.cn

Abstract. With the increasing exploitation of marine gas, electric submersible pump technology has attracted more and more attention. Under the influence of the bad underground environment, the insulation performance of the electric submersible pump will gradually deteriorate during its operation, resulting in current leakage, causing phase-to-phase open circuit, fire or electric leakage to ground and other dangerous phenomena, which brings destructive damage to the electric submersible pump equipment. In order to solve this problem, this paper designs a real-time monitoring system of the leakage current for the electric submersible pump based on the principle of current mutual inductance and transducing, and constructs the indoor well condition simulation experiment. The experimental data shows that the system can monitor and display the current leakage of the electric submersible pump in real time to ensure the safe operation of the electric submersible pump.

1. Introduction

In recent years, with the decrease of easy-to-produce crude oil reserves, the well condition of oil well is more and more complicated. As an important artificial lifting equipment, electric submersible pump plays an irreplaceable role in the secondary and tertiary oil recovery, which puts forward higher requirements for the application reliability of electric submersible pump[1]. However, the insulation of the electric submersible pump is affected by various factors such as electricity, heat, machinery and bad environment during its operation. Its performance will gradually deteriorate[2], resulting in current leakage, causing dangerous phenomena such as phase-to-phase open circuit, fire or electric leakage to ground, which brings destructive damage to the electric submersible pump equipment. At present, in order to ensure the safe use of the electric submersible pump[3], the insulation test of the electric submersible pump system is carried out by megger before going downhole according to the regulations. If the insulation performance is to be tested after going downhole, the whole electric

IOP Conf. Series: Materials Science and Engineering **799** (2020) 012008 doi:10.1088/1757-899X/799/1/012008

submersible pump well must be powered off and stop operating. It is impossible to monitor the insulation damage of electric submersible pump in real time. In order to solve these problems, a real-time monitoring system of leakage current for electric submersible pump is designed based on the existing monitoring system of electric submersible pump and downhole condition. Finally, the well simulation experiment shows that the system can accurately monitor the current leakage of the electric submersible pump in real time, evaluate the insulation damage of the electric submersible pump well correctly, and ensure the safe operation of the electric submersible pump unit.

2. Monitoring system of electric submersible pump and downhole condition

Schematic diagram of monitoring system of electric submersible pump and downhole condition is shown in Figure 1.

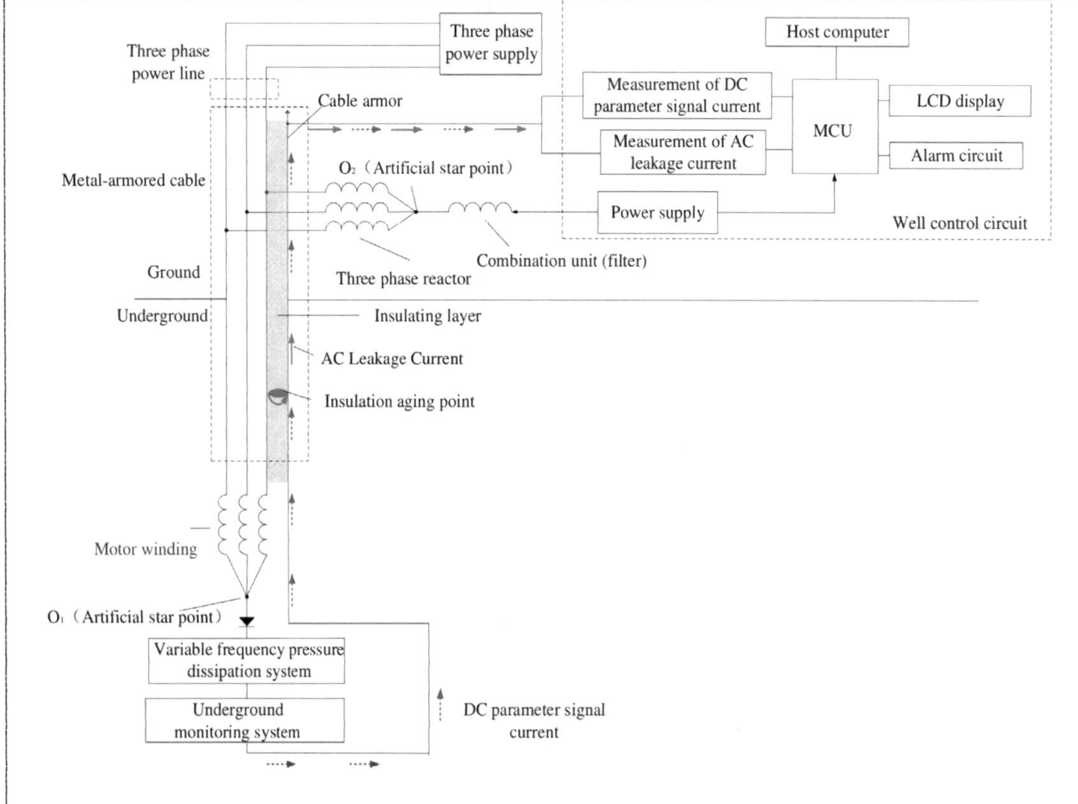

Figure 1. Schematic diagram of monitoring system of electric submersible pump and downhole condition

The ground system is mainly composed of ground artificial star point, combination unit, ground control circuit, etc. The artificial star point is constructed by a three-phase electric reactor with the same three-phase winding parameters as the downhole motor, which supplies power for the downhole monitoring circuit based on the principle of star point equipotential. The underground part is mainly composed of submersible motor part, metal armored cable, frequency conversion and pressure suppression system, and downhole monitoring circuit. The metal armored cable is used as the power supply channel. The three-phase power supply on the ground with variable frequency supplies power to the downhole motor through the three-phase power line of the metal armored cable. The motor is responsible for driving the whole set of electric submersible pump equipment to start operation[3]. The outer side of the three-phase power line is wrapped with an insulating layer to isolate the power and protect the cable. The outermost layer of armored cable is cable armor, which is used as the

IOP Conf. Series: Materials Science and Engineering **799** (2020) 012008 doi:10.1088/1757-899X/799/1/012008

information carrier to transmit the DC signal corresponding to each measured parameter, such as downhole temperature, pressure, vibration, etc. monitored by the downhole monitoring circuit to the ground.

Due to the harsh underground environment, the insulation performance of the electric submersible pump will gradually deteriorate during its operation[3], which is a potential threat to the operation of the electric submersible pump. In view of this, the armor shown in Figure 1 is used as the transmission channel of measurement information to monitor the leakage current that may be generated during the operation of the electric submersible pump in real time on the ground[4].

3. The principle of real-time monitoring for leakage current of electric submersible pump

3.1. Working principle of current mutual inductance transducer

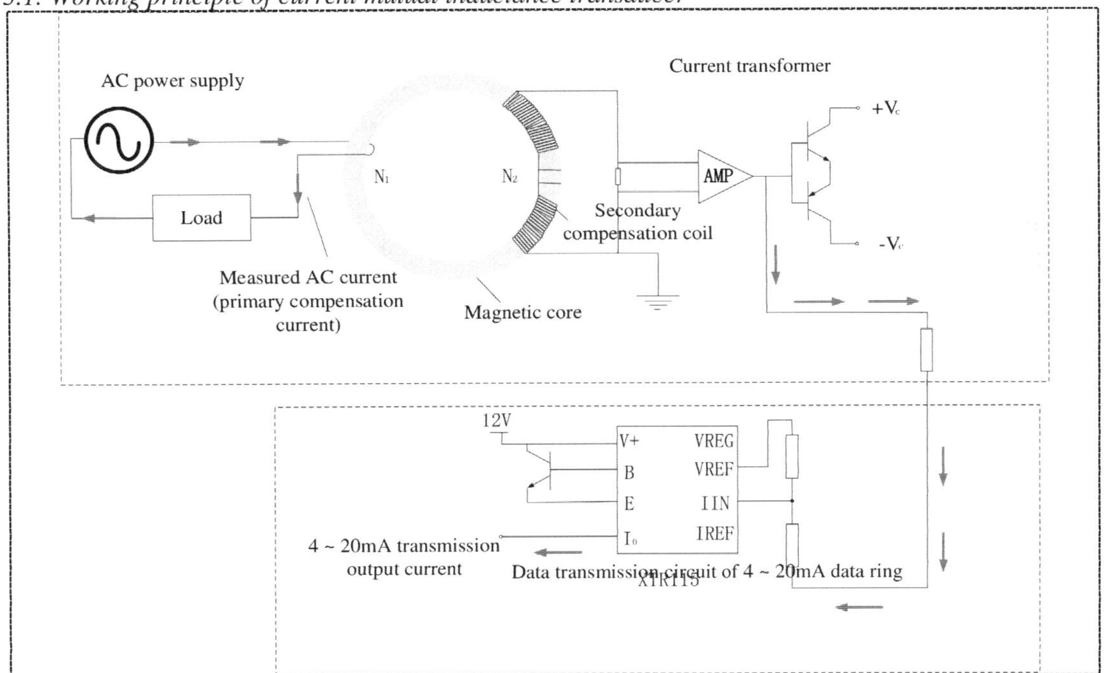

Figure 2. Schematic diagram of current mutual inductance transducer measurement

The current transformer is an instrument which converts the large primary current into the small secondary current according to the electromagnetic induction principle to measure the current. Current transformer is generally used in power system, which plays an important role in current transformation and electrical isolation. It is composed of a closed iron core and two windings of the primary and secondary sides[5]. As shown in Figure 2, when the circuit-under-test passes through the core ring, a magnetic field will be generated around the circuit-under-test, which will generate alternating magnetic flux through electromagnetic induction, and the secondary winding will generate a current decreased by the ratio of turns.

Current transformer is based on the principle of electromagnetic induction, that is, only when the measured current is alternating current, the corresponding current value can be measured. Therefore, the current transformer can only measure the AC current value, which will not be affected by the DC signal transmitted from the downhole by the cable armor. In order to solve the influence of field interference of industrial control[6], this paper selects two-wire current transformer chip XTR115 to transform the current signal received by current transformer into 4 ~ 20mA standard current signal.

In this paper, the selection of the standard rated value of the current mutual inductance transducer refers to the measurement index of leakage current[3] of Phoenix xt150, the electric submersible pump monitoring equipment of Schlumberger, an oilfield technology company, as shown in Table 1.

IOP Conf. Series: Materials Science and Engineering **799** (2020) 012008 doi:10.1088/1757-899X/799/1/012008

Table 1. Measurement parameters and indexes of Phoenix xt150

Measurement parameters	Measuring range	Accuracy	Resolution	Drift	Rate
Inlet pressure	0~5800psi	± 5psi	0.1psi	5psi/year	4s
Outlet pressure	0~5800psi	± 5psi	0.1psi	5psi/year	4s
Inlet temperature	0~150°C	1.33%	0.1°C	--	4s
Winding temperature	0~409°C	1%	0.1°C	--	36s
Vibration	0~30g	3.33%	0.1	--	Variable
Leakage current	0~25mA	0.20%	0.001mA	--	Variable

3.2. Calibration experiment of current mutual inductance transducer

Construct a DC circuit (Loop 1) powered by 5V constant current source and an AC circuit (Loop 2) powered by 12.78V AC source, as shown in Figure 3. Loop 2 includes adjustable load, precision ammeter 1 and precision ammeter 2. Loop 2 can get multiple sets of measurement data by adjusting the adjustable load resistance, precision ammeter 1 can measure current value of Loop 2 in real time, and precision ammeter 2 can measure 4 ~ 20mA standard current signal output by current mutual inductance transducer in real time.

Pass the two circuits through the air core in the middle of the current mutual inductance transducer. At this time, there are both DC current and AC current flowing through the air core of the current mutual inductance transducer. The current data of double circuit experiment under different loads are measured by precision ammeter 1 and 2. If the constant current source is turned off, at this time, only AC current flows through the core of current mutual inductance transducer. The experimental platform is as shown in Figure 3 (b). The current data of AC circuit experimental under different loads is measured by precision ammeter 1 and 2.

Under different loads, the experimental data of double circuit and AC circuit are shown in Table 2. The load of Loop 1 is 510, and the current is always 9.891mA.

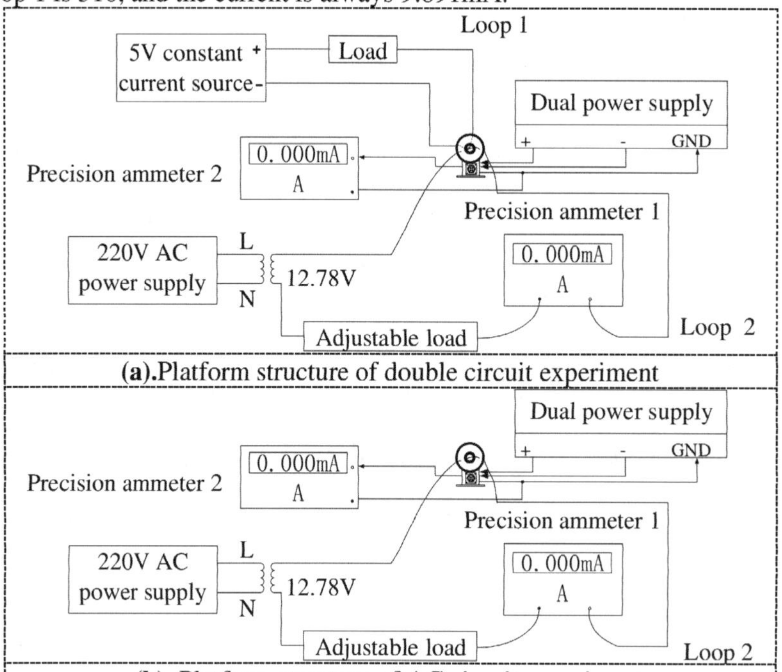

(b). Platform structure of AC circuit experiment

Figure 3 Platform structure of calibration experiment

IOP Conf. Series: Materials Science and Engineering **799** (2020) 012008 doi:10.1088/1757-899X/799/1/012008

Table 2. Experimental data of current mutual inductance transducer under different loads

Loop 2 resistance/Ω	Loop 2 current/mA	Transmission current of loop 2/mA	Double circuit transmission current/mA	relative error/%
∞	0	4.046	4.041	0.12
935	13.766	8.478	8.440	0.45
795	16.179	9.251	9.205	0.49
634	20.300	10.615	10.545	0.66
539	23.847	11.726	11.676	0.42
491	26.181	12.987	12.427	0.45
415	30.988	14.024	14.008	0.11
356	36.090	15.659	15.658	0.006
327	39.288	16.740	16.664	0.45
303	42.441	17.799	17.894	-0.53
271	47.384	19.429	19.453	-0.12
264	48.695	19.802	19.903	-0.05

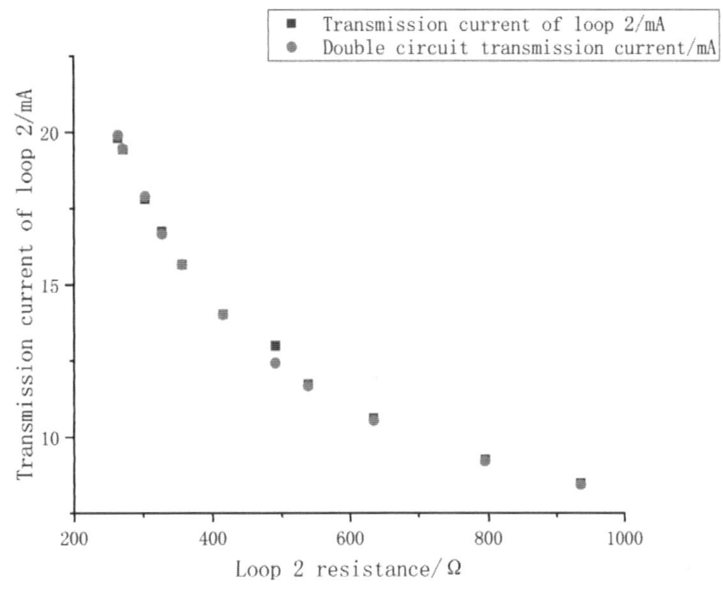

Figure 4. Comparison of current transducing value of AC circuit and double circuit

It can be seen from Figure 4 that the output current value of the current transducer has almost no change when the AC current and the AC&DC current pass through its core. Therefore, the current mutual inductance transducer can only measure the AC current, and the presence of DC in the loop does not affect the measurement results.

4. Real-time monitoring system for leakage current of electric submersible pump based on the principle of current mutual inductance

4.1. System structure

The structure of the real-time monitoring system for the leakage current of the electric submersible pump is shown in Figure 5.

When the electric submersible pump and downhole multi-parameter monitoring system work normally, the downhole monitoring circuit transmits the DC current corresponding to the monitored multi-parameter to the multi-channel AD acquisition module in sequence along the cable armor. At the same time, the cable armor goes through the central hole of the current mutual inductance transducer, so as to monitor the leakage current of the electric submersible pump, then the current will be output as a 4 ~ 20mA standard current signal to the multi-channel AD acquisition module. After being processed by MCU, it will be displayed on LCD screen in real-time, and the corresponding insulation curve is drawn in the upper computer, and the historical data is stored as the basic data to evaluate the working condition of electric submersible pump.

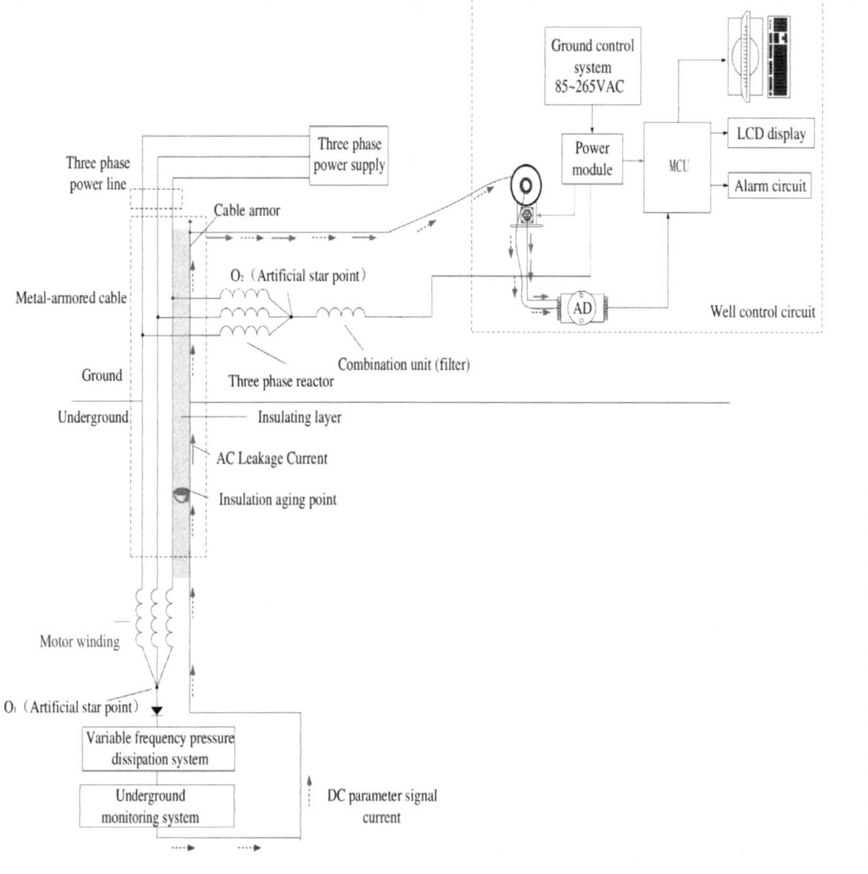

Figure 5. Structure of monitoring system

4.2. Measurement circuit

The structure diagram of hardware circuit system is shown in Figure 6. STM32F103 is used as the main control chip. The system mainly includes current mutual inductance transducer, RS232 communication circuit, multi-channel AD acquisition module, LCD screen, SD memory card, alarm circuit, start button and upper computer.

After the system is powered on, press the start button to start the whole system. The main control chip controls the multi-channel AD acquisition module through the RS232 communication port to receive the current signal transmitted by the downhole equipment and current mutual inductance transducer. The multi-channel AD acquisition module sends the received current signal to the main control chip for processing and to calculate the insulation condition of the downhole electric submersible pump, which will be displayed on the LCD screen and sent to the upper computer at the same time, and the SD card stores the signal data processed by MCU.

IOP Conf. Series: Materials Science and Engineering **799** (2020) 012008 doi:10.1088/1757-899X/799/1/012008

When the monitoring system detects the insulation leakage of the electric submersible pump, the main control chip controls the alarm to ring and light the alarm indicator.

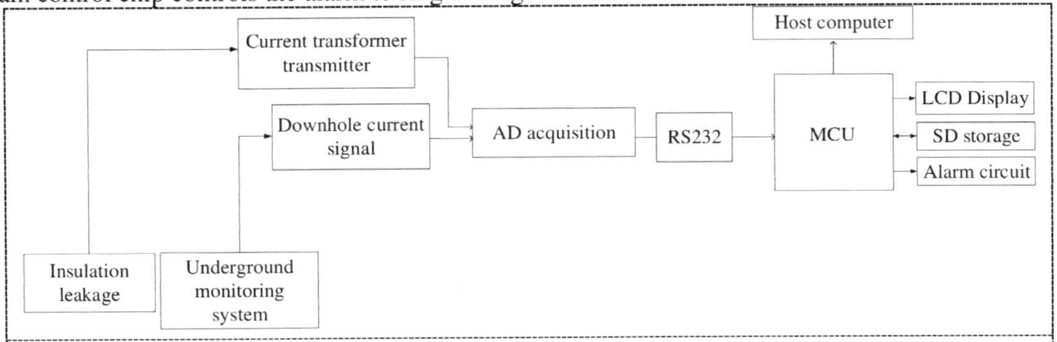

Figure 6. Hardware circuit structure

5. Well condition simulation experiment and test results

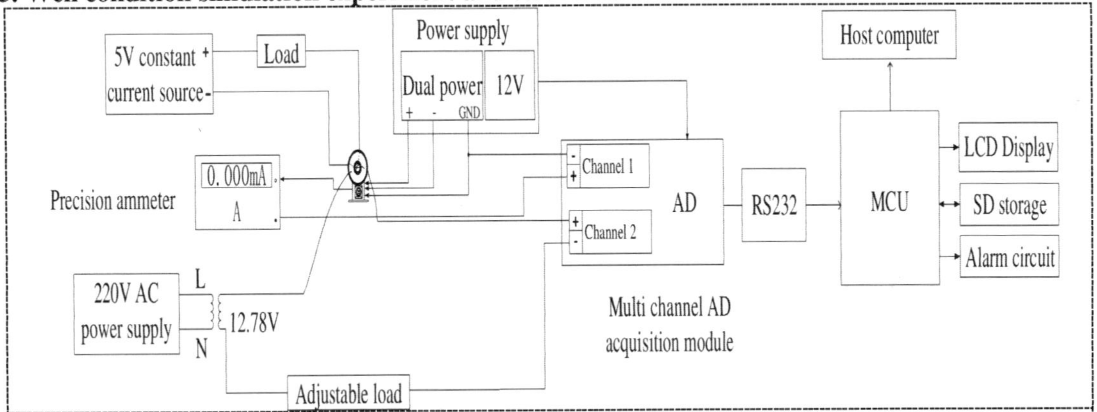

Figure 7. Indoor experiment platform

The well condition simulation experiment is shown in Figure 7. The AC circuit representing the insulation leakage of the electric submersible pump and the DC circuit representing the working state of the downhole unit are respectively built, and the two circuits pass through the air core in the middle of the current mutual inductance transducer. Then connect the relevant hardware circuit and the upper computer, and adjust the AC circuit load to get multiple sets of measurement values.

Table 3. Monitoring data of real-time monitoring system for leakage current of electric submersible pump

Ammeter reading/mA	Module acquisition current/mA	relative error/%	LCD display/mA	relative error/%	Upper computer interface/mA	relative error/%
8.154	8.152	-0.024	8.151	-0.036	8.152	-0.024
9.772	9.771	-0.01	9.771	-0.010	9.783	0.112
10.586	10.585	-0.009	10.585	-0.009	10.585	-0.009
11.624	11.628	0.034	11.628	0.034	11.628	0.034
12.724	12.722	-0.015	12.732	0.062	12.743	0.149
13.735	13.732	-0.021	13.742	0.050	13.735	0.000
14.589	14.578	-0.075	14.579	-0.069	14.569	-0.137
15.258	15.265	0.045	15.266	0.052	15.274	0.103
16.302	16.299	-0.018	16.298	-0.024	16.299	-0.018
17.669	17.672	0.016	17.671	0.011	17.671	0.011

From the measurement results in Table 3, it can be seen that the measurement error of the designed real-time monitoring system for leakage current of electric submersible pump is kept within 0.15%, which can accurately monitor the current leakage of electric submersible pump in real time.

6. Conclusion

Based on the real-time monitoring system of electric submersible pump and downhole state, this paper designs a real-time monitoring system for leakage current of electric submersible pump based on the principle of current mutual inductance. In addition, an indoor verification experiment for this real-time monitoring system is carried out. The results show that the system can accurately monitor the leakage current of the electric submersible pump in real time, and ensure the safe operation of the electric submersible pump unit.

Acknowledgments

This paper was supported by Postgraduate Innovation and practical ability training program of Xi'an Shiyou University (YCS19113052); National Natural Science Foundation of China (51974250); National Natural Science Foundation of China (41874158); Natural Science Basic Research Program of Shaanxi(2018JQ5133); Key laboratory project of Shaanxi Provincial Department of Education(17JS106).

References

[1] Liang XD. (2003) High Voltage Engineering. tsinghua university press. Beijing.
[2] Cao L. (2018) Research On Monitoring Technology of Operating Condition Of Electrical Submersible Pump. Northeast Petroleum University.
[3] Wu YF. Research on multi-parameter monitoring system of Electrical Submersible Pump .Xi'an Petroleum University.2018.
[4] Han JQ. (2015) Research on 150°C ESP Working Condition on Monitoring. Harbin Industrial University.
[5] Xia YP, Wang M. (2015) Design of oil level monitoring system for oil immersed current transformer. Electronic technology and software engineering.25:225~226.
[6] Li C.(2014)Development of Pressure Transmitter Based on INA333 and XTR115.Instrument Technique and Sensor,25:8~10.

IOP Conf. Series: Materials Science and Engineering **799** (2020) 012009 doi:10.1088/1757-899X/799/1/012009

Design and Implementation of Launching Subsystem of a Radar Parameter Measurement Simulation Training Platform

Jun Wang[1,2*], Bin Deng[3], Chuanqi Li[4] and Yiyuan Li[1]

[1]Graduate team 1, Air Force Early Warning Academy, Wuhan, Hubei, 430019, China

[2]The 1 team, No.32382 Unit, the PLA, Wuhan, Hubei, 430311, China

[3]Air defense early warning equipment department, Air Force Early Warning Academy, Wuhan, Hubei, 430019, China

[4]Graduate team 2, Air Force Early Warning Academy, Wuhan, Hubei, 430019, China

[*]Corresponding author's e-mail: 278923898@qq.com

Abstract. This paper introduces the design and implementation of the launch subsystem of a radar performance measurement simulation training platform. First, the main functions of the launch subsystem of the platform are determined according to the training requirements of the outline for measuring the radar performance parameters of the troops, including test items and technical indicators; then, comb the launch subsystem analog circuit of work process; finally, the design method of each functional circuit is introduced in detail. The practical application results are given. The launch subsystem of the simulation training platform can realistically simulate the performance parameter measurement training of the radar real launch subsystem. In addition, it can compare the feedback measurement results with the designed indexes, meet the training requirements of radar equipment support personnel, and increase the means of training and integration. Ensuring training time and enhancing support capability can effectively solve the problems of too many personnel and planes and equipment damage in actual training.

1. Introduction

Parameter test is a necessary means to master the technical status of equipment[1], and it is an important skill that radar maintenance personnel must master. For this reason, equipment support training (Air Force military training program) puts forward clear training and assessment requirements for radar equipment support personnel, requiring radar technicians to master the test methods of radar main performance parameters. However, at present, the radar technical support team lacks the real training conditions on the one hand, and the corresponding simulation training platform on the other hand, so it is difficult to carry out test skill training[2-3]. In order to solve the practical problems of the training of parameter testing skills, it is urgent to develop the corresponding performance testing simulation training platform. In order to meet the actual needs of the technical support team of Radar Brigade to carry out the test and training of the performance parameters of the main subsystems of radar, a set of simulation training platform that can provide the test of the main parameters of the subsystems such as antenna feed, launch and reception is proposed to improve the test ability of maintenance technicians.

In this paper, the design and implementation of the simulation training system for the measurement of radar performance parameters are described in terms of its main functions, working process, circuit

IOP Conf. Series: Materials Science and Engineering **799** (2020) 012009 doi:10.1088/1757-899X/799/1/012009

design, etc.

2. Main functions

According to the general requirements of the outline, the test items and performance indexes of the platform simulation are designed considering the test items and performance indexes that the army needs to carry out for a certain radar launching subsystem.

2.1. test items

Through the reserved test port on the panel of the platform, using spectrum analyzer, microwave power meter, oscilloscope and other measuring instruments, the simulation training of the main performance parameters of the transmitter, the main performance parameters of the local oscillator module and the main performance parameters of the transmission signal branch is completed. The specific test items are as follows:

- Main performance parameters test of transmitter: radar transmitting power, working frequency, transmitting pulse repetition frequency, transmitting pulse envelope parameter, transmitting pulse spectrum, transmitter improvement factor and other parameters.
- Main performance parameters test of local module: local signal frequency, local signal strength, local signal clutter suppression system, local signal stability and other parameters.
- Main parameters test of transmission signal branch: DDS output signal, pre excitation output power, power amplifier module input power, power amplifier module output power, power synthesis output power and other parameters.

2.2. Performance index

According to the test items, the performance parameters of DDS signal generation, up conversion, power synthesis and other functional modules are required. After the software and hardware joint debugging of the training platform is completed, the function of the platform is tested and trained, and the test and training results can be compared and analyzed with the design indexes.

2.2.1. DDS signal source output frequency and power

- Center frequency f_0 30MHz
- Output power P_0 0-3dBm

2.2.2. Output signal type and index of DDS signal source

2.2.2.1. Fixed carrier signal

- Output frequency f_0 30MHz
- Pulse width 100, 200, 300, 400, 500μs
- Pulse repetition period T_r 100, 200, 300, 400, 500Hz

2.2.2.2. Linear (nonlinear) FM signal

- Signal bandwidth ΔB 1, 2, 5MHz
- Pulse width τ 100, 200, 300, 400, 500μs
- PRF T_r 100, 200, 300, 400, 500Hz

2.2.3. Frequency synthesizer output

- The first coherent local oscillator 250-300MHz (11 frequency points in total, one frequency point 5MHz apart)
- The second coherent local oscillator 70MHz
- Coherent IF 30MHz

2.2.4. Upconverter assembly

- Output frequency 150-200MHz (11 frequency points in total, one frequency point 5MHz apart)

- Input power ≥10dBm
- Output power ≤15dBm
- Attenuation range of attenuator 31dB (step 1dB)

2.2.5. Power synthesizer assembly

- Output power ≥300mW
- Digital phase shifter 4 bits
- Digital phase shifter step 22.5°

2.2.6. Transmitter load

- Load type matching and mismatching load (open circuit, short circuit, 20Ω, 80Ω resistance load)
- Characteristic impedance 50 Ω

3. Working process

The basic working process of the analog circuit of the transmission subsystem is as follows.

3.1. DDS signal generation and filter amplification

DDS is a kind of if radar signal which transforms a series of digital signals into analog signals through a digital to analog converter. Through the parameter selection function on the platform, the signal type (fixed carrier frequency, linear frequency modulation, nonlinear frequency modulation), local frequency, pulse width, repetition frequency and other parameters can be set. If radar signals are set by "switch filter selection", and "1MHz" (i.e. filter amplifier 1) or "5MHz" (i.e. filter amplifier 2) amplification channels are selected respectively. The bandwidth of filter amplifier 1 is 1MHz and that of filter amplifier 2 is 5MHz.

3.2. Up conversion

If radar signal output by filter amplifier 1 or filter amplifier 2 is sent to the first frequency converter through panel jumper connection. The if radar signal is mixed with the first coherent local frequency to get the difference frequency signal, which is connected to the filter amplifier 3 through the internal circuit of the transmitter, and then sent to the second frequency converter to get the difference frequency of the second coherent local frequency, so as to get the RF radar signal, and then connected to the filter amplifier 4 and sent to the 4-bit digital attenuator through the internal circuit of the transmitter.

3.3. Power amplification synthesis

The RF radar signal output by the 4-bit digital attenuator is sent to the power distributor of the power amplifier module by the way of panel jumper connection. The power distributor divides the input RF radar signal power into three equal parts according to the proportion of 1:3, which are respectively sent to the first, second and third power amplification branches for power amplification through the internal line connection of the transmitter. The RF radar signal output by each power´amplification branch is connected to the 3:1 power synthesis input through the internal line of the transmitter, and the power synthesizer synthesizes the output power of the three power amplifier branches into one The bidirectional directional coupler connected through the internal line is output to the panel. The "forward coupling" output of the bidirectional directional coupler is a part of the incident wave power transmitted from the transmitter to the load direction, and the "reverse coupling" output is a part of the reflected wave power transmitted from the load to the transmitter direction. The bidirectional directional coupler can measure the power and standing wave ratio output from the transmitter to the load.

The RF radar signal is directly connected to the load on the panel through "power synthesis", and the load impedance is equal to the antenna impedance.

4. design of main circuit

The main functional circuits of radar transmitter simulator include DDS signal source (waveform generation) module, first and second filter amplifier modules, local frequency source module, up conversion module and power amplifier module.

4.1. Design of DDS source (waveform generation) module

Firstly, get the type information of signals input by the trainees on the panel, such as fixed carrier frequency signal, linear (non-linear) frequency modulation signal, pulse width, repetition frequency, etc., send these input information to the MCU for processing after being buffered by 74hct245 input buffer, and the SCM generates DDS amplitude, phase, frequency control words and other register control words according to the input of the trainees According to the register specified by A1-A5 address bus, write the data into the register through d0-d7 data bus, complete the control of DDS working mode, make it output the specified signal according to the input of trainees, and the schematic diagram is shown in Figure 1.

Figure 1. Schematic diagram of DDS (waveform generation)

4.2. First and second filter amplifier circuits

Firstly, the input signal work is divided into two channels, i.e. a filter with a center frequency of 30MHz and a bandwidth of 1MHz and 5MHz. A matching circuit is added at the input end and output end of the filter to realize better energy transmission of the signal. After selection by a RF SPDT Switch, the switch is selected by the trainees on the panel as a filter with "1MHz" or "5MHz". After selection, the signal is after over amplification and power division, it is output to the panel as the test signal, in addition, it is output to the rear stage as the baseband signal, and the schematic diagram is shown in Figure 2.

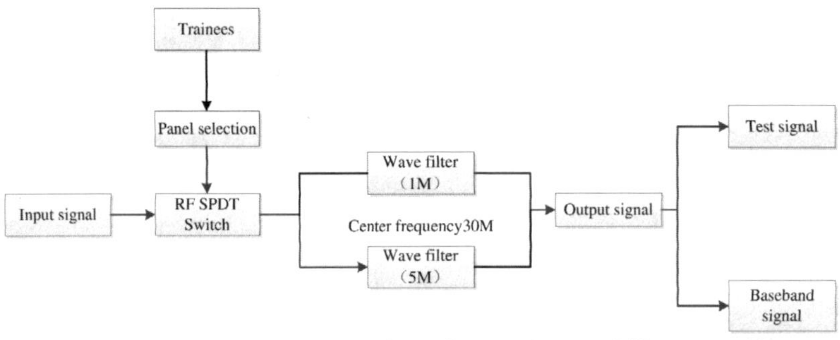

Figure 2. Schematic diagram of the first and second filter amplifiers

4.3. Local frequency source circuit

In order to achieve coherent signal output, the frequency reference of the same 40MHz is used. After 4-way drive, amplification and isolation, one channel is used as the clock signal, and the other 3 channels are used through the PLL circuit to output one local oscillator and two local oscillators, as shown in Figure 3.

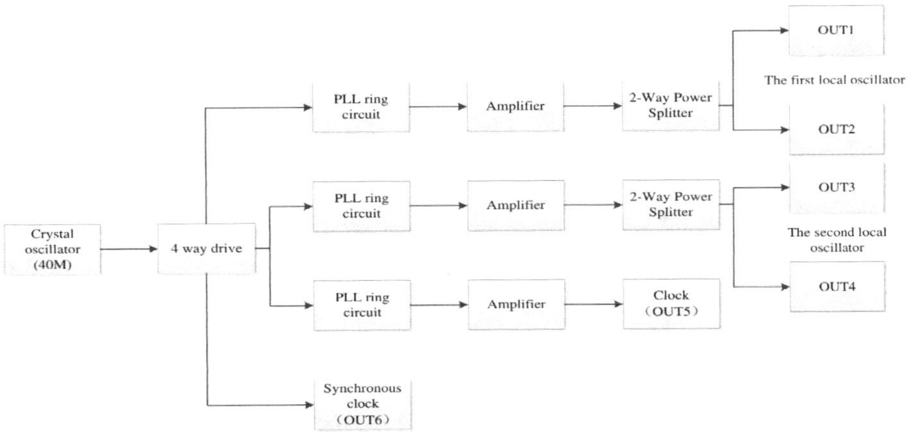

Figure 3. Schematic diagram of local frequency source

4.4. Up conversion module

The input local oscillator signal and if signal arrive at the mixer after being filtered and amplified for mixing, and then filter Z3 and Z4 are used to select the frequency of the generated RF signal. Since the passive mixer is used, the insertion loss of the signal is large, so N10 and N12 amplifiers are added to improve the amplitude of the output RF signal, as shown in Figure 4.

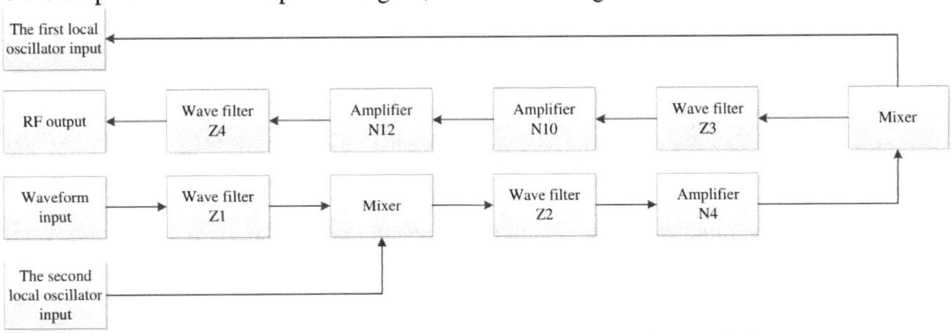

Figure 4. Schematic diagram of upconversion module

4.5. Power amplifier module

In this module, the input signal is divided into four channels through two-stage Wilkinson power divider. One channel is used as the input of power amplifier circuit after the digital control phase shift, attenuator P1 phase shift and attenuation. The phase shift and attenuation can be selected by the trainees on the panel. The other three channels are directly input to the power amplifier circuit. The output of four channels of power amplifier circuit is synthesized into one channel through two-stage Wilkinson power divider A small part of the coupler is coupled as the detection branch signal, and the other part is output through the isolator. The circuit diagram is shown in Figure 5.

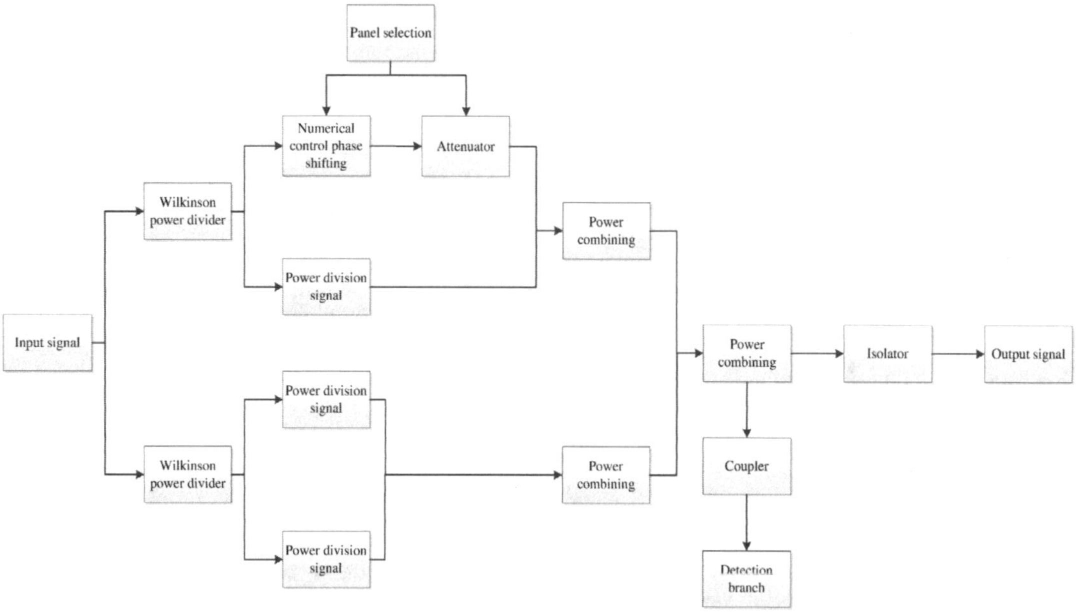

Figure 5. Schematic diagram of power amplifier module

5. Concluding remarks

The test results show that the radar performance parameters measurement simulation training platform can carry out the simulation measurement training of the main performance parameters of the radar launch subsystem, and meet the training requirements of the training program for radar support personnel. It can effectively solve the practical difficulties of the current radar technical support team, such as the lack of real training conditions and the corresponding simulation training platform. It can also enable the operators to complete the test experiment of the main performance parameters of the launch subsystem in a safe environment, better understand the working principle of the radar launch subsystem, and reduce the damage to the equipment caused by repeated operation in the actual equipment training.

References

[1] Liu, X.X., Wang, Y., Shao, X.(2010)Design and realization of training simulation system based on radar. Ship Electronic Engineering,12:84-86.
[2] Wang, J.J., Deng, B., Hao, Y., et al.(2016)Research and practice of "radar parameter measurement" network course construction. Contemporary Education Research and Teaching Practice,6:202-203.
[3] Shu, T., Yang, D.X., Chen, H.B., et al.(2010)CTL-713C weather radar parameter survey question and processing method. Meteorological and Environmental Sciences,33:80-83.
[4] Deng, B.(2008)Electronic Measuring Instrument. National Defence Industry Press, Beijing.
[5] Wang, F.,Qin, W., Wei, X.H., et al.(2013)Measurement technology of radar performance. TECHNOLOGICAL PIONEERS,5:172.
[6] Huang, Y.L.(2010)ADS RF Circuit Design Foundation and Typical Application. People's Post and Telecommunications Press, Beijing.
[7] Gu, T.(2013)Easy to Learn C#. Electronic Industry Press, Beijing.
[8] Han, S.C.(2018)Development of simulation training system for engineering vehicle maintenance based on CAN bus. MECHANICAL MANAGEMENT AND DEVELOPMENT,184:178-181.

IOP Conf. Series: Materials Science and Engineering **799** (2020) 012010 doi:10.1088/1757-899X/799/1/012010

Research on Multi-points Measurement of a Novel Alternate arrangement dynamometer Model

DANAISH [1*], REN ZONGJIN[1], ZHANG JUN[2], MUHAMMAD AYAZ AKBAR[3]

[1*,1,2,3] School of Mechanical Engineering, Dalian University of Technology, Dalian 116024, China

*Corresponding author's email: danaish@mail.dlut.edu.cn,

Abstract. The basis of this research is to design and fabricate a new type of alternating arrangement of three-axis piezoelectric sensors dynamometer for measuring six-component force/torque measurement. The system consists of a long dynamometer with eight tri-axial piezoelectric sensors alternatively distributed. The design theory of the piezoelectric six-axis force dynamometer is established, including a finite element model (FEM: ANSYS analysis) and experimental model (static and dynamic calibrations). Structural design is constructed for analyzing the operating principle and performing the experimental measurements. The multi-points measurement (MP-M) model is fabricated based on the established alternate arrangement pattern and force application of multi-points allocation to apply the force/torque. Therefore, this work describes linearity error, repeatability error and cross-talk error of the constructed model. However, the designed model is not limited only to an alternate arrangement pattern of tri-axial sensors. But in the future, other types of arrangements will be researched and will be experimentally performed.

1. Introduction

The multi-points measurement model (MP-M) concept can be implemented in the manufacturing process, robot design, Rocket thrust measurement and space station. The model is designed for measuring the external forces and collecting the spatial force information. The model has been designed, constructed and tested based on using tri-axial piezoelectric sensors technology. Therefore, this study requires a detailed review of the literature and an overall design approach based on the static characteristics of the fabricated model.

There are various ways to measure spatial force/moments by arranging sensors in different locations and using a different types of arrangements. A lot of research has been done on the selection of sensors and their arrangement geometry which supports the measurement of six components force/moments[1]. Literature can be found on a different combination of sensors arrangement as Zhenyuan Jia et al, investigated a multi-point and a square arrangement with 3 and 4 sensors mounted respectively and their maximum error is up to 5.3%[2]. Moreover, Hexagonal and octahedral arrangements with six and eight tri-axial piezoelectric sensors installed respectively and their cross-talk error is within 4.5%[3]-[4].

A six-component force sensor may be a base, a table dynamometer, or a wrist type according to its location of application, although the force-sensing principles are basically the same[5]. However, table dynamometers offer precise and effective force measurement, usually typical dynamometer consists of force sensors that are mounted in between two plates[6]. The different types of combinations of the tri-axial sensors allow for six degrees of measurement as Brimhall Z.N. used a combination of tri-axial sensors on the forward structure and in the back to measure six degrees of freedom of rocket motor test

IOP Conf. Series: Materials Science and Engineering **799** (2020) 012010 doi:10.1088/1757-899X/799/1/012010

stand[7]. Alexander Winkler and Jozef Suchy used a 12DOF sensor for dynamic force/torque measurement[8]. Another researcher discussed strain-gauge balances as high-precision instruments that are capable of measuring up to six degrees of freedom of aerodynamic loads[9].

A Multi-point piezoelectric balance enabling the measurements of 6DOFs is designed and is developed for the experimental calibration. The Tri-axial piezoelectric load cells are developed to be used in this designed arrangement pattern as they offer an increased stiffness and fast response. At the first stage, all the sensors are statically calibrated one by one and their sensitivity and errors have been recorded. Secondly, the design arrangement has been implemented in the MP-M model during the experimental setup system. And in the last static and dynamic calibrations have been performed by applying designed load for developed model.

2. Model structure and measurement principle

According to the structural characteristics of the tri-axial piezoelectric force, sensors can measure the three components of force in X, Y and Z directions. But for measuring the moments such as pitching, yawing and rolling the single sensor is not enough, therefore, a combination of piezoelectric sensors by a specific arrangement has been designed and implemented for measuring force/torque measurements. The model structure is based on the designed distance between the alternate allocated sensors and assigned multi-directions force output to the sensors.

2.1. Model structure

The analysis of the structure is defined as the following assumptions are made; the rigidity of all the tri-axial sensors regarding the quartz crystals are identical, with almost equal sensitivities and symmetric output ratios. The base plate and cover plate for sensors installation of the long dynamometer are rigid bodies.

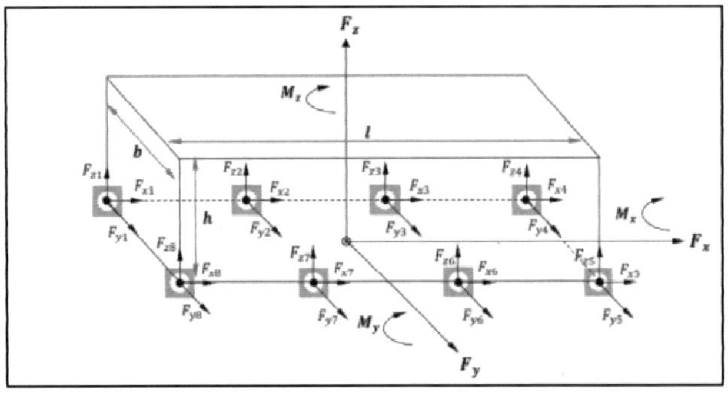

Fig 1. Multi-points measurement model structure

In Fig 1, ‘l’ is the total length of arrangement pattern in which the eight sensors are installed in an alternate method. The same way, breadth ‘b’ is considered as the horizontal distance between installed sensors in the Y direction and height ‘h’ is measured as the vertical height of sensors in the Z direction. However, eight mode supports of piezoelectric sensors offer high rigidity and a wide measuring range. Because of the flexibility in the model, the pattern can be changed during the experimental work or in future research.

2.2. Measurement principle

The measurement principle of the MP-M model mainly includes a base plate, a cover plate, and eight triaxle piezoelectric sensors alternate installation as a long dynamometer. The main measuring principle is using piezoelectric technology for measuring the external forces. Piezoelectric quartz crystal chips are used as an active element inside the tri-axial piezoelectric sensors and tri-axial sensors are used together to design and construct a dynamometer as shown in below Fig 3. And the alternate arrangement shape pattern of the designed model is shown in Fig 2 with eight tri-axial sensors alternatively mounted.

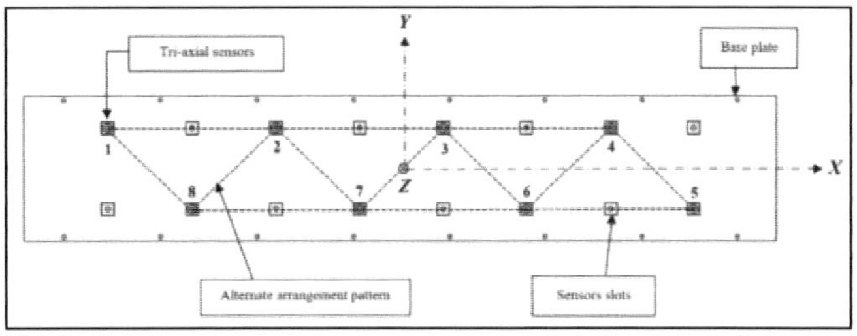

Fig 2. Alternate arrangement pattern of eight tri-axial sensors

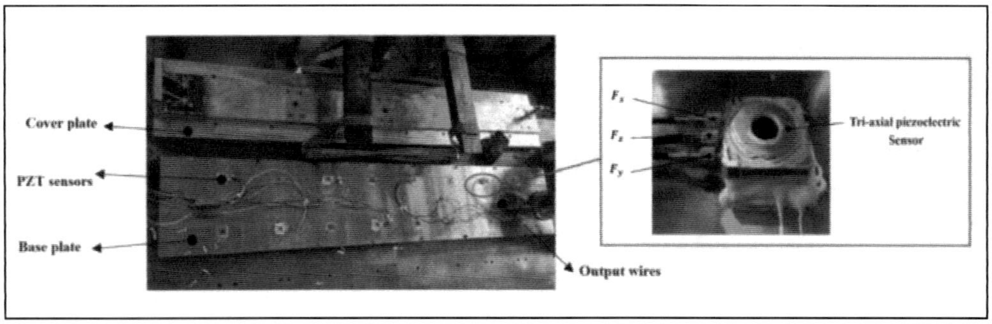

Fig 3. Allocation of tri-axial sensors in alternate arrangement pattern

3. Numerical model

ANSYS software is used to evaluate the static characteristics of a multi-points dynamometer model in which total deformation and stress analyses have been performed. A designed modeling approach and static load methods application are used in FEM analysis to verify the effectiveness of the designed model.

3.1. Modeling and Meshing

The design of all components of the main model is built with SolidWorks and all parts are assembled according to the operating and measurement principle to construct the MP-M model. Then the designed geometry model is imported into ANSYS software and a mesh is generated. The dimensions of the main parts of the designed model and the material are stated in above Table 1. The ANSYS Workbench is used for meshing the MP-M model, the model has meshed into 564894 elements and 851447 nodes as shown on the right side of Fig 4.

Table 1. Main structural parameters of Multi-points measurements model

Component's name	Length (mm)	Width (mm)	Height (mm)	Material	Poisson's ratio (pa)	Elastic modulus (pa)	Density (Kg/m^3)
Base plate	1800	340	29	C45	0.3	2^{11}	7850
Cover plate	1800	300	55	C45	0.3	2^{11}	7850
Tri-axial sensor	30	30	12	304	0.3	2^{11}	7850
Small plate	150	150	100	C45	0.3	2^{11}	7850
Preload bolts	75	M10	-	304	0.3	2^{11}	7850

IOP Conf. Series: Materials Science and Engineering **799** (2020) 012010 doi:10.1088/1757-899X/799/1/012010

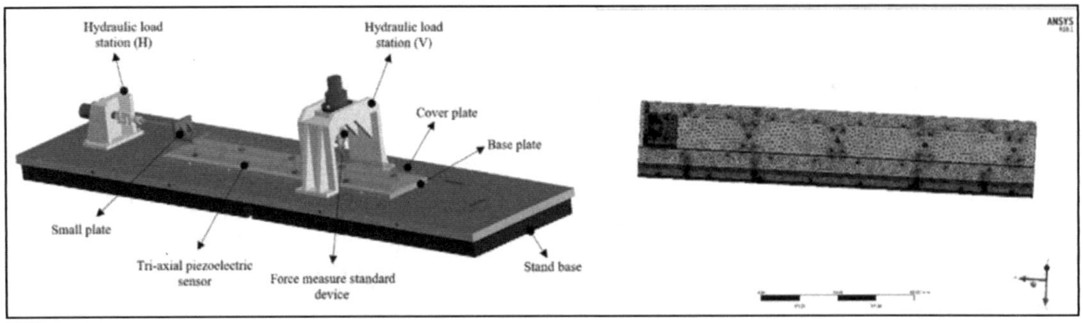

Fig 4. 3D modeling of MP-M model and meshing coordinates

The 3D modeling of the MP-M model highlight all components which are assembled together to construct the main model. The cover plate is shown transparent to easily distinguish the position of tri-axial sensors as shown in above Fig 4. The total length of the balance plate model is 180cm, breadth is 39cm and height is 9.6cm. The vertical distance and horizontal distance between mounted sensors are 19cm and 40.25cm respectively.

3.2. Total deformation and force reactant output

The following Fig 5, illustrates the deformation when a maximum force 12kN is applied on the allocated point and the area with maximum deformation can be assessed nearer the force applied point. The load applied in the Z direction is push force and is taken as positive and load in X-direction can be considered as pulling force. The force reactant output of applied load in X- and Z-direction are discussed in below Table 5.

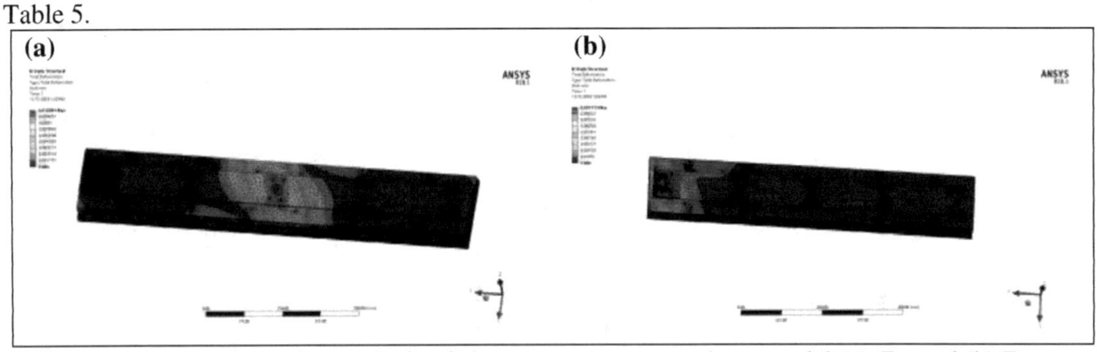

Fig 5. Total deformation analysis of alternate arrangement shape model (a) F_Z, and (b) F_X.

4. CALIBRATION EXPERIMENT AND DISCUSSION

The calibration of the multi-points measurement model is very important in designing and constructing the six-axis force/torque measurement model since the use of the model is directed to calibration results. Two types of experimental calibrations are designed for MP-M model one is static calibration and the second one is dynamic calibration.

4.1. Static calibration

The static calibration model mainly focuses on such indexes as linearity, repeatability and cross-talk error of the MP-M model. The fabricated model as shown in Fig 6 is calibrated for a single force up to 12kN applied on each assigned point individually, and the data output in all three directions recorded for every 3kN. This step is repeated three times to get the average values and to calculate the linearity and repeatability errors for the allocated points.

IOP Conf. Series: Materials Science and Engineering **799** (2020) 012010 doi:10.1088/1757-899X/799/1/012010

Fig 6. Diagram of experimental work (b) horizontal force load method F_x in X-direction (c) vertical force load method F_z in Z -direction.

4.1.1. Static calibration in Z- and Y-directions

The multi-load range force is applied in Z-direction on the marked point on the top of the cover plate of the designed model. The same way load is applied on one selected point on side of the cover plate in X-direction. $\sum x_i$, $\sum y_i$, and $\sum z_i$ where i=1,2,3,...8, is the summation of the output of eight sensors in the respective directions respectively. The calibration analysis of both points is discussed in below Table 2 and Table 3.

Table 2. Calibration experiment analysis in Z-direction

Located Point	Standard force (N) <u>Direction</u>	Measured voltage (V) <u>Co-ordinates</u>			Nonlinear Error (%)	Repeatability Error (%)	Sensitivity (pC/N)	Cross-talk error (%)
F_z	F_z	$\sum x_i$	$\sum y_i$	$\sum z_i$				
Push force in Z-direction	0	0.00	0.00	0.00	4.32	0.75	4.4	1.18
	3000	0.01	0.04	3.13				
	6000	0.01	0.08	6.26				
	9000	0.01	0.11	9.41				
	12000	0.01	0.15	12.61				

Table 3. Calibration experiment analysis in X-direction

Located Point	Standard force (N) *Direction*	Measured voltage (V) *Co-ordinates*			Nonlinear Error (%)	Repeatability Error (%)	Sensitivity (pC/N)	Cross-talk error (%)
F_x	**F_x**	$\sum x_i$	$\sum y_i$	$\sum z_i$				
	0	0.07	0.01	0.03				
Pull	3000	2.84	0.01	0.01				
force in	6000	6.11	0.04	0.02	2.65	0.51	8.8	0.97
X-	9000	9.43	0.08	0.02				
direction	12000	12.7	0.10	0.12				

In above Table 2 and Table 3, in their first column, the located point has been mentioned and the direction of force is highlighted whether the force is pulling or pushing force. The second column shows the applied standard forces in every 3kN and maximum up to 12kN. Furthermore, in the static calibration experiment, the linearity, repeatability, sensitivity and cross talk error are calculated for the constructed model.

4.2. Dynamic calibration

The dynamic calibration of the MP-M model as the response characteristics to the input signal that varies as a function of time. To calibrate dynamic characteristics of fabricated model an impact technology is used as shown in Fig7 and an FFT analysis carried out using DEWESoft software to find out the natural frequency (ω_n). The dynamic response values for each coordinate are listed in Table 4.

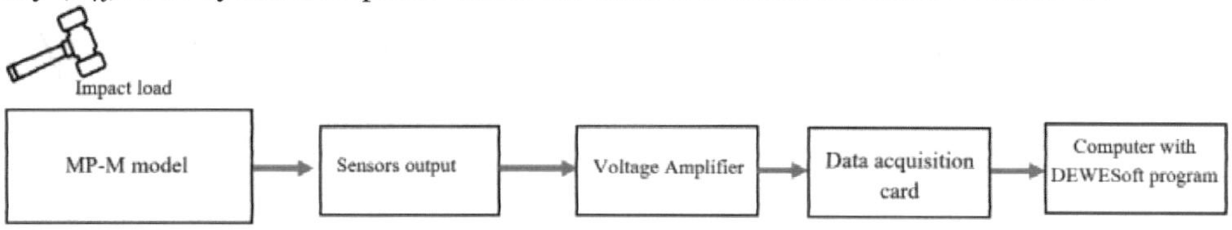

Fig 7. Schematic diagram of dynamic calibration

Table 4. Natural frequencies in each direction

Natural frequency	F_x	F_y	F_z
ω_n (Hz)	>1000	>400	>350

The model has a good response characteristic, frequency curve and can gratify measurement requirement as the three-dimensional natural frequency is 1000Hz, 400Hz, and 350Hz in X, Y and Z directions respectively.

4.3. Comparison of load output (%)

According to analysis discussed in above Table 2 and Table 3, the load-out ratios of FEM and Experimental analyses are obtained of multi-points are shown in the below Table 5.

Table 5. Comparison of Load output percent values of FEM and Experimental analysis

Multi-load points	Load output (%)	
	FEM analysis	Experimental analysis
F_z	99.9	105.1
F_x	99.9	105.9

Load output percent values of the MP-M model are in a range from 99.9 to 100 percent in FEM analysis and approximately 105.1 to 105.9 percent in experimental analysis. It can be observed, the percent output difference of FEM analysis is lower than the experimental analysis, because of the symmetry in structure and rigid body analysis in FEM. The load output percent differences between the FEM and Experimental analyses of F_z is 5%, in Z-direction and on point F_x is 6% in X-direction respectively.

The experimental results can be compared to the known rigid body FEM analysis result, an agreement is within approximately 1-6% error. The multi-points load output percent values of FEM and experimental analyses are close which proves the rationality of the theoretical model and supports the calibration experiment results of the novel designed alternate arrangement model.

5. CONCLUSION

The proposed research is based on the theoretical and experimental calibration of an established novel alternate arrangement shape dynamometer. A structure model is constructed to highlight the proposed arrangement of tri-axial sensors and to confirm the designed coordinates for sensors. The measurement principle of the established model is discussed and a mathematical model is derived. The developed model is verified by theoretical and experimental analyses. The experimental analysis of the fabricated model is in agreement with the theoretical analysis results of the designed model. The experimental results are compared to the known rigid body FEM analysis result and the consistency is in the range of about 1-6%. The static calibration verifies that the cross-talk error of the studied model is under 4 percent. However, the developed model needs more research for experimental calibration to minimize the occurred errors.

6. DECLARATION OF CONFLICTING INTERESTS

The author(s) declared no potential conflicts of interest with respect to the research, authorship, and/or publication of this article.

ACKNOWLEDGMENT

The author(s) gratefully acknowledges the following financial support for the research, authorship, and/or publication of this article: This project was supported by the National Natural Science Foundation of China (No. 51675084) and Aeronautical Science Foundation of China (20160163001).

REFERENCES

[1] A. M. Wright, A. B. Wright, T. Born, and R. Strickland, "A six degree-of-freedom thrust sensor for a labscale hybrid rocket," *Meas. Sci. Technol.*, vol. 24, no. 12, pp. 1–10, 2013.

[2] Z. Jia, Y. Gao, Z. Ren, S. Gao, and Y. Shang, "Design and calibration method for a novel six-component piezoelectric balance," *Proc. Inst. Mech. Eng. Part C J. Mech. Eng. Sci.*, vol. 227, no. 8, pp. 1841–1852, 2013.

[3] Z. Y. Jia, S. Lin, and W. Liu, "Measurement method of six-axis load sharing based on the Stewart platform," *Meas. J. Int. Meas. Confed.*, vol. 43, no. 3, pp. 329–335, 2010.

[4] L. Qin, C. Jiang, J. Liu, and Y. Duan, "Design and calibration of a novel piezoelectric six-axis force/torque sensor," *Seventh Int. Symp. Precis. Eng. Meas. Instrum.*, vol. 8321, pp. 83210–83219, 2011.

[5] S. A. Liu and H. L. Tzo, "A novel six-component force sensor of good measurement isotropy and sensitivities," *Sensors Actuators, A Phys.*, vol. 100, no. 2–3, pp. 223–230, 2002.

[6] A. Albrecht, S. S. Park, Y. Altintas, and G. Pritschow, "High frequency bandwidth cutting force measurement in milling using capacitance displacement sensors," *Int. J. Mach. Tools Manuf.*, vol. 45, no. 9, pp. 993–1008, 2005.

[7] Z. Brimhall, N. Divitotawela, J. Atkinson, D. Kirk, and H. Peebles, "Design and Validation of a Six Degree of Freedom Rocket Motor Thrust Stand," in *44th AIAA/ASME/SAE/ASEE Joint Propulsion Conference & Exhibit*, 2008, no. July, pp. 1–15.

IOP Conf. Series: Materials Science and Engineering **799** (2020) 012010 doi:10.1088/1757-899X/799/1/012010

[8] A. Winkler and J. Suchý, "Dynamic force/torque measurement using a 12DOF sensor," *IEEE Int. Conf. Intell. Robot. Syst.*, pp. 1870–1875, 2007.

[9] K. C. Lynn, S. A. Commo, N. M. Ulbrich, and C. P. Harris, "Experimental design considerations for calibration of semispan force measurement systems," *J. Aircr.*, vol. 52, no. 2, pp. 638–649, 2015.

IOP Conf. Series: Materials Science and Engineering **799** (2020) 012011 doi:10.1088/1757-899X/799/1/012011

The Positioning Accuracy and Performance Simulation Analysis of BeiDou-3 in Cote d'Ivoire

Sub-Lieutenant Brou Eugene Krou[1], Houpu Li[1*], Yue Song[1]

[1]Department of Navigation, Naval University of Engineering, Wuhan, Hubei, 430033, China

[*]Corresponding author's e-mail: lihoupu1985@126.com

Abstract: BeiDou Navigation Satellite System will operate world widely in 2020. Therefore, it is very important to master the accuracy and positioning performance of BeiDou-3 in West Africa. The number of visible satellite, the navigation accuracy and Geometric Dilution of Precision (GDOP) of BeiDou-3 (3GEO-3IGSO-24MEO) are simulated using STK software in Cote d'Ivoire. Using data of five stations in Cote d'Ivoire, the simulation results are compared for BeiDou-3 and GPS. Thus, the results are allowed to possess a database about the positioning performance of BeiDou-3 not only for Cote d'Ivoire but also benefit for surrounding Regions of the Gulf of Guinea.

1. Introduction

The three steps strategy of BeiDou Navigation Satellite System (BDS) development opted by China is almost accomplished. The experimental system (BeiDou-1, 2000-2003) has provided Radio Determination Satellite Service, two ways of positioning, timing and short message service in China[1-3]. In addition, the second step has established a regional service system (BeiDou-2, 2004-2012) with more satellites orbiting around the earth and provided more applications in Asia-Pacific region[4-6].

After the two compatible steps hereinafter, the interoperable BeiDou-3, which combines the qualities of BeiDou-1 and BeiDou-2 including short message service, will be revealed to the world in 2020 and join GPS (USA) and GLONASS (RUSSIA) in the exclusive club of the GNSS in full operation. A great number of researchers have done considerable works about the satellite visibility and GDOP (Geometric Dilution of Precision) of BeiDou-3 in China and other Asia Pacific region [7-9], but little work is done about the positioning performance analysis of BeiDou-3 in Cote d'Ivoire.

In order to show positioning performance of BeiDou-3 from five specific locations in Cote d'Ivoire, the STK10.0 software is used to simulate BeiDou-3 in term of the number of access of MEO satellites, navigation accuracy, and geometric dilution of precision. The results are compared with those of GPS in these locations.

2. BeiDou-3 constellation and simulation parameters

2.1. BeiDou-3 constellation

BeiDou-3 is composed of three (3) GEO (Geostationary orbit), three (3) IGSO (Inclined Geosynchronous Orbit, 55 degrees) and twenty-four (24) MEO (Medium Earth Orbit). 3-Dimension and 2-Dimension scenes of this hybrid constellation are shown by STK software in Figure1 and Figure2.

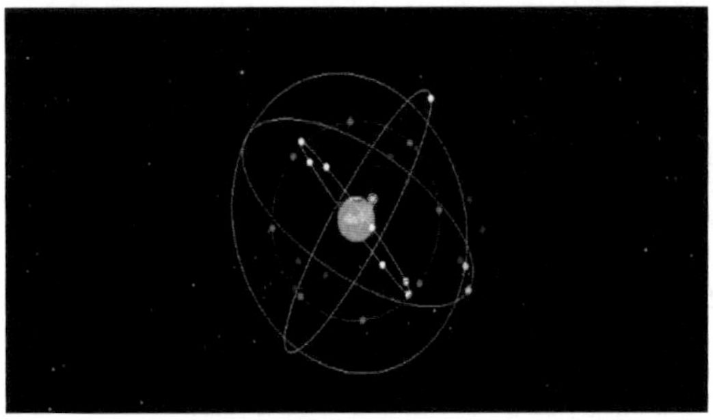
Figure 1. 3D BeiDou- 3 constellation

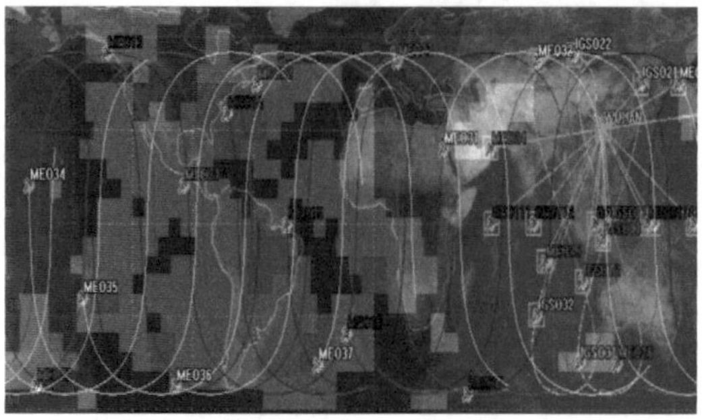
Figure 2. 2D BeiDou-3 contellation

2.2. Simulation parameters

The Positioning performance is observed by using the number of access of BeiDou-3 MEO satellites. STK software provides data, graphs of the above types of parameters in real time. The maximum, minimum, average and standard deviation values are calculated.

The minimum elevation angle is 15 degrees (angle between the horizontal plane and the line of sight as show in Figure3, measured in the vertical plane and referring to the horizontal plane). The simulation time is set from 04 o'clock 02 May 2019 to 04 o'clock 03 May 2019.

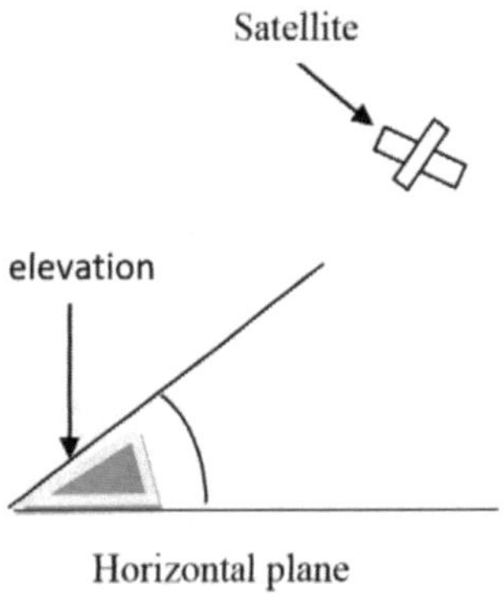

Figure 3. Elevation angle of satellite observation

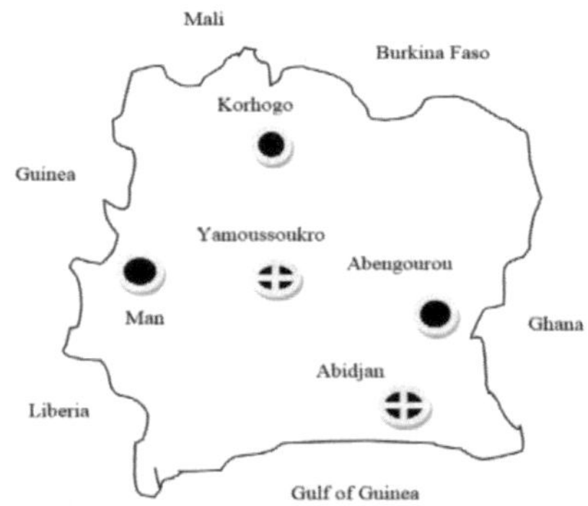

Figure 4. Location points in Cote d'Ivoire

3. Observation stations analysis

Cote d'Ivoire is a country of western sub-Saharan Africa. Located in the northern hemisphere close to the equator, it borders Liberia and Guinea in the west, Mali and Burkina-Faso in the north, Ghana in the east, and the Gulf of Guinea (Atlantic Ocean) in the south. Because of its geographic position and the geometric shape of this coastal country, the choice of Abidjan in the south, Yamoussoukro in the center, Korhogo in the north, Man in the west and Abengourou in the east can replace the distribution characteristics of the entire country. The distribution of Cote d'Ivoire is displayed in Figure4.

3.1. BeiDou-3 simulation results

Table 1. Mean values of BeiDou-3

	Number of Access	GDOP	Navigation Accuracy(m)
Abidjan	6.9	2.91	14.53
Yamoussoukro	6.7	3.27	16.36
Korhogo	9.6	1.73	8.84
Abengourou	9.7	1.69	8.46
Man	9.6	1.73	8.66

The main conclusions about BeiDou-3 are as follows:

- In general, 9 MEO satellites are observed at the same time, but 6 MEO satellites are observed at the same time for Abidjan and Yamoussoukro, it means that BeiDou-3 can calculate the users' position accurately in Abidjan and Yamoussoukro.
- GDOP is around 2.27, it can meet the requirements of trilateral measurement
- The navigation accuracy value is between 8.5 and 16.5 meters.

3.2. Constellation and simulation results of GPS

3.2.1. GPS constellation

GPS Constellation consists of twenty-four (24) satellites are uniformly distributed on six (6) medium orbital plane inclined by 55 degrees with respect to the equator:

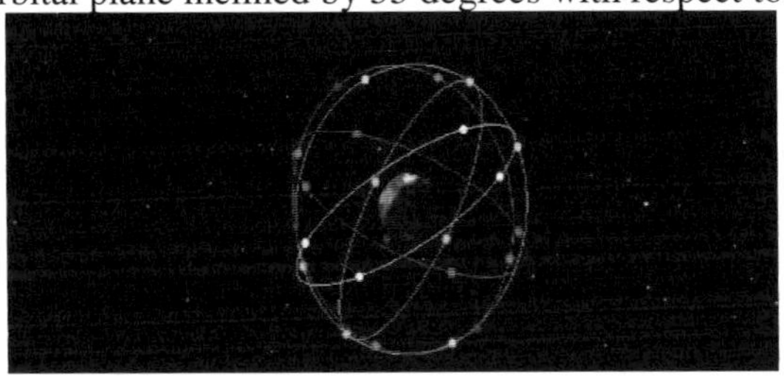

Figure 5. 3D GPS constellation

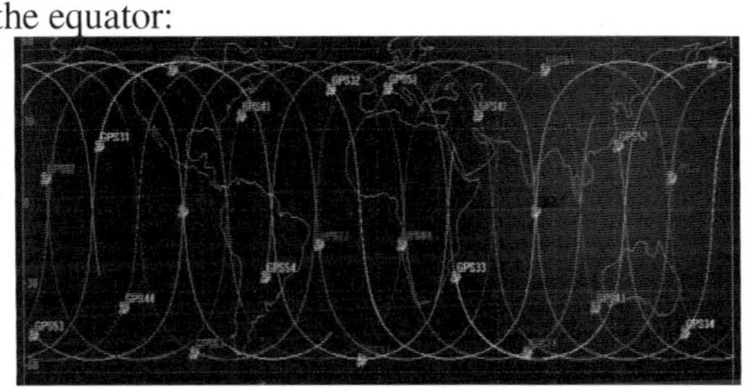

Figure 6. 2D GPS constellation

3.2.2. Simulation results of GPS

Table 2. Mean values of GPS

	Number of Access	GDOP	Navigation Accuracy
Abidjan	9.7	1.62	8.12
Yamoussoukro	9.7	1.62	8.09
Korhogo	9.6	1.62	8.1
Abengourou	6.6	3.07	15.33
Man	9.5	1.62	8.24

Using the mean values of the three analysis parameters, the conclusions are as follows:

- Taking Abengourou as an example, where there are 6 satellites in views, 9 GPS satellites are in views in the others cities. GPS can calculate position accurately.
- Except for this Abengourou , GDOP is stable around 1.62. thus geometrics distribution of GPS satellites are excellent in most areas.
- The navigation accuracy is less than 8.5 meters in most areas.

4. Comparison of BeiDou-3 and GPS

According to the above results for BeiDou-3 and GPS, the number of access, GDOP and the navigation accuracy are compared in the five stations of Cote d'Ivoire.

4.1. Comparison of the number of access

The number of access or the visible satellite of BeiDou-3 and GPS can be seen in the following statistics Table 3.

IOP Conf. Series: Materials Science and Engineering **799** (2020) 012011 doi:10.1088/1757-899X/799/1/012011

Table 3. Number of access BDS-3 VS GPS

	BeiDou-3	**GPS**
Abidjan	6.9	9.7
Yamoussoukro	6.7	9.7
Korhogo	9.6	9.6
Abengourou	9.7	6.6
Man	9.6	9.5

Table 3 shows that the number of GPS and BeiDou-3 satellites are both good enough for trilateration in the five cities, but GPS users can capture more satellites than BeiDou-3 in Abidjan and Yamoussoukro, while BeiDou-3 captures more in Abengourou.

4.2. Comparison of GDOP

The statistics of geometric dilution of precision of BeiDou-3 and those of GPS are shown in table 4.

Table 4. GDOP BeiDou-3 VS GPS

	BeiDou-3	GPS
Abidjan	2.91	1.62
Yamoussoukro	3.27	1.62
Korhogo	1.73	1.62
Abengourou	1.69	3.07
Man	1.73	1.62

Mean value of geometric dilution of precision of BeiDou-3 is around 2.27 while that of GPS is stable at 1.62. Both are excellent, but the GPS' GDOP is less than BeiDou-3' GDOP, excepted in Abengourou where BeiDou-3' GDOP is a little over half of GPS.

4.3. Comparison of navigation accuracy

At least, the navigation accuracy are obtained in the locations and the Simulation results are summarized in Table 5.

Table 5. Navigation accuracy BeiDou-3 VS GPS

	BeiDou-3(m)	**GPS(m)**
Abidjan	14.53	8.12
Yamoussoukro	16.36	8.09
Korhogo	8.84	8.1
Abengourou	8.46	15.33
Man	8.66	8.24

The navigation accuracy value of BeiDou-3 is more than 8.5 meters in general, the minimum value is obtained in Abengourou where the accuracy of BeiDou-3 is higher than GPS (Table5). Otherwise, the navigation accuracy value of GPS is less than 8.5 meters in the four other cities.

5. Conclusion

(1) Using STK software to simulate BeiDou-3 constellation from 4 o'clock 02 may to 4 o'clock 03 may 2019 in five representative locations (Abidjan, Yamoussoukro, Korhogo, Abengourou and Man) situated in south, center, north, east and west respectively in Cote d'Ivoire, the results are conclusive for BeiDou-3.The BeiDou-3 is accurate with less than 16.5 meters position errors, and this value can decrease to 8.5 meters in the east part of Cote d'Ivoire.

(2) Navigation accuracy of GPS at the same moment shows better accuracy in the different parts of Cote d'Ivoire with less than 8.5 meters position errors, but this performance of GPS is less than that of BeiDou-3 in the east part of Cote d'Ivoire.

(3) BeiDou-3 can provide continuous and reliable navigation services in Cote d'Ivoire, indirectly in the surrounding region of the Golf of Guinea. In general, GPS positioning is better in the most important cities in Abidjan and Yamoussoukro. The output positioning performance of BeiDou-3 in the east part of Cote d'Ivoire is better than GPS. Simply, using some BeiDou-3 specific receivers and products of enhancement should improve the performance of BeiDou-3 in Cote d'Ivoire.

Acknowledgements

This work is supported by National Natural Science Foundation of China under Grant Nos. 41771487, 41631072 and Natural Science Foundation for Distinguished Young Scholars of Hubei Province of China under Grant No. 2019CFA086.

References

[1] Chen B. (2019) "BeiDou" opens the curtain of high-density networking in 2019. International Space, 5: 4-7.
[2] Zhang H, Wu W. (2019) Research status and overview of BeiDou satellite navigation industry standard. Time Automotive, 2: 52-53.
[3] Yun M, Sun B. (2013) BeiDou satellite navigation system officially provides regional services. Satellite Applications, 1: 5-8.
[4] Li G, Zhu X. (2013) Overview of BeiDou system application, opportunities and challenges. Journal of Navigation and Positioning, 3: 22-27.
[5] Su X. (2019) Analysis of the current situation and development prospect of BeiDou satellite navigation system. Journal of Guangxi Radio and Television University, 3: 89-92.
[6] Li D, Li S, Zhao J, Niu W, Liu J. (2016) Review of research on signal acquisition methods of BeiDou System. Computer Engineering and Science, 5: 905-913.
[7] Li H, Tang Q, Zhang H. (2018) The positioning performance simulation analysis of BeiDou-2 and GPS in the arctic region. 2018 International Conference on Computational, Modeling, Simulation and Mathematical Statistics,11-15.
[8] Yang Y X, Li J L, Wang A B, et al. (2014) Preliminary assessment of the navigation and positioning performance of BeiDou regional navigation satellite system. Science China: Earth Sciences, 57: 144-152.
[9] Montenbruck O, Hauschild A, Steigenberger P, et al. (2013) Initial assessment of the COMPASS/BeiDou-2 regional navigation satellite system. GPS Solutions, 17: 211-222.

Study on acoustic transmission characteristics of BHA in directional well

Haiming Xie[1], Jing Zhou[1], Peifen Zhang[2]

[1]National Engineering Laboratory for oil and gas drilling technology, Xian shiyou university, Xian, China

[2]Xi'an Quan Feng Zhi Kong Technology Co., Ltd. Xi'an, China

Abstract: Acoustic telemetry technology while drilling may effectively solve the practical and speed bottleneck of downhole information transmission, but its channel characteristics are subject to the influence of drill string structure in the channel. According to the requirements of reservoir exploitation and drilling technology, the directional well technology is becoming more complex, and the changeable bottom hole assembly (BHA) in the well causes the aperiodicity of acoustic channel. Through the modeling and analysis of BHA under different drilling technology, the characteristics of various aperiodic channels are obtained, and compared with the characteristics of ideal periodic channels, the acoustic transmission characteristics of drill string in directional well are obtained. The aperiodicity brought by BHA makes the high frequency section of the channel seriously attenuated, so the number of aperiodic drilling tools used in the acoustic remote transmission channel needs to be strictly controlled. It can be excluded from the channel if necessary.

1. Introduction

The telemetry while drilling technology is used to transmit downhole instruments and surface data, which is one of the key technologies of intelligent drilling. At present, the most widely used measurements while drilling (MWD) technology is based on mud medium, and data transmission based on electromagnetic wave is also used in oil fields, but both technologies have limitations, and the data transmission rate is difficult to meet the transmission requirements of a large number of data while drilling[1]. The high-speed remote transmission of data while drilling has become a bottleneck to limit the intellectualization of downhole drilling instruments.

In theory, the data transmission speed of acoustic wave in drill string is 1-2 orders of magnitude higher than that of mud pulse and electromagnetic wave. In 1948, sun oil company of the United States began to study the technology of acoustic wave transmission in drill string. In 1983, Drumheller used the one-dimensional sound propagation theory to establish the sound propagation characteristics of an ideal drill string system[2-3]. As of 2011, Xact has served more than 400 wells, of which the maximum well depth is 4000m under the use of two transponders[4]. In 1991, relevant research began in China, and Liu Qingyou established the mechanical model of drill string vibration[5]. At present, the channel characteristics of the periodic drill string and the attenuation of the acoustic signal by the environment have been fully studied in China, and the influence of different drill string structures on the channel has also been studied.

In thin reservoir, fault block reservoir, marginal reservoir, remaining reservoir and other hard to develop reserves, the difficulty and cost of oil and gas production continue to increase[6]. To further improve oil recovery and minimize formation damage, directional drilling techniques such as

IOP Conf. Series: Materials Science and Engineering **799** (2020) 012012 doi:10.1088/1757-899X/799/1/012012

horizontal wells and extended reach wells are needed [7-9].

The structure of complex bottom hole assembly (BHA) and drill string used in directional well is not consistent, which will lead to the aperiodicity of the acoustic channel of drill string. The analysis of the channel characteristics of drill string in directional well can improve the design method of MWD when complex BHA is used.

2. Channel analysis of BHA in directional drilling

In the process of directional drilling, according to the design requirements of different drilling technology, the BHA with bent joint and downhole screw power are often used.With the use of stabilizers, non-magnetic drill collars, weighted drill string, flexible short joints, conversion joints and other drilling tools, the bottom drilling tool generates guiding force by using the bending angle of the bending joint itself with the aid of the fulcrum of the well wall, and the lateral cutting of the formation causes the deviation of the newly drilled well axis, so as to achieve the purpose of directional drilling[10,11]. However, the structural change brings about the non periodic change of the drill string channel.

The structure of drill collars in the BHA is a cylindrical pipe with constant cross-section, and there is no abrupt change of cross-section. According to the analysis of the transmissive film model, the transmission coefficient is 1 and the reflection coefficient is 0. Therefore, the influence of stabilizer on the transmission characteristics of drill string is mainly considered in the BHA of directional drilling. It is assumed that the number of drill strings to be analyzed is 100 and the transmission distance is 948m.The structural parameters of different drilling tools in the drill string channel are shown in Table 1 [12].

Table 1 structural parameters of different drilling tools

Type	ID(mm)	OD(mm)	Cross-sectional area(m^2)	Length(m)
Drill string body	121.3	139.7	0.0038	8.6848
Drill string joint	88.9	177.8	0.0186	0.4572
Stabilizer joint	71.4	177.8	0.0208	1.50
Stabilizer working face	71.4	215.9	0.0326	0.5

1) Channel analysis of BHA with single stabilizer

Only by changing the position of a single stabilizer in the BHA and fully considering the well parameters, drilling parameters and formation factors, can the well deviation of the well trajectory be adjusted[13]. As shown in Fig. 1, sound waves are emitted from the left side of the string. A stabilizer is used next to the acoustic generator, and the rest are periodic drill string structures composed of pipe body and joint. The stabilizer is used to calculate the propagation characteristics of the acoustic wave in the drill string structure, which is equivalent to the drill string periodic sequence by using the penetrating anti boundary mask model. Through the equivalent reflection film model, the stabilizer is included in the combination of drill string period, and the propagation characteristics of acoustic wave in drill string structure are analyzed as a whole.

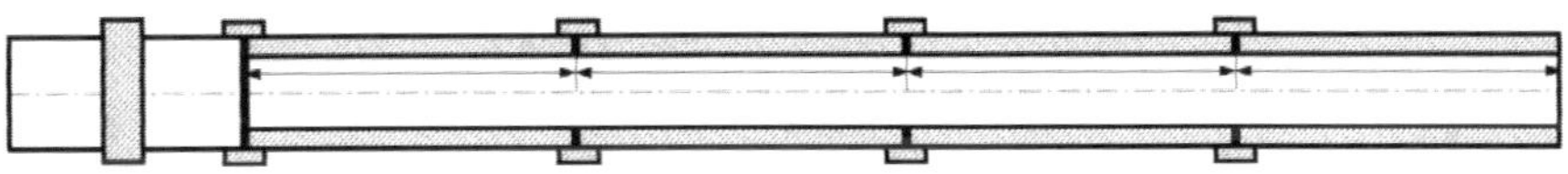

Figure 1 the equivalent structure of acoustic transmission of BHA with single stabilizer

IOP Conf. Series: Materials Science and Engineering **799** (2020) 012012 doi:10.1088/1757-899X/799/1/012012

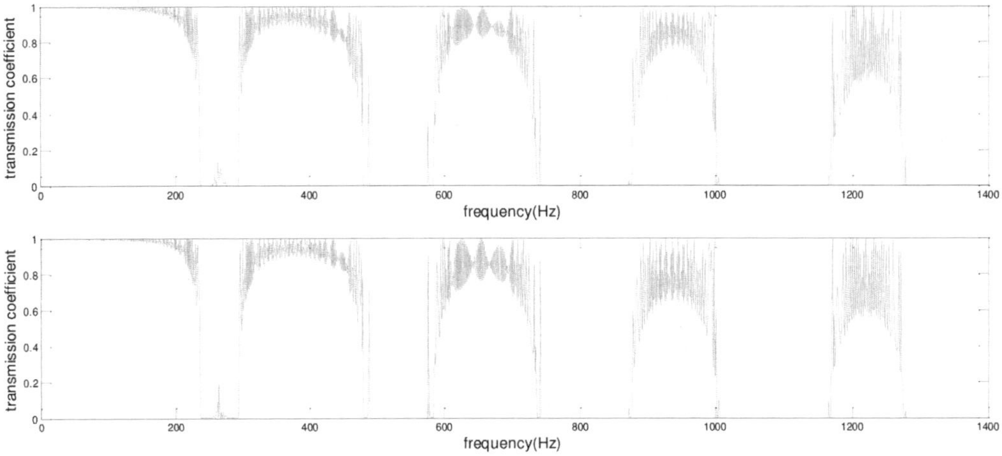

Figure 2 Comparison of influence of single stabilizer on acoustic transmission characteristics

In Figure 2, the bottom figure shows the ideal acoustic transmission characteristics of 100 periodic drill strings without stabilizer, and the top figure shows the acoustic transmission characteristics of the BHA with stabilizer.

It is found that a single stabilizer has a certain influence on the acoustic wave transmission, and the overall transmission characteristics have little influence.

2) Channel analysis of BHA with double stabilizers

The single stabilizer can only reduce and increase the deviation of the well trajectory. When directional drilling, the double stabilizer is often used in different positions of the drill string combination to achieve the operation requirements of stabilizing, reducing and increasing the deviation. The transmission characteristics of the double stabilizers are analyzed here.

Figure 3 the equivalent structure of acoustic transmission of BHA with double stabilizers

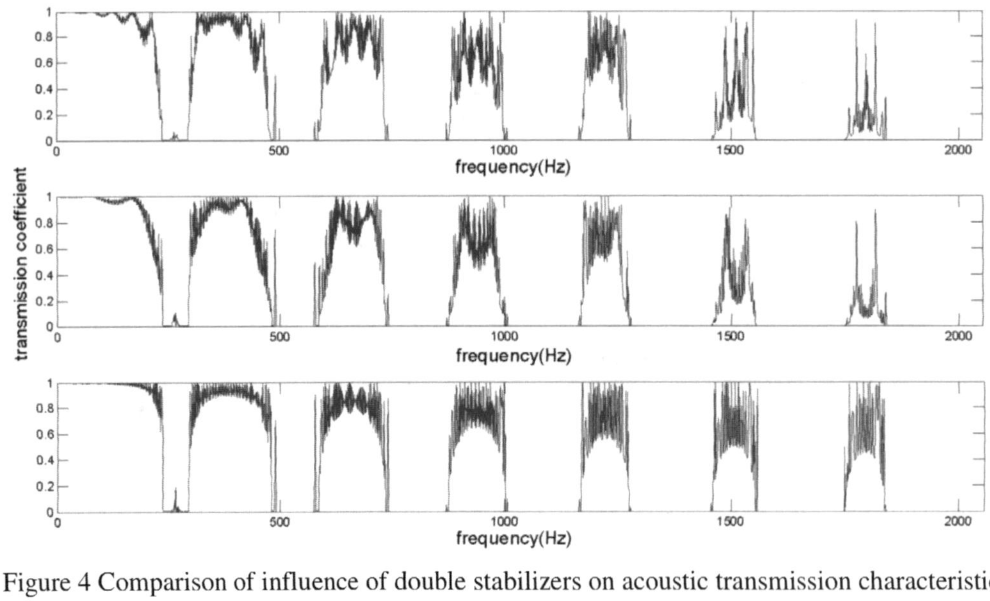

Figure 4 Comparison of influence of double stabilizers on acoustic transmission characteristics

The figure below in Figure 4 shows the acoustic transmission characteristics of 100 periodic drill strings without stabilizer. The figure above shows the acoustic transmission characteristics of the BHA when the second stabilizer is 50m away from the acoustic generator. The figure in the middle shows the transmission characteristics of the second stabilizer 20m away from the acoustic generator.

It can be seen from the comparison that the different positions of the stabilizers lead to the great fading of the acoustic transmission characteristics of the drill string channel, especially for the passband with the frequency above 1300hz. However, the position of the second stabilizer in the double stabilizer BHA has little influence on the acoustic transmission characteristics.

3) Channel analysis of BHA with multiple stabilizers

Under the influence of formation dip angle and geological structure, it is difficult to achieve directional drilling with stable well deviation. In order to achieve better results, three or more stabilizers are often used in the actual drilling process to increase the rigidity of downhole BHA and achieve the purpose of better stable well deviation, as shown in Figure 5. Generally, the distance between two stabilizers in the BHA is between 10-20m. According to different process requirements, specific matching parameters are shown in Table 2.

Table 2 matching dimensions of BHA with stable inclination

BHA	L1(m)	L2(m)	L3(m)	L4(m)	L5(m)
Strong	0.8~1.2	4.5~6.0	9.0	9.0	9.0
Medium	1.0~1.8	3.0~6.0	9.0~18.0	9.0~27.0	——
Weak	1.0~1.8	4.5	9.0	——	——

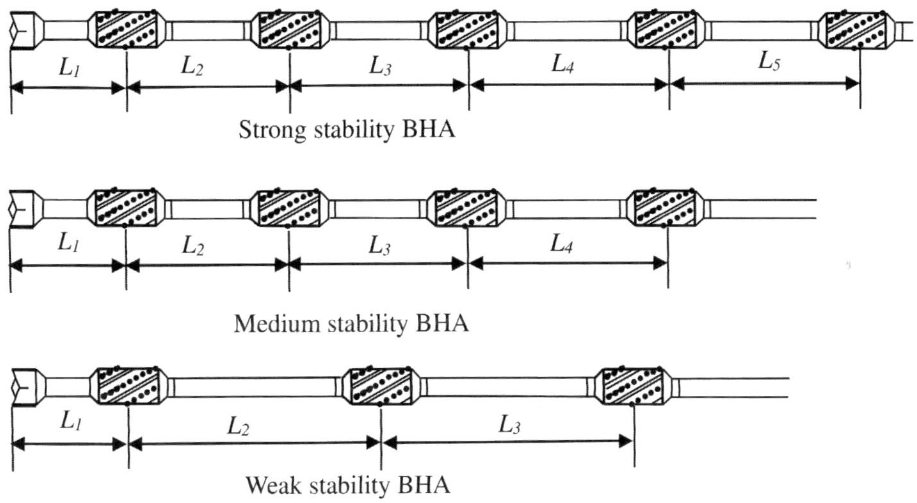

Figure 5 the structural of BHA with multiple stabilizers

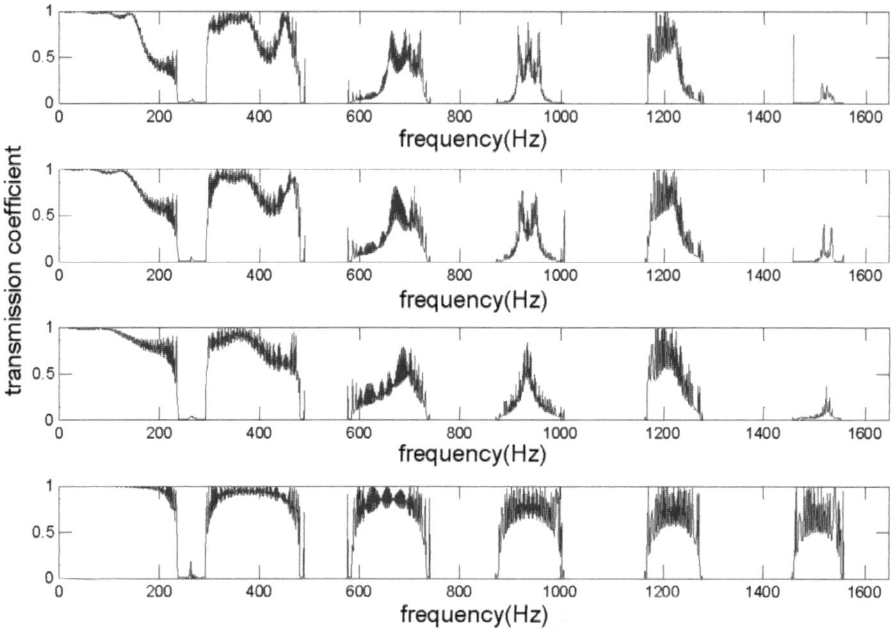

Figure 6 Comparison of influence of multiple stabilizers on acoustic transmission characteristics

In Figure 6, the acoustic transmission characteristics of drill string channel with 5, 4 and 3 stabilizers are respectively shown from top to bottom. The bottom figure shows the ideal acoustic transmission characteristics of 100 periodic drill strings without stabilizers. It can be seen that the use of multiple stabilizers in the channel seriously attenuates the acoustic transmission signal, and has relatively small impact on the passband below 500Hz, but the passband attenuation above 1400hz is almost zero. In the combination of multiple stabilizers, the distance between the second connected stabilizers is equal, so the number of multiple stabilizers has little effect on the acoustic transmission.

3. Conclusion

The structure of stabilizer in BHA destroys the periodicity of the drill string channel and affects the channel characteristics of the periodic drill string. Three stabilizers attenuate the frequency band above 500Hz seriously, and the number of stabilizers has a great influence on the acoustic transmission characteristics. If multiple stabilizers appear in the BHA aperiodically, the situation will be more serious.

Therefore, the aperiodic drill string structure in the drill string channel has a significant impact on the channel characteristics. Although the directional requirements are considered in the actual drilling process, it is necessary to avoid the occurrence of similar structure drilling tools in the transmission channel as much as possible to avoid the occurrence of various abnormal structures in the channel. If not, the position of the transducer must be moved up.

Acknowledgements

The authors acknowledge the Class A Strategic Leading Science and Technology Subproject of the Chinese Academy of Sciences (Grant: XDA14030103), the National Natural Science Foundation of China (Grant: 51874238), the PhD Start-up Foundation of Xian Shiyou University.

Reference

[1] ZHOU Xiaohui, SONG Guiqiao, ZHANG Weihua, NI Yao, CAI Jiexiong. Current research progress of seismic while drilling technology. Geophysical Prospecting for Petroleum, 2016, 55(6): 913-923.

[2] Yan Z, Wei C, Geng Y, et al. Design of a rotary valve orifice for a continuous wave mud pulse

generator[J]. Precision Engineering, 2015, 41:111-118.
[3] Suh A. Innovative Instrumented Motor with Near-bit Azimuthal Gamma Ray and Inclination Improves Geosteering in Thin-bedded Formations[J]. Oil Gas European Magazine, 2013, 39(1):22-24.
[4] Barnes T G, Kirkwood B R. Passbands for Acoustic Transmission in an Idealized Drill String[J]. Journal of the Acoustical Society of America, 2005, 51(5B):1606-1608.
[5] Lee H Y. Drillstring axial vibration and wave propagation in boreholes[D]. 1991.
[6] Rodríguez-Rozas A, Pardo D, Torres-Verdín C. Fast Simulation of 2.5D LWD Resistivity Tools[C] Eage Conference and Exhibition. 2017.
[7] Arata F, Mele M, Tarchiani C, et al. Look Ahead Geosteering via Real Time Integration of Logging While Drilling Measurements with Surface Seismic[C] Spe Technical Conference and Exhibition. 2017.
[8] Zhang Q, Zhang F, Wang Y, et al. Dual-spacing hydrogen index correction method for neutron gamma density measuring in LWD[J]. Zhongguo Shiyou Daxue Xuebao, 2017, 41(4):78-84.
[9] Kumar L S, Han W K, Guan Y L, et al. Downhole pipe selection and arrangement for acoustic drillstring telemetry[C] Industrial Electronics and Applications. IEEE, 2011:1539-1542.
[10] Kumar L S, Han W K, Guan Y L, et al. Optimization of acoustic communication for industrial drilling[C] Information & Communication Technologies. IEEE, 2013:1060-1063.
[11] Gutierrez-Estevez M A, Krueger U, Krueger K A, et al. Acoustic broadband communications over deep drill strings using adaptive OFDM[C] Wireless Communications and NETWORKING Conference. IEEE, 2013:4089-4094.
[12] Reeves, M., Smith, D. G., Groves, D., Brehm, A., Rovira, S. A., & Armagost, K.Unique Acoustic Telemetry Network With Distributed Pressure Measurement Nodes Enables Accurate Real-Time Analysis of Sweep Effectiveness. Society of Petroleum Engineers[C]. SPE Annual Technical Conference and Exhibition , 2013.
[13] Zhou L, Annamalai M, Koh J, et al. A Crystal-Less Temperature-Independent Reconfigurable Transmitter Targeted for High-Temperature Wireless Acoustic Telemetry Applications[J]. IEEE Transactions on Circuits & Systems II Analog & Digital Signal Processing, 2013, 60(9):542-546.

IOP Conf. Series: Materials Science and Engineering **799** (2020) 012013 doi:10.1088/1757-899X/799/1/012013

Development and Current Situation of Radar Performance Parameter Measurement Simulation Training Platform

Jun Wang[1,2*], Bin Deng[3] and Chuanqi Li[4]

[1]Graduate team 1, Air Force Early Warning Academy, Wuhan, Hubei, 430019, China

[2]The 1 team, No.32382 Unit, the PLA, Wuhan, Hubei, 430311, China

[3]Air defense early warning equipment department, Air Force Early Warning Academy, Wuhan, Hubei, 430019, China

[4]Graduate team 2, Air Force Early Warning Academy, Wuhan, Hubei, 430019, China

[*]Corresponding author's e-mail: 278923898@qq.com

Abstracts. The measurement of radar performance parameters is the basic work of radar support forces, and the development of simulation training platform has a very good role in promoting this work. This paper describes the current situation of radar parameter measurement, introduces the development of equipment simulation training platform, and combs the development process of radar performance parameter measurement simulation training platform. At last, the practical requirements of the simulation training platform for radar performance parameter measurement are put forward, that is, by building the simulation training platform for radar main parameter measurement, to meet the requirements of the new two-level maintenance system for the parameter measurement of military maintenance personnel.

1. Introduction

In the modern battle command system, radar is a kind of sensor that can collect all kinds of military information, and it can detect the target's distance, measure the target's parameter information quickly, and work around the clock. It becomes the multiplier of aircraft, missile, ship and other battle platforms to play their combat performance and the "eye of a thousand" of land air defense alert. The measurement of radar performance parameters runs through all stages of the life cycle of radar equipment. It is a basic technical work to evaluate the quality of equipment and check whether the radar performance meets the design requirements. It is also a technical means to master the technical status of equipment and isolate and locate equipment faults. It is very important not only to ensure the design of radar equipment is reasonable in the development and production stage, but also to save the cost of production and debugging. At the same time, it is also of great significance to improve the availability of radar equipment in the use stage, to ensure that radar equipment is always in good working condition, and to give full play to the operational efficiency of radar equipment [1].

2. Current situation of radar parameter measurement

In the work of radar equipment support, parameter measurement is an important technical means for radar equipment maintenance technicians to understand the technical status of equipment, analyze and judge equipment faults, implement equipment maintenance and carry out technical level identification

[2]. However, with the wide application of various new technologies and systems in the new radar equipment, there are many items of equipment performance parameter measurement and high precision requirements. Most of them need to use microwave power meter, spectrum analyzer, network analyzer and other radio-frequency instruments to complete the test, which requires more and more technicians [3]. The radar equipment of the grass-roots units has been on duty for a long time, so it is difficult to make time for targeted parameter measurement training. In the training stage of colleges and universities, the teaching of radar parameter measurement also faces the contradiction of more people and less machines. During the course, the students have less opportunities to get on the computer, so it is difficult to ensure that everyone can master the testing skills and can not fully meet the equipment support posts of the grass-roots units Competency requirements.

After the adjustment of the military establishment system, most of the basic radar stations no longer allocate cadres and technicians, so they have higher and higher requirements for the comprehensive maintenance ability of the sergeant technicians, especially for the maintenance ability combined with the test parameters of instruments and meters. On the one hand, from the requirements of practical combat training, higher requirements are put forward for radar equipment maintenance support and parameter test. Technical support personnel must strengthen radar parameter test training and radar maintenance support training. On the other hand, there is no platform dedicated to parameter measurement training for grass-roots radar units, and traditional training methods and means cannot meet the requirements of military training program in the new situation The test and maintenance ability of technical support personnel cannot be improved rapidly and effectively [4].

3. Development of equipment simulation training platform

3.1. Development of foreign forces

In the late 20th century, the world's major military powers vigorously carried out simulation training in the field of military training, the essence of which is to use modern simulation technology based on computer to realize the leap of military training. The simulation training of the US army began in the early 1970s, and experienced different stages such as artificial simulation, semi-automatic simulation and computer simulation[5]; in the early 1980s, the US army began to use simulation technology for simulation training, paid attention to the development of combat simulation system suitable for actual combat requirements, and successively organized the development of "Contract Tactical Simulation Training System" and "Army Combat Simulation System", etc.; in the 1990s, the U.S. military established the U.S. National simulation center, which is specially responsible for the development, development and management of simulation training system, to support the simulation training of the U.S. military, especially the application of virtual reality technology and distributed interactive simulation technology, which greatly promoted the extensive application of computer simulation training in the U.S. military[6].

With the development of virtual reality technology, the U.S. Army has developed a large number of training simulation systems which are closer to the actual combat. At present, the simulation training system of the U.S. Army has been widely used in college teaching, weapon equipment operation training, complex professional technology training, command and decision-making training, campaign and tactical training and even strategic training [7]. In recent years, the simulation training of the U.S. Army has integrated various high and new technologies. Based on the training base, the training simulation system has been used to provide simulation training similar to the actual combat for all kinds of military and theater forces. Some of the results have been tested in the actual combat, and the simulation equipment development has been modified in combination with the actual combat [8].

3.2. Development of our army

On the basis of learning from the advanced experience of foreign troops, combined with the characteristics of our army, a batch of command simulation training systems for different levels and arms are developed, forming the model design specifications and unified data interaction standards

IOP Conf. Series: Materials Science and Engineering **799** (2020) 012013 doi:10.1088/1757-899X/799/1/012013

suitable for the actual command simulation training system of our army. There are many theoretical and practical researches on the equipment simulation training system. In reference [9], a general solution to the control and display problem of weapon equipment simulation training system is proposed by establishing a CDLD model. In reference [10], the overall framework, logical structure, functional modules and operation process of the simulation training system for vehicle equipment support are designed, and the theoretical framework of the simulation training system for vehicle equipment support is established. According to the requirement of military training in the new situation, the paper puts forward the conception and Countermeasures of the construction of flight simulation training equipment system from the top level. According to the requirement of integrated training of all elements of equipment support, a simulation training system of integrated training of all elements of equipment support is designed in reference [12]. In reference [13], the data integration mode of data warehouse is adopted to solve the problem of data sharing and exchange of simulation training subsystem, which provides data support for the integration of simulation training system of ordnance equipment support. In reference [14], the realization of driving simulation training system for amphibious armored equipment is studied. Literature [15] focuses on improving the use efficiency of equipment simulation training system in combat training, and summarizes the practical application value of simulation training system through comparative analysis, principle analysis and function analysis. In reference [16], a design method and idea of radar simulation training system console is proposed, and a console of radar simulation training system is designed and implemented accordingly. In reference [17], according to the operation and training characteristics of group vehicles, the overall scheme and implementation method of simulation training system based on hardware in the loop simulation are proposed. In reference [18], in order to improve the electrical maintenance support ability of a certain type of equipment, the design and development of electrical simulation maintenance training platform for a certain type of equipment is realized by the comprehensive application of virtual reality, modeling and simulation, human-computer interaction and other technologies. In reference [19], advanced computer simulation technology and satellite communication and navigation technology are adopted, and the design concept of communication equipment support simulation training system adapted to the future war is proposed. Literature [20] discusses the application of virtual reality technology in the development of equipment command simulation training system by studying the development process of editing and analysis system of virtual battlefield environment in a certain equipment command simulation training system.

4. The development of radar parameter measurement simulation training platform

Radar troops have always attached great importance to military training. In order to improve training efficiency, they have paid special attention to the development and application of simulation training software since 1980s. In the revised military training program in 2008, the research results of simulation training software were firstly absorbed. At that time, 20 mature radar simulation training system software was included in the program, which opened the precedent of radar simulation training and opened the prelude of radar simulation training. It not only adapts to the new situation of rapid development of radar equipment, but also improves the quality, efficiency and combat effectiveness of radar training Provide training platform and means. At present, the system platform for training and development of radar equipment performance parameter testing can be divided into two categories, one is simulation training software developed by computer software, the other is semi physical simulation training platform.

4.1. Simulation training software for performance parameter measurement

Since 2008, the development of radar operation training platform has become more and more perfect, but it is mainly reflected in the training of radar operators, mainly for the tactical operation instruction training of equipment. However, after comparing the training of equipment performance parameter testing skills, it can not meet the needs of maintenance technicians for the new system equipment performance parameter measurement training in a period of time, so it is urgent to develop the corresponding simulation training software. In this context, in the construction of the air force simulation training center project, the test and training software for the performance parameters of the active main combat radar equipment was approved, and the task was assigned to develop the "24 type new radar performance parameter test and training simulation system" under the leadership of Professor Yang Jiangping of the air force early warning college. The system is aimed at the current main combat radar equipment (24 types in total) The test of performance parameters is designed by using three-dimensional virtual technology. The virtual test scene is carried out around the test process of main performance parameters. The test connection and instrument operation of the whole process of real simulation equipment test provide a realistic training environment for the trainees, which alleviates the urgent need of parameter performance test training software to a certain extent.

In addition, some radar units have developed the corresponding parameter test training software for some types of radar equipment. The most successful and mature training software is the 408e radar parameter test multimedia operation training system developed by the 24th brigade of the radar corps, which consists of four functional modules: test principle, test video, test training and test assessment. Among them, the "test principle" function module teaches the basic knowledge and test methods of performance parameter test through micro course; the "test video" module provides the demonstration of main performance parameter test steps and the explanation and demonstration of specific test process by video recording. The "test training" module is to provide training scenarios and guidance for test connection and instrument operation. The "test assessment" module is to comprehensively assess the basic knowledge, method steps, test skills, precautions and other contents of the trainees in the way of question bank.

4.2. Semi physical training platform for performance parameter measurement

The hardware in the loop simulation platform for performance parameter measurement is actually an important part of the equipment simulation training platform, and its development is basically consistent with the development of the equipment hardware in the loop simulation platform.

In order to promote the students to study and consolidate the radar principle, radar testing principle and other knowledge, the air force early warning Institute independently or jointly developed multiple sets of radar parameter measurement teaching equipment to ensure professional teaching. The main function of the platform is to complete the comprehensive test of each subsystem, which is helpful to enhance the students' memory of radar basic principle knowledge and improve maintenance support ability.

- In 1995, we first developed a test platform for non-solid state transmitter, which opened the way for radar parameter measurement simulation training;
- In 2000, the air force early warning Institute and 38 institutes jointly developed the receiver experimental platform, which can simulate the traditional receiver test (excluding A/D sampling test), including the P-segment radar receiver, etc. The P-segment radar receiver as shown in Figure 1;

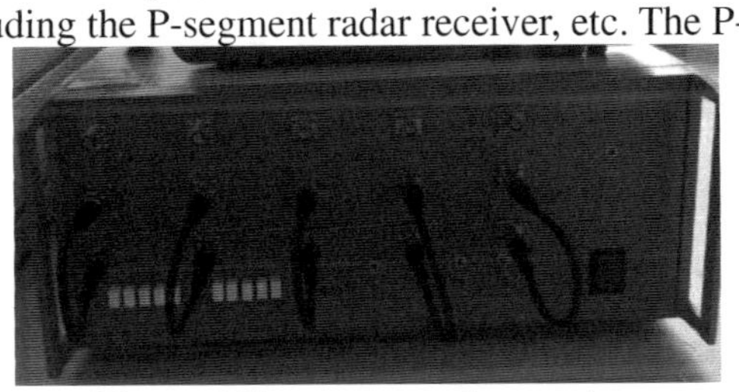

Figure 1 P-segment radar receiver

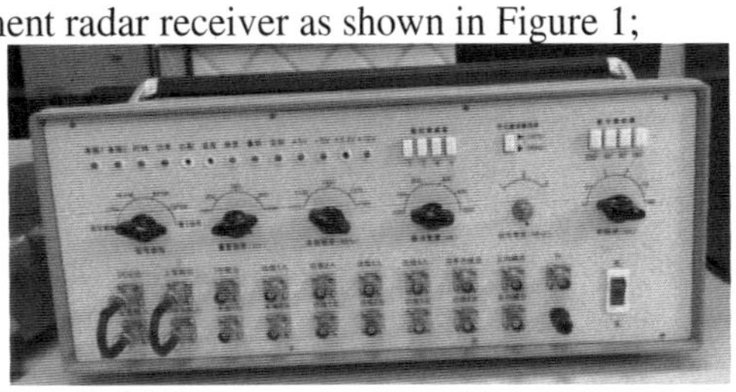

Figure 2 Modern radar launching test chamber

- From 2009 to 2010, the simulation experiment platform of launch vehicle was developed, which realized solid-state transmitter test. However, the selection of platform parameters and repetition frequency were preset by analog switch, such as modern radar launch test box, L-band radar launch parameter test and training assessment system, radar transmitter performance parameter test platform, etc. As shown in Figure 2, the modern radar launch test chamber;

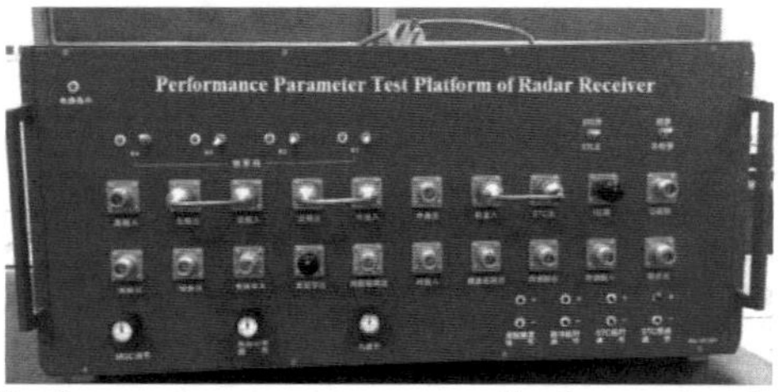

Figure 3 Performance parameter test platform of radar receiver

- At the same time, a digital receiver test platform was developed in 2010 to realize A/D sampling test, such as radar receiver performance parameter test platform, etc. The radar receiver performance parameter test platform as shown in Figure 3;
- In 2015, the air force early warning college and Radar Brigade 5 jointly developed a digital transmitter test platform, and realized the functions of parameter selection and repetition frequency through single-chip microcomputer and computer software; in 2016, combined with the construction of 2110, phase 3 of the air force early warning college, Huangpi non military academy developed and equipped a number of launch and receive test platforms. However, there is no special testing platform in the College of antenna feeder system testing, and the laboratory has not done relevant research and development, only the standing wave ratio and loss testing equipment of feeder.

In recent years, other colleges and universities have also developed some radar performance parameter measurement simulation training platforms. For example, in 2017, the ordnance Sergeant School of the Army Engineering University and Anhui egret Electronic Technology Co., Ltd. jointly developed a microwave antenna test system. Designed according to the measurement requirements of microwave antenna, it is composed of spectrum analyzer, signal source, transmitting antenna, receiving antenna, turntable and system software. It can be compatible with traditional measurement methods and automatic measurement methods based on human-computer interaction. It has the functions of antenna standing wave ratio measurement, antenna pattern measurement, polarization characteristic measurement, antenna gain measurement, etc. it can carry out multiple tests at the same time To help the students understand the antenna test principle and test equipment, and master the basic test methods of the antenna.

4.3. Existing problems

Colleges and universities have developed and distributed simulation training platforms for equipment parameter measurement and radar parameter measurement. The functions of these platforms are mainly focused on teaching use. The test items of each subsystem are comprehensive, basically covering all parameter tests. When there are many people, they can teach separately, understand the mechanism thoroughly, and meet the teaching requirements. However, according to the requirements of the new military training program for the technical support personnel of the grass-roots radar force, the force needs to test the main performance parameters that affect the radar, and to test the parameters required in the daily maintenance and repair. The existing test items of the simulation training platform are too comprehensive and cumbersome to be suitable for the maintenance and test training of the force. For example:

- Antenna feeder subsystem: the antenna pattern is not paid much attention to in the army, only focusing on the low power mode of VSWR measurement through network analyzer, and focusing on testing this mode with different instruments, which has not been tested before;
- Launch subsystem: the army mainly tests the output power, plus the signal-to-noise ratio at most, and does not pay too much attention to the power amplification of the front stage and the last stage. At the same time, the up conversion, the first stage mixing, the second stage mixing and other parameters have been packaged in the modern radar;
- Receiving subsystem: the technical support personnel of the army only test the parameters such as frequency source performance and launch excitation signal performance, while according to

the requirements of the new program, the parameters such as sensitivity and noise coefficient are tested by the base level repair organization. As for the bandwidth, dynamic range, channel consistency and other indicators, they are only accurate in the design and production process, and they are not required in the maintenance of the army to pay too much attention.

5. Real needs

In order to further improve the technical support ability of grass-roots radar forces, in the new "Air Force military training program" equipment support training, the training subjects and assessment standards of radar equipment support specialty are clearly required. Radar technicians need to be familiar with the contents of parameter test and data accumulation, master the methods and applications of test and data analysis. However, there is no special simulation training platform for radar technicians to carry out parameter test. By building a simulation training platform for radar main parameters measurement, which has the same test interface with the radar real installation, the simulation real installation training can adapt to the requirements of the new two-level maintenance system for the parameter measurement of the maintenance personnel in the army level (basic level, relay level), that is, radar station and repair Institute, solve the problems of lack of platform and less actual mobile phones in the practical training, and improve the military technology Equipment support ability of personnel, especially performance parameter test ability.

Through mass production, it can also solve the contradiction between radar units' training of parameter measurement and radar equipment's long-term duty task of combat readiness. It can provide important means and platform for radar units to carry out instrument operation, parameter test, maintenance skill training, combat assessment and other activities. Through continuous parameter measurement training on the platform, operators can understand the testing mechanism, master the testing methods, improve the testing skills, and finally achieve the purpose of improving the equipment capability of radar equipment technicians. Through the platform to carry out test training, training benefits can be greatly improved, training costs can be saved, and more importantly, the presence rate of technical cadres in grass-roots units can be guaranteed. The test and training system is also suitable for radar overhaul plant, brigade technical support team, grass-roots radar station and other repair levels. It can be used for radar maintenance technicians to carry out radar launch parameter test training and assessment, and can effectively and rapidly improve the level of parameter test of technical personnel.

6. Concluding remarks

With the development of electronic technology, communication technology, computer technology and other cutting-edge science and technology, radar equipment is constantly updated, and the means of parameter measurement are gradually innovated. It is of certain practical significance to develop and develop a new parameter measurement simulation training platform to meet the requirements of the new program, promote the rapid improvement of the support ability of the army, and improve the combat effectiveness.

References

[1] Deng, B.(2008)Electronic Measuring Instrument. National Defence Industry Press, Beijing.
[2] Wang, J.J.,Deng, B.,Hao, Y.,et al.(2016)Research and practice of “radar parameter measurement” network course construction. Contemporary Education Research and Teaching Practice,6:202-203.
[3] Shu, T., Yang, D.X., Chen, H.B., et al. (2010) CTL-713C weather radar parameter survey question and processing method. Meteorological and Environmental Sciences,33:80-83.
[4] Wang, F., Qin, W., Wei, X.H., et al. (2013) Measurement technology of radar performance. TECHNOLOGICAL PIONEERS,5:172
[5] Wang, H.C., Zhao, X.T., Teng, Y.A. (2010) The main trend of us military simulation equipment training. SCIENCE & TECHNOLOGY INFORMATION,9:39.

[6] Liang, Z.P., Tang, C.(2010)Experience analysis of U.S. military simulation training. World Military Review,12:70-71.
[7] Wu, M.W., Meng, X.C.(2005)Research on the development of simulation training in the U.S. military. World Military Review,1:48-51.
[8] Liu, X.X., Wang, Y., Shao, X.(2010)Design and realization of training simulation system based on radar. Ship Electronic Engineering,12:84-86.
[9] Xu, L., Wu, X.X., Zhang, C.J. (2011) CDLD: an introduction to the development model of weapon equipment simulation training system. SCIENCE & TECHNOLOGY INFORMATION,3:64-65.
[10] Chen, X.B., Shi, Y., Liu, Y.D.et al.(2015)Vehicle equipment support simulation training system.Journal of Military Transportation University,17:13-17.
[11] Jiang,M.J., Liu,H.J.(2012)Construction of flight simulation training equipment system.Journal of Sichuan Ordnance,33:67-69.
[12] Su,X.J.(2014)Design of all-factor integrated simulation training system for equipment support.Modern Electronics Technique,37:60-63.
[13] Lian,Y.F.,Lu,Y.,Chen,L.Y.,et al.(2015)Research on data integration of ordnance equipment training simulation system. Information Technology,5:8-11.
[14] Sun,Y.F.,Jia,Q.,Huang,Q.M.,et al.(2013)Implementation of driving simulation training system of amphibious armored equipment. Ordnance Industry Automation,32:12-16.
[15] Ma,Q.,Wang,W.S.,Chen,Z.(2016)Simulative training equipment application battle-focused training.Journal of Equipment Academy,27:108-111.
[16] Ma,H.P.,Nie,C.L.,Hou,S.L.,et al.(2012)Engineering Design and Realization of the Control Platform of a Radar Training Simulator System. Electronic Sci. & Tech.,25:94-96.
[17] Jiang,J.Z.,Zhang,W.M.,He,D.A.(2009)Hardware-in-the-Loop simulation of certain refueling equipment. Ordnance Industry Automation,28:43-45.
[18] Lv,X.P.,Yan,Y.M.,Liu,Z.C.,et al.(2013)Research on electrical simulation maintenance training platform of a certain equipment. Computer Measurement & Control,31:2486-2489.
[19] Xie,X.H.,Yang,J.T.,Yu,W.C.(2011)Research of imitation and training system of correspondence instruments guaranteeing under the information-based conditions. National Defense Science & Technology,1:29-32.
[20] Xi, X.Q., Wang,Y.C.,Qian,H.L.,et al.(2012)Application of virtual reality technology in the development of equipment command training simulated system. Journal of Equipment Academy,23:111-115.

IOP Conf. Series: Materials Science and Engineering **799** (2020) 012014 doi:10.1088/1757-899X/799/1/012014

Numerical Simulation of the Plane Electromagnetic Wave Transient Propagation in Lossy Medium

Li Gao, Mingjun Li, Feng Liu, Lulu Zhang

Departments of Electronic Engineering, University of Xi'an Shiyou, 710065, Xi'an

Email: 12162635@qq.com, 563962694@qq.com , liufeng1904@163.com, 1770764759@qq.com

Abstract: In the field of geophysical exploration, it is of great value to study the transient propagation characteristics of electromagnetic waves in lossy medium. The equations of electric field and magnetic field are derived. The absorbing boundary conditions and numerical stability conditions are analyzed. The two-layer medium model is established to simulate the propagation characteristics of plane waves. The simulation results show that the transient analysis of electromagnetic wave propagation can reflect the time fluctuation characteristics of electromagnetic field in multilayer media, and provide the indication of the boundary position.

1. Introduction

With the wide application of electromagnetic wave technology in communication, microwave imaging, geological exploration and many other fields, it is very important to study the propagation characteristics of plane electromagnetic wave. In the field of geophysical exploration, most of the geology is lossy medium, so the study of the propagation characteristics of plane electromagnetic wave in lossy medium is of great significance. Due to the layered structure of the earth medium, especially in the detection of underground targets, the targets buried in the multi-layered medium are often encountered. Therefore, it is necessary to study the reflection and transmission of plane electromagnetic waves in the multi-layered medium. The analysis of the transient characteristics of electromagnetic wave propagation can directly reflect the wave process in the electric field and magnetic fields of the plane electromagnetic wave with the passage of time, and can identify the position of the interface [1-4].

In order to analyze the transient characteristic of electric field and magnetic field, two layer inhomogeneous formation models are set up and the plane electromagnetic waves propagating characteristics numerical modeling are simulated.

2. Equations for solving 3-D electromagnetic fields

For the time harmonic field, substituting into the constitutive relation $\boldsymbol{D}=\varepsilon\boldsymbol{E}$ 、$\boldsymbol{B}=\mu\boldsymbol{H}$ 、$\boldsymbol{J}=\sigma\boldsymbol{E}$, the equations of electric and magnetic fields can be obtain[5-9].

$$\nabla\times\left(\frac{1}{\mu_r}\nabla\times\boldsymbol{E}\right)-\omega^2\varepsilon_r\boldsymbol{E}=-i\omega\boldsymbol{J} \tag{1}$$

$$\nabla\times\left(\frac{1}{\varepsilon_r}\nabla\times\boldsymbol{H}\right)-\omega^2\mu_r\boldsymbol{H}=\nabla\times\left(\frac{1}{\varepsilon_r}\boldsymbol{J}\right) \tag{2}$$

IOP Conf. Series: Materials Science and Engineering **799** (2020) 012014 doi:10.1088/1757-899X/799/1/012014

Where, $\boldsymbol{J}$ is the impressed current, $\boldsymbol{E}$ is the electric field strength, $\boldsymbol{H}$ is the magnetic field strength, $\boldsymbol{D}$ is the electric displacement vector, $\boldsymbol{B}$ is the magnetic induction intensity, ε_r is the dielectric constant, and μ_r is the magnetic permeability. When the boundary conditions is considered, we can get

$$\begin{aligned} F(\boldsymbol{E}) = & \frac{1}{2}\iiint_V \left[\frac{1}{\mu_r}(\nabla \times \boldsymbol{E})\bullet(\nabla \times \boldsymbol{E}) - k_0^2 \varepsilon_r \boldsymbol{E}\bullet\boldsymbol{E} \right] dV \\ & + \iint_{S_2} \left[\frac{\gamma_e}{2}(\hat{\boldsymbol{n}} \times \boldsymbol{E})\bullet(\hat{\boldsymbol{n}} \times \boldsymbol{E}) + \boldsymbol{E}\bullet\boldsymbol{U} \right] dS \\ & + ik_0 Z_0 \iiint_V \boldsymbol{E}\bullet\boldsymbol{J} dV \end{aligned} \tag{3}$$

$$\begin{aligned} F(\boldsymbol{H}) = & \frac{1}{2}\iiint_V \left[\frac{1}{\varepsilon_r}(\nabla \times \boldsymbol{H})\bullet(\nabla \times \boldsymbol{H}) - k_0^2 \mu_r \boldsymbol{H}\bullet\boldsymbol{H} \right] dV \\ & + \iint_{S_2} \left[\frac{\gamma_h}{2}(\hat{\boldsymbol{n}} \times \boldsymbol{H})\bullet(\hat{\boldsymbol{n}} \times \boldsymbol{H}) + \boldsymbol{H}\bullet\boldsymbol{V} \right] dS \\ & - \iiint_V \boldsymbol{H}\bullet\left(\nabla \times \frac{1}{\varepsilon_r}\boldsymbol{J} \right) dV \end{aligned} \tag{4}$$

They are the equations of solving electric field and magnetic field.

A single value can be assigned to each node when the solve area contains discontinuous interface. Because the boundary condition of normal field continuity is also imposed in this process, the grid at the boundary surface must be encrypted. However, grid refinement near the interface will increase the number of unknowns, and may make grid generation difficult, especially when the interface is irregular. It is also possible to introduce twice as many nodes on both sides of the discontinuity boundary.

3. Numerical simulation

In numerical simulation, the frequency is 2MHz, the space step is $\Delta z = 0.03\text{m}$ and the stability condition is $\Delta t = {}^{\Delta z}\!/_{c} \approx 10^{-10}(s)$ .The field on the boundary meet the absorbing boundary condition.

3.1. Normal incidence to the ideal conductor from vacuum

According to the theory of electromagnetic field, when the plane wave is incident from vacuum to the surface of ideal conductor, it will produce total reflection. The superposition of incident wave and reflected wave on the same side of the medium produces standing wave. For the pure standing wave, the anti-node and the node are fixed, and the electric field amplitude of the wave node is zero and constant. The maximum value of electric field amplitude is in the wave inverse node. The wave i anti-node of magnetic field and the node of electric field intensity are in the same position, and vice versa. The electromagnetic wave propagates along the z-direction, the anti-node and node is determined respectively Equation (5), (6)

$$z_n = \frac{n\lambda}{2} \quad (n = 0,1,2,\cdots) \tag{5}$$

$$z_m = \frac{(2m+1)\lambda}{4} \quad (m = 0,1,2,\cdots) \tag{6}$$

In numerical simulation, the conductivity of vacuum and metal is 0.0S/m and $1\times10^7 S/m$ respectively .The transient characteristics of electromagnetic wave propagation of $\frac{1}{8}\lambda$ 、

$\frac{1}{4}\lambda$、$\frac{3}{8}\lambda$ and $\frac{1}{2}\lambda$ far from interface is shown in Figure 1.

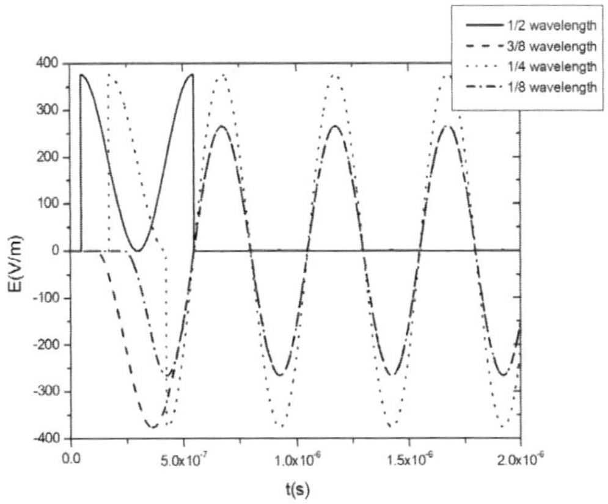

(a) Electric field

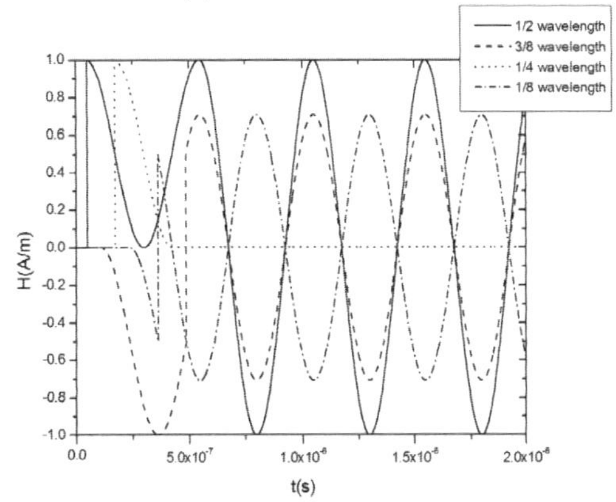

(b)Magnetic field

Figure 1. Electromagnetic propagating characteristic of total reflection.

3.2. The wave is normal incident to different resistivity medium

The study region is 0～1m, the sampling interval is 0.001m. There is the interface at 0.55m and the source is located at 0.2m. Two observation points are located at 0.35m and 0.75m respectively.

1) Plane wave is incident to the high resistivity medium from the low resistivity medium

Uniform plane wave is incident from medium 1($\sigma = 0.1S/m$) to medium 2($\sigma = 1.0S/m$) and the plane electromagnetic wave propagation transient characteristics is shown in Figure 2.

IOP Conf. Series: Materials Science and Engineering **799** (2020) 012014 doi:10.1088/1757-899X/799/1/012014

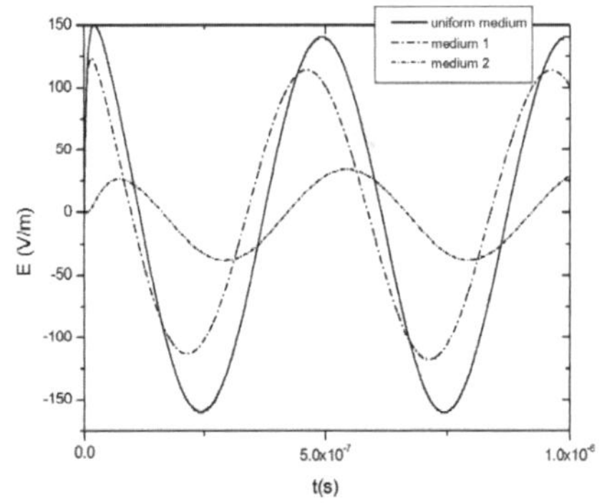

(a) Electric field

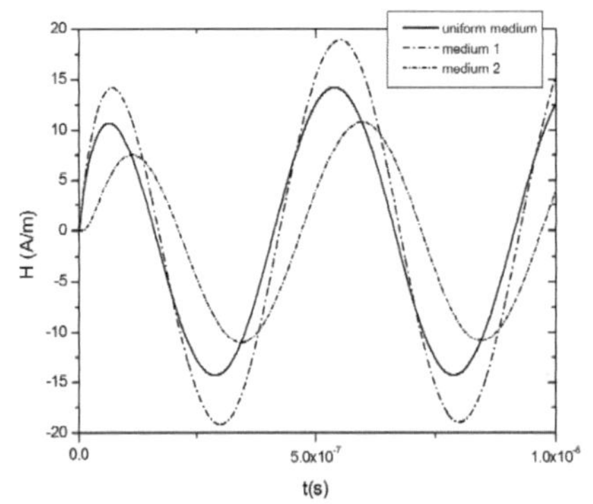

(b) Magnetic field

Figure 2. Transient characteristic when the wave is normal incident from high to low resistivity medium.

The amplitude of electric field in homogeneous medium (the amplitude of electric field when plane wave is incident from medium 1 to medium 2) is compared with that in homogeneous medium (the amplitude of electric field when plane wave is incident from medium 2 to medium 1), the former is larger than the latter. So it is obvious that the reflection phenomenon will not happen at the interface. Part of the energy of the wave is converted into heat, and the other part is the transmission wave through the interface. Wave impedance, reflection coefficient and transmission coefficient are all complex numbers, so there is phase shift between reflection field and transmission field.

2) Plane wave is incident to the low resistivity medium from the high resistivity medium

When a uniform plane wave is incident from medium 1($\sigma = 1.0S/m$) to medium 2($\sigma = 0.1S/m$), the transient characteristics of plane electromagnetic wave propagation are shown in Figure 3. When plane wave incident from medium 1 ($\sigma = 0.1S/m$)to medium 2, the amplitude of electric field in homogeneous medium is a little smaller than that in homogeneous medium. So it's clear that the reflection is very small at the interface.

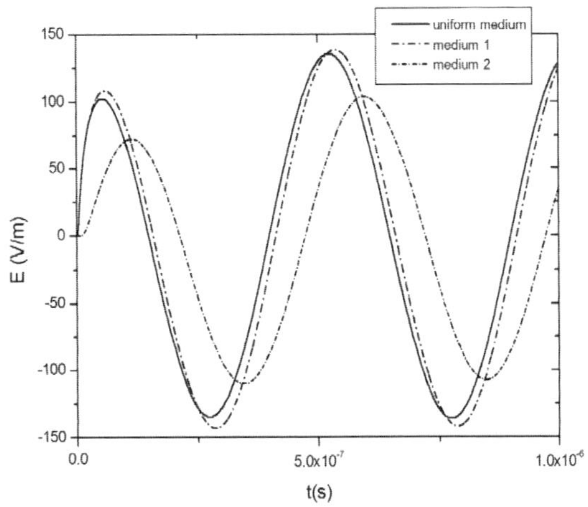

(a) Electric field

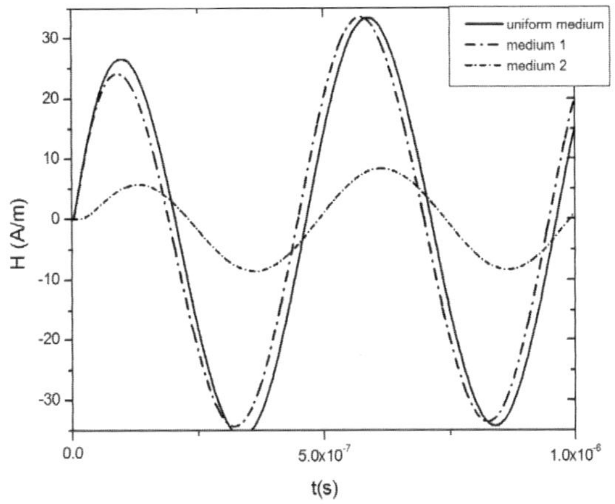

(b)Magnetic field

Figure 3. The transient characteristic when the wave is normal incident from low to high resistivity medium.

4. Conclusions

1) When the plane electromagnetic wave is directly incident from vacuum to the surface of the ideal conductor, the total reflection occurs, and the incident wave and the reflected wave are superposed on one side of the medium to produce the pure standing wave. When a high resistivity medium is incident on an ideal conductor, there are discontinuities in the transient curves of the electric and magnetic fields, which can identify the location of the interface.

2) When the plane electromagnetic wave is directly incident into the low resistance medium from the high resistance medium, there is no reflection phenomenon, and the reflection coefficient is negative; when it is incident into the high resistance medium from the low resistance medium, there is a reflection phenomenon, the reflection coefficient is positive, part of the incident wave energy is converted into heat energy, one part into reflected wave energy, the other part into transmitted wave energy, the reflection field and transmitted field There is phase shift between them.

Acknowledgements
The research was supported by Major National S&T Program of China No. 2011ZX05020-004-05.

References

[1] Hue Y K, TEIXEIRA F L. (2005) Modeling of EM Logging Tools in Arbitrary 3-D Borehole Geometries Using PML-FDTD. IEEE Geoscience and Remote Sensing Letters, ,2(1):78-81.
[2] Hue Y K, TEIXEIRA F L. (2005) Numerical mode-matching method for tilted-coil antennas in cylindrically layered anisotropic media with multiple horizontal beds. Transactions on Geoscience and Romote Sensing, 45(8):2451-2462.
[3] Chen H, Zhang L, Wang X M. (2012) Numerical modeling of responses to a titled-coil antenna in a Transversely Isotropic Formation. Chinese Journal of Geophysics, vol.55, no.10:3493-3505.
[4] Ou Y F, Wei B J. (2012) Simulation of titled coil's response in cylindrically stratified media and its application for electromagnetic propagation measurement-while-drilling. Journal of China University of Petroleum, vol. 36, no.5:72-79.
[5] Lin N, Wei B J, Yang J Z, et al. (2009) Basic theory of electromagnetic wave resistivity measurement while drilling tool with titled antennae and its application for geo-steering. Journal of China University of Petroleum, vol. 33, no. 1:44-49.
[6] WEI X C, LI E P. (2005) Efficient Solution to the Large Scattering and Radiation Problem Using the Improved Finite-Element Fast Multiple Method. Transactions on Magnetics, 41(5):1684-1687.
[7] Robert K. (2009) 3D numerical simulation of borehole sonic measurements acquired in dipping, anisotropic, and invaded formation. In: SPWLA 50th Annual Logging Symposium, The Woodlands, Texas, United States. 21-24 June.
[8] Lei Y Z. (2010) Analytical solution to time-harmonic electromagnetic field of titled circular coil by the coordinate transformation method. Transactions of China Electrotechnical Society, vol. 25, no. 4:15-18.
[9] Cui F Y, Wang G L, Zhang G J , et al. (2003) Application of staggered grid finite difference method to the computation of 3-D induction logging response. Chinese Journal of Geophysics, vol.46, no.4:561-567.

IOP Conf. Series: Materials Science and Engineering **799** (2020) 012015 doi:10.1088/1757-899X/799/1/012015

Air-Ground Rotary Scanning Imaging System for Ground Vehicles Based on Passive Infrared Detection

Bingshan Lei[1,2*], Weina Hao [3], Wenbin Zhu[2], Keding Yan[2]

[1]School of Mechanical and Electrical Engineering, Xi'an Technological University, Xi'an, 710021, China

[2]School of Electronic and Information Engineering, Xi'an Technological University, Xi'an, 710021, China

[3]Technological Center, Xi'an Technological University, Xi'an, 710021, China

[*]Corresponding author's e-mail: leibingshanxatu@163.com

Abstract. General target detection system is based on active detection method, which is usually of poor concealment and bulky. In order to improve the concealment and dexterity of airborne detection, an airborne infrared scanning imaging system for ground vehicles based on passive infrared detection is proposed. Under the condition of guaranteeing the minimum effective detection interval, a rotary scanning dynamic detection model is established based on the principle of single-point passive infrared detection, and the relationships between the detection distance, the rotating angular velocity, the scanning angle, the relative motion velocity and the size of detection area are obtained. According to the expression of infrared characteristic of ground vehicles, the relationship between the response voltage of radiation signal and the relationships among the detection distance is obtained. Based on the projection distribution of the infrared feature area received by the infrared detector, the signal response expression including the size information of the target is obtained at a certain rotating angular velocity. Finally, through simulation and experiment, the system can effectively acquire the signal intensity and signal response width information in the infrared feature area of ground vehicles between the detection distance of 5.4 m and 33.4 m, which verifies the effectiveness of the system.

1. Introduction

Passive infrared detection are useful for many applications in both civil and military areas [1]. Most traditional approaches adopt infrared camera, millimeter wave radar and other detection technology to detect and identify targets. It is widely used in military target recognition [2], aircraft target search [3-4], deep space target detection [5], etc. It has incomparable advantages in the field of anti-terrorism equipment requiring a large number of low-cost detection devices [6-7]. On air-ground infrared scanning imaging platform, the detection of the armored vehicle is uncertain. The infrared photoelectric detection system with rotating scanning in the air receives the radiation energy of the vehicle in the infrared characteristic region and the radiation energy of the vehicle scattering in complex environment. Therefore, the combined action of multiple factors makes the contribution of the vehicle to the energy of the airborne infrared scanning imaging system present a random state. Sharma [8] proposed a new threshold of background subtraction based on linear discriminant ratio (LDR) in the detection of moving human body under the condition of in-depth study of background dynamic change and illumination change. Chengyi Xu [9] proposed a new method to adjust the infrared radiation characteristics of the target itself inspired by cephalopod skin. Hanlin Qin [10] proposed a template matched filtering target detection method to improve the detection probability of dim and small targets. Chen Zhong [11] proposed a new spatio-temporal detection method based on two-dimensional empirical mode decomposition and time-domain difference to solve the problem that small moving infrared targets are difficult to detect due to low signal-to-noise ratio and large fluctuation of background. Chengming Sun [12] proposed an accurate modeling method for infrared radiation characteristics of space targets according to the law of light radiation and scattering. Peipei Yan [13] proposed a method for calculating the illumination of the reflected light from the surface of the space target. These researches are based on the infrared radiation characteristics of the target in the static state of the detection system. The model does not consider the influence of the movement state of the detection system and the change of its attitude on the infrared radiation signal characteristics. In the aspect of detection and recognition model and spectral analysis of armored vehicles under unknown motion state, there exists a mechanism of passive detection and recognition from air to ground and infrared radiation analysis under this model has not been studied in depth. The existing detection model and infrared radiation theory are also difficult to quantitatively and qualitatively give the calculation model of passive scanning detection of armored vehicles and infrared radiation analysis of targets. The infrared radiation characteristics expression of ground armored targets with uncertain altitude, uncertain detection distance and uncertain scanning angle for the airborne rotary infrared scanning imaging detection system is establish.

2. Air-ground rotary infrared scanning imaging model for ground vehicles

The principle of the rotary infrared scanning detection system is shown in Fig. 1. The ground vehicle is taken as the reference coordinate system, the forward direction of the ground vehicle is X-axis, and the moving speed is v_t. The unmanned aerial vehicle rotary infrared scanning detection platform, it is detection distance of L, infrared detector field center axis and the angle between the ground plumb axis θ, scanning angular velocity is ω, the unmanned aerial vehicle platform moves in the negative direction of the X-axis and the velocity is v_r. The rotary infrared scanning detector receives the infrared radiation characteristics of the reflected environment from the body of the ground vehicle or the surface of the vehicle, which is S_1, S_2 and S_3.

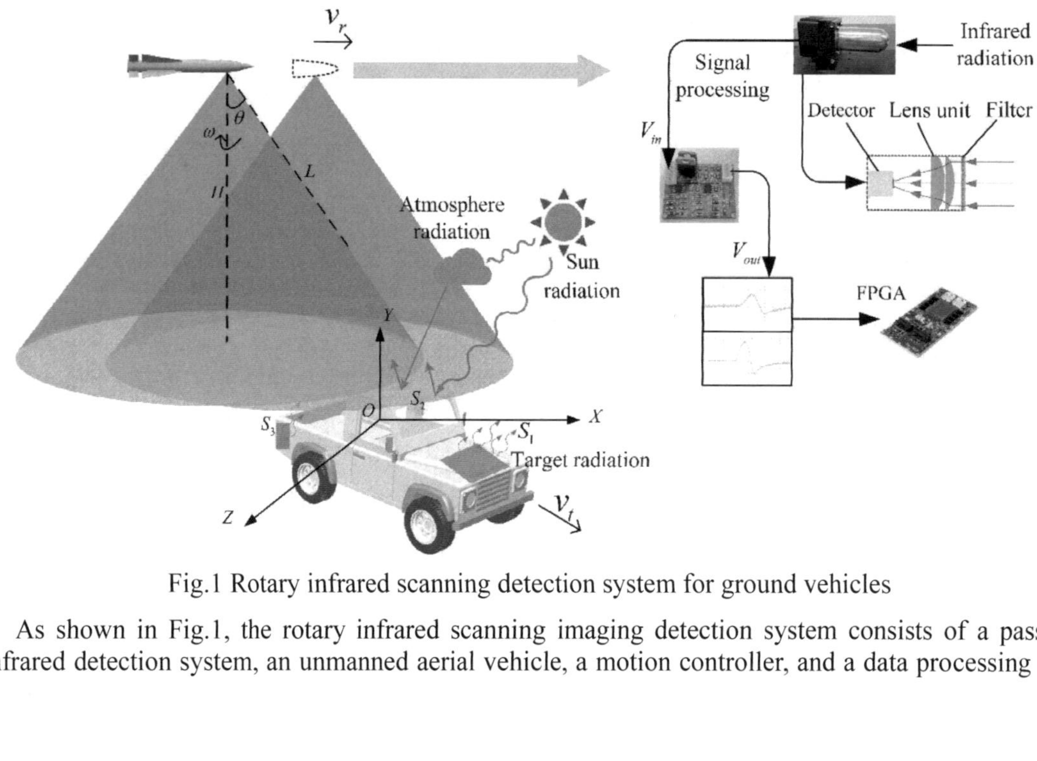

Fig.1 Rotary infrared scanning detection system for ground vehicles

As shown in Fig.1, the rotary infrared scanning imaging detection system consists of a passive infrared detection system, an unmanned aerial vehicle, a motion controller, and a data processing and analysis system. Passive infrared detection system is mainly composed of optical lens group, multi-element infrared sensor, signal processing circuit and lithium battery power supply system.

With the rotation of the scanning system, the detection scanning line is spiral shape to scan the ground. In order to ensure the validity of the scanning system, the scanning angular velocity, the detection distance, the scanning angle and the flight speed are matched, and the minimum effective detection interval can not be greater than the width of the ground vehicle. Taking the ground vehicle coordinate system as the reference coordinate, the ground scanning trajectory parameter equation of the rotary infrared scanning detection system is Eq. (1).

$$\begin{cases} x = L \times \sin\theta \times \cos(\omega t) + \left(\vec{v_r} - \vec{v_t}\right)t \\ y = L \times \sin\theta \times \sin(\omega t) \end{cases} \tag{1}$$

In Eq. (1), relative velocity is $V = \vec{v_r} - \vec{v_t}$, detection time is t.

3. Effect of infrared radiation characteristics of ground vehicles on signal amplitude

When the intrinsic radiation of ground vehicles and the radiation of reflected background reach the detector through atmospheric transmission and optical system conversion and generate the response voltage of radiation signal can be obtained as,

$$V_S = \tau_p^*(\lambda)\tau_{air}^*(\lambda)\int_{\lambda_1}^{\lambda_2}\left[\frac{I^*}{L^2} - \Phi^* \times \omega\right]A_0 R(\lambda)d\lambda \tag{2}$$

In Eq. (2), V_S is the signal voltage, the detection band of the infrared detection system is $\lambda_1 \sim \lambda_2$, A_0 is the area on which the target radiation area is mapped onto the detector sensitive surface. the detection distance between ground vehicles and infrared detection system based on air platform can be obtained as,

$$L^2 = \frac{A_0 V_N \tau_p^*(\lambda)\tau_{air}^*(\lambda)}{V_S/V_N\sqrt{A_d \Delta f_N}}\int_{\lambda_1}^{\lambda_2}\left[I^* - \Phi^* \times S\right]D^*(\lambda)d\lambda \tag{3}$$

In Eq. (3), V_N is the effective value of noise voltage, A_d is the area of detector sensitive surface, $D^*(\lambda)$ is the spectral detection rate of infrared photodetector, Δf_N is equivalent noise bandwidth.

4. Response analysis of ground vehicles radiometric projection size to infrared signal

The area size of the infrared radiation received by the infrared photodetector from the ground vehicles is a variable, resulting in the uncertainty of the energy contribution provided by the infrared feature area of the target to the infrared photodetector, it is closely related to the scanning angle of the detector, the height of the detecting system from the ground, the scanning speed of the rotating system of the detecting system and the change of the moving speed of the target. The pulse broadening feature of the signal received by the infrared photodetector contains the geometric size information of the target, which is a main reference index for target recognition and processing. Fig. 2, gives a schematic analysis of infrared radiation signal broadening under uncertain conditions.

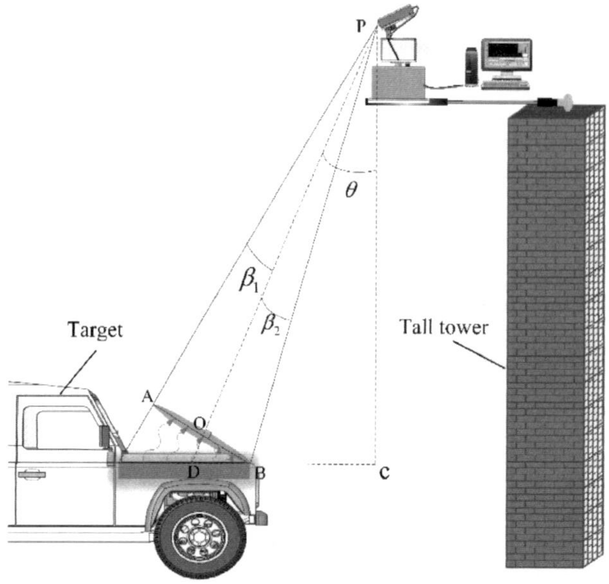

Fig. 2 Analysis schematic diagram of infrared radiation signal broadening under uncertain conditions

In Fig.2,D is the theoretical diameter of target radiation region, $\overline{AB}$ is the chord length of the radiation area projected from the target radiation area to the detector field of view, O is the central point on the chord length of the projected area, $\angle APB$ is the field of view of infrared photodetector, OP is the central line of view of the detector, The angle between PA and the center line of field of view of the detector is β_1, PB and the center line of field of view of the detector is β_2, and $\beta_1 = \beta_2$, θ is the scanning angle of infrared photodetector. The pulse width response of infrared signal of infrared photodetector sweeping through the target heat source region is,

$$t = \frac{\alpha}{\omega} = \arcsin\left(\frac{\sin\beta_2}{\sin\theta\bullet\cos(\theta-\beta_2)}\right)\overline{PC}\Big/\pi f\,\overline{PO} \tag{4}$$

In Eq. (4), $v = \overline{CD}\times\omega$, $V = 2\pi f\,\overline{PO}\sin\theta$, $\overline{PO} = \sqrt{\left(\overline{PC}/\cos(\theta-\beta_2)\right)^2 - \left(\overline{AB}/2\right)^2}$

5. The simulation and experimental analysis

5.1. Simulation analysis

According to the rotating scanning detection mechanism, when the infrared detector flies in one direction with velocity along one direction through the flight platform, its scanning line intersects with the ground. Suppose the maximum detection interval is $\Delta R \le 2m$, the diameter of detection area is $40m$, the detection altitude is $H = 30m$, the detection angle is $\theta = 30^\circ$, the rotary speed is $\omega = 4r/s$, and the flight speed is about $v_r = 12m/s$. The scanning trajectory of an independent single element detector is obtained as shown in Fig. 3.

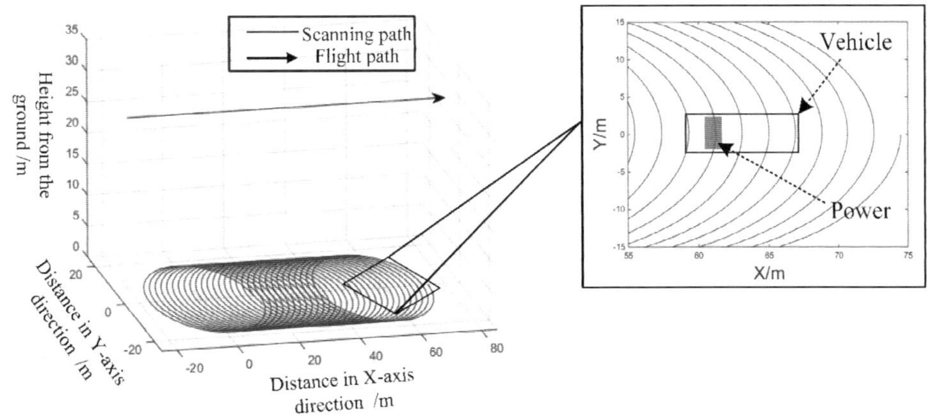

Fig. 3 Rotary scanning trajectory of single element detector on the ground

In Fig. 3, the black arrow represents the flight direction and the blue track represents the scan trajectory. When the armored vehicle exists in the main detection area of the front end indicated by the enlarged image, the detector will capture the infrared characteristic area of the armored vehicle. When the angular velocity and relative velocity are constant, the relationship between the response time and the response voltage of the signal under different detection distances is analyzed, as shown in Fig. 4.

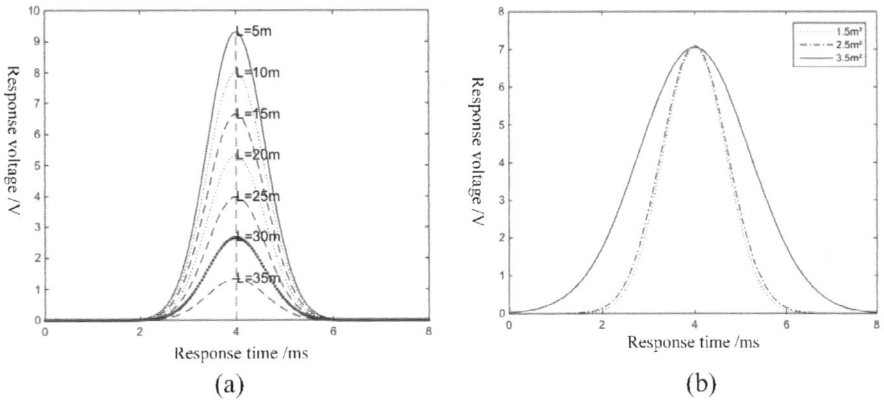

Fig.4. Change of detection distance, relationship between signal response time and response voltage

In Fig.4, it can be concluded that when the detection distance decreases, the amplitude of the signal response voltage increases and the response time of the signal increases. the change of response time corresponding to the change of target infrared feature area profile size is assumed as shown in Fig.4(a). In Fig. 4(b), it can be seen that when the radiation characteristic region changes, when the angular velocity is constant, the speed of the detector sweeping through the infrared characteristic region is constant, and the signal response time will increase with the increase of the radiation characteristic region area.

5.2. Experimental analysis

In order to facilitate the acquisition of measurement data, using the experimental platform of rotating infrared scanning system, the civilian cars of 4.7 m long and 1.9 m wide is tested. Detection system verification experiment is shown in Fig. 5.

IOP Conf. Series: Materials Science and Engineering **799** (2020) 012015 doi:10.1088/1757-899X/799/1/012015

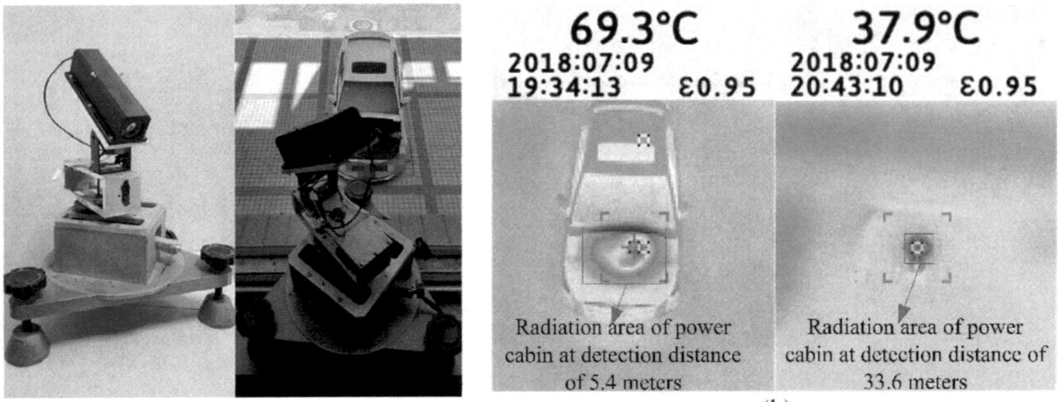

(a) (b)

Fig. 5 Detection system verification experiment

When the scanning angle is 30 °, the radiation area of the power cabin is detected. With the increase of the detection distance, the area of the radiation area decreases gradually with the increase of the distance, and the radiation energy decreases accordingly. Fig.5(b) shows the energy attenuation caused by the change of radiation area when the detection distance is 5.4m and 33.6m.According to the detection performance of the rotary infrared scanning detection system, the radiation intensity at different detection distances and angles is detected by the rotary infrared scanning system, and the detection probability of ground vehicles is studied by setting the signal threshold. The signal amplitude and pulse width at the detection distance of 5.4 m and 33.6 m are analyzed, as shown in fig.6.

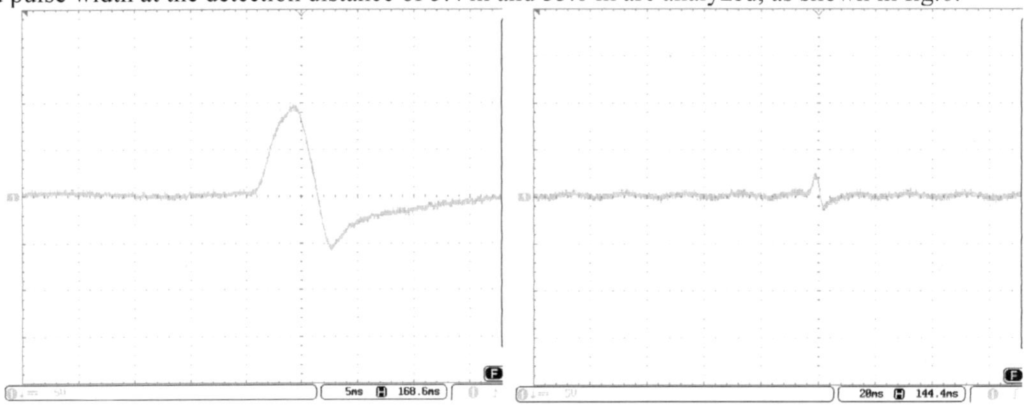

(a) (b)

Fig.6 Signal characteristics of target radiation area and detection distance

According to the comparative analysis of the signal characteristics in the radiation areas of figs.12 and 13 the target detection threshold can be set according to the voltage value and pulse width of the signal. When the detection distance is 5.4 m, the signal size is 9.3V, the pulse width of the signal is 6.3ms, and when the detection distance is 33.6 m, the signal size is 2.8V and the pulse width of the signal is 1.8ms. By studying the changing rules of these two eigenvalues, the detection rate of ground targets can be effectively improved.

6. conclusions

In this paper, the airborne infrared scanning imaging system for ground vehicles based on passive infrared detection is established. Through the simulation and experiment analysis, the energy released from the infrared feature area of ground vehicles decreases with the increase of detection distance when the parameters such as detection angle and rotation angular velocity are constant, but the characteristics of signal amplitude and pulse width received by the detector are consistent. Establishing the system and theoretical model, the energy and size information of the target can be obtained effectively, which provides a scientific basis for distinguishing different targets with template matching recognition algorithm and improves the recognition ability of passive infrared detection system.

Acknowledgments
This work has been supported by Project of the National Natural Science Foundation of China (No. 11804263).

References

[1] Piñeiro-Ave, José, et al. "Target detection for low cost uncooled MWIR cameras based on empirical mode decomposition." Infrared Physics & Technology 63.3(2014):222-231.
[2] Li, K. M. , et al. "Review of ground vehicles recognition." Acta Electronica Sinica 42.3(2014):538-546.
[3] Wang, Chuanyun , and S. Qin . "Adaptive detection method of infrared small target based on target-background separation via robust principal component analysis." Infrared Physics & Technology 69(2015):123-135.
[4] "A real-time infrared radiation imaging simulation method of aircraft skin with aerodynamic heating effect." Infrared Physics & Technology 71(2015):533-541.
[5] Huang, Fuyu, et al. "Analysis of space target detection range based on space-borne fisheye imaging system in deep space background." Infrared Physics & Technology 55.6(2012):475-480.
[6] Kattnig, Alain, et al. " Sieleters, an airborne infrared dual-band spectro-imaging system for measurement of scene spectral signatures." Optics Express 23.12(2015):16164-76.
[7] Rousset-Rouviere , Laurent , et al. "System design of an airborne infrared measurement system." Proceedings of SPIE -The International Society for Optical Engineering 5787(2005):29-37.
[8] Sharma, Lavanya , D. K. Yadav , and A. Singh . "Fisher's linear discriminant ratio based threshold for moving human detection in thermal video." Infrared Physics & Technology 78(2016):118-128.
[9] Xu, C., G. T. Stiubianu, and A. A. Gorodetsky. "Adaptive infrared-reflecting systems inspired by cephalopods." Science 359.6383(2018):1495-1500.
[10] Qin, Hanlin , et al. "Infrared small moving target detection using sparse representation-based image decomposition." Infrared Physics & Technology 76(2016):148-156.
[11] Chen, Zhong , et al. "A novel spatial–temporal detection method of dim infrared moving small target." Infrared Physics & Technology 66(2014):84-96.
[12] Sun, C. M. , Y. Yuan , and X. B. Zhang . "Modeling of infrared characteristics of deep space target." Acta Physica Sinica 59.10(2010):7523-7530.
[13] Yan, P. P. , C. W. Ma , and W. J. She . "Influence of earth's reflective radiation on space target for space based imaging." Acta Physica Sinica 64.16(2015).

IOP Conf. Series: Materials Science and Engineering **799** (2020) 012016 doi:10.1088/1757-899X/799/1/012016

Numerical simulation of reflected wave characteristics in oil wells measured by Acoustic method

Ke Wang, Yanping Liu*, Jie Wu, and Jianshen Gao

Departments of Electronic Engineering, University of Xi'an Shiyou, Xi'an, Shaanxi, 710000, China

*Corresponding author's e-mail: liuyp@xsyu.edu.cn

Abstract. In order to analyze and identify the corrosion of the inner interface of the coupling, perforating and casing, or the collapse and depression of the casing, as well as the reflection form of the dynamic liquid level, the COMSOL finite element analysis software was used to carry out numerical simulation research on the above situation, and its response characteristics were obtained through calculation. In this paper, three-dimensional propagation characteristics of acoustic waves in the annulus of the oil jacket were studied using signal sources of different polarities. The results showed that when the polarity of the signal source was positive (negative), the coupling reflection signal was first positive (negative) and then negative (positive), and the perforation reflection signal was first negative and then positive (positive then negative). The waveform of the reflection signal in the case of corrosion or casing collapse was consistent with the waveform of the perforation reflection signal; the reflection signal of the moving surface was of the same polarity as the emission signal source. When the signal source was a positive-negative superimposed source, the conclusion obtained by applying the above-mentioned positive (negative) polarized signal source was based on the polarity of the first peak of the superposed source. The research results can provide strong support for the popularization and improvement of related theories, and correct reference standards for the performance verification of echo processing methods, as well as important theoretical basis for the correct identification of echo positions on the Dynamic fluid interface.

1. Introduction

In recent years, many scholars have done a lot of research work on the detection of Dynamic fluid interface in oil wells, but lack of in-depth research on the mechanism and characteristics of echo generation. Li Xiangyu et al. [1-6] used different methods to establish a hydro-mechanical surface soft-sensing model. However, these are only the research of measurement methods, and there is no in-depth study on the generation mechanism of measurement objects, that is, no systematic research has been made on the characteristics of the echo generated when the sound waves propagate in the air of the oil collar. Zhang Peng [7] combined the relevant references and a large amount of field data from oil wells, using mathematical expressions to simulate the coupling signal and the liquid level echo signal. This is only a simulation of waveform shape, which can not truly reflect the propagation characteristics of sound waves in tube-casing annulus. Airborne propagation characteristics. Liu Yingxin et al. [8] used the pseudorandom code amplitude modulation signal of the cosine carrier as the detection signal to perform oil well echo simulation, and this method of simulating a wideband signal source in the field with a single frequency signal is only a simplified simulation process. Research on the process is still lacking in guidance. Zhou Wei et al. [9,10] proposed a method for detecting the

IOP Conf. Series: Materials Science and Engineering **799** (2020) 012016 doi:10.1088/1757-899X/799/1/012016

fluid level of an oil well based on a string sound field model, and made a lot of analysis of the sound field characteristics, but did not further study the reflection characteristics of sound waves when propagating in the sound field model of the string. Up to now, the related research of sonic oil wells is mainly for the study of hydrodynamic surface measurement methods. There is no systematic research on the reflection mechanism and echo characteristics of acoustic waves when propagating in the air of the oil jacket ring. In addition, most of the methods for estimating the sound velocity and the method for detecting the hydrodynamic surface are directly applied to the measured data to draw conclusions. To identify the performance of a method requires a standard as a reference. This standard is the result of processing the ideal signal. However, the ideal signal needs to be obtained by analog methods as close to the real situation as possible. Based on this, the author uses COMSOL multiphysics simulation software to construct an annulus model of the oil jacket to simulate the reflection waves generated by the corrosion of the coupling, perforation, the inner interface of the casing or the collapse of the casing when the different signal sources are excited. Coupling waves and hydrodynamic surface waves. Conduct systematic research on the reflection characteristics of sound waves, analyze the mechanism and morphological characteristics of sound waves when they encounter different obstacles when they propagate in the air, and the results can provide strong support for the popularization and improvement of related theories. The performance verification of the echo processing method provides a correct reference standard, and provides an important theoretical basis for correctly identifying the echo position of the liquid surface.

2. Acoustic reflection theory of oil and gas wells

Figure 1 shows the physical model for oil and gas well measurement. Oil and gas wells consist of tubing connected by casing and couplings. The moving fluid level measuring instrument is installed at the wellhead. The pulsed sound source generated by the instrument propagates to the moving fluid level through the annulus between the casing and the tubing. The sound wave receiving sensor collects the sound waves of the annulus, the moving fluid level and the surrounding oil well at a certain sampling rate. Reflected signal. Because the length of the tubing is fixed, by measuring the periodic reflection signal time of the coupling, the propagation speed of the sound waves in the annulus can be determined. The speed multiplied by the single-pass reflection time of the moving liquid surface is equal to the depth of the moving liquid surface. This is the basic principle of moving liquid surface measurement. The core is to determine the position of the hydrodynamic surface and calculate the speed of sound. The basic theory of acoustic wave propagation in oil and gas wells is analyzed below.

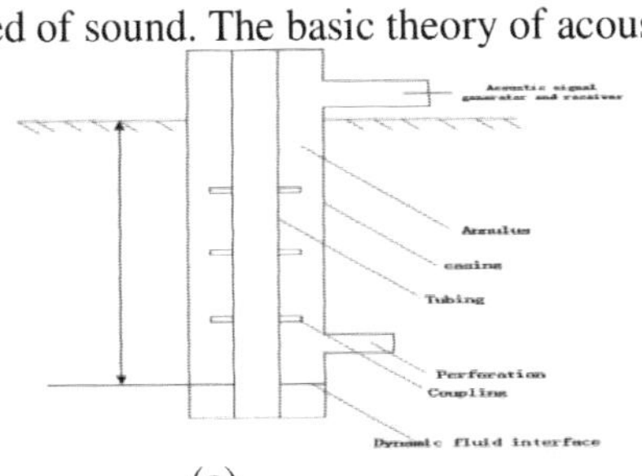

(a)

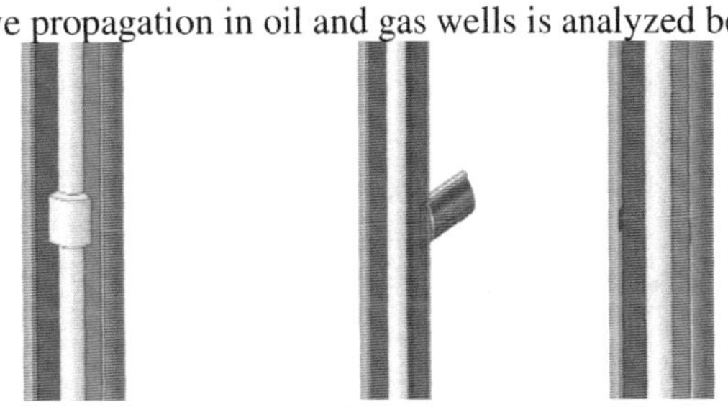

(b) (c) (d)

Figure 1. Schematic diagram of oil and gas well level measurement（a）

Figure 2. Three-dimensional perspective view of the coupling (b), perforation (c), and depression of the inner interface of the casing (d)

2.1. *Principle of reflection of acoustic signals in annulus of oil jacket*

The reflection of a hydrodynamic surface acoustic wave signal is equivalent to the normal incidence of the acoustic wave signal at the interface of two different impedances. At this time, the law of sound pressure and sound velocity reflection and transmission [11] is

$$r_p = \frac{R_2 - R_1}{R_1 + R_2} \quad (1) \qquad r_v = \frac{R_1 - R_2}{R_1 + R_2} \quad (2) \qquad t_p = \frac{2R_2}{R_1 + R_2} \quad (3) \qquad t_v = \frac{2R_1}{R_1 + R_2} \quad (4)$$

IOP Conf. Series: Materials Science and Engineering **799** (2020) 012016 doi:10.1088/1757-899X/799/1/012016

In equations (1) and (2),$r_p(r_v)$and$t_p(t_v)$ are the reflection and transmission coefficients of sound pressure (velocity of sound),$R_1$and$R_2$are the acoustic impedances of medium 1 (annulus) and medium 2 (oil-gas-water).$R_1 = \rho_1 c_1$,$R_2 = \rho_2 c_2$,ρ and $c$are the density and sound velocity of the medium (annulus, oil-gas-water), respectively. Equations (1) and (2) show:

1) In normal oil and gas wells, the speed of sound in the oil and water is much greater than the speed of sound in the annulus, therefore $R_2 \gg R_1$, Substituting (1) to (4) gives 1, -1, 2 and 0, It shows that the total speed of sound occurs on the interface, and the sound pressure continues to propagate in a static pressure manner.

2) When sound waves pass through the perforated water outlet layer of the oil and gas surface to reach the water outlet layer,$R_2 > R_1$, The sound pressure reflection coefficient is greater than 0 and the velocity reflection coefficient is less than zero. Conversely, when passing through the water layer,$R_1 > R_2$, negative sound pressure, the sound pressure of the reflected wave is opposite to the sound pressure of the incident wave, and the sound velocity is greater than 0.

2.2. Coupling, perforating, and sinking signal reflection principles

When a sound wave passes through a coupling with a changed cross section, the reflection coefficient of the sound pressure can be described by reflection formulas with different cross-sectional area parameters [11]:

$$r_p = \frac{S_1 - S_2}{S_1 + S_2} \qquad (5)$$

In formula (5),S_1 and S_2 are the cross-sectional areas of the upper and lower surfaces of the coupling, perforation, and depression. Equation (5) shows:

1) On the upper surface of the coupling, the perforation or the recessed lower disk, the area changes from large to small,that is $S_1 > S_2$, the reflection coefficient is greater than 1, and the reflected wave signal is in the same phase (same polarity) as the incident;

2) On the lower surface of the coupling, the perforation or the recessed upper disk, the area changes from small to large,that is$S_1 < S_2$, the reflection coefficient is less than 1, and the reflected wave signal is opposite to the incident (opposite polarity).

3. Forward model construction and result analysis

In order to clarify the corrosion of couplings, perforations, casing internal interfaces or casing collapse depressions and liquid surface reflection wave shapes, corresponding three-dimensional models were constructed and their acoustic reflection characteristics were studied, as shown in Figure 2.

3.1. Numerical simulation and characteristic analysis of coupling reflection wave using COMSOL software

In order to clarify the coupling wave formation mechanism, the upper and lower surfaces of the coupling are divided into two aspects to study, and a three-dimensional three-dimensional model of the abrupt cross-sectional area is constructed: ① The cross-sectional area of the tubing remains unchanged, and the cross-sectional area of the casing changes from large to small (Figure 3a); ②The cross-sectional area of the tubing remains unchanged, and the cross-sectional area of the casing changes from small to large(Figure 3b). The red area is the acoustic wave propagation medium, and the yellow area is the oil pipe.

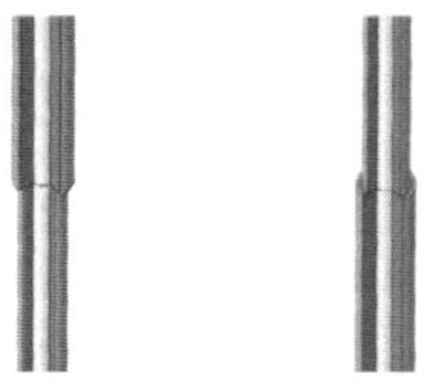

(a) (b)

Figure 3. Casing cross-sectional area from large to small (a) and small to large (b)

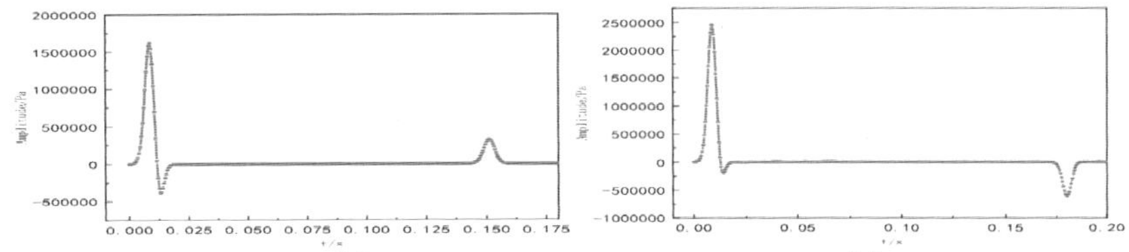

(a) (b)

Figure 4. Acoustic response of the casing cross-sectional area from large to small (a) and small to large (b)

When the cross-sectional area of the tubing remains unchanged, the cross-sectional area of the casing changes from large to small, and the shape of the reflected signal at the interface is consistent with the incident signal, that is, the two have the same polarity (Figure 4a). At this time, this interface generates a negative reflected acoustic wave signal, which is opposite to the polarity of the incident signal (Figure 4b).

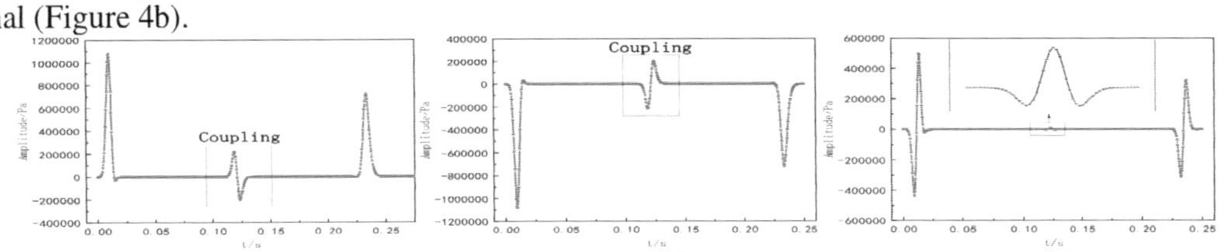

(a) (b) (c)

Figure 5. Coupling sound wave response under positive excitation (a), negative (b), and positive and negative superimposed (c) excitation

Figure 5 shows: the positive signal source is excited, the coupling reflection signal is positive (rise) and then negative (falling) (Figure 5a); the negative signal source is excited, the coupling reflection signal is negative (falling) first and then positive (rising) (picture 5b); positive and negative superimposed signal source excitation, the polarity of the first peak of the coupling reflection signal is consistent with the signal source, and its reflection signal is superimposed by the responses generated by the negative and positive signal sources respectively (Figure 5c). When the sound wave reaches the coupling, it is equivalent to the cross-sectional area from large to small, and the polarity of the reflected signal is the same as that of the incident; when the sound wave signal leaves the lower interface of the coupling, the cross-sectional area is from small to large, and the polarity of the reflected signal is opposite to the incident. The available coupling signals are obtained by superimposing the two signals.

3.2. Numerical simulation and characteristic analysis of perforated reflected waves using COMSOL software

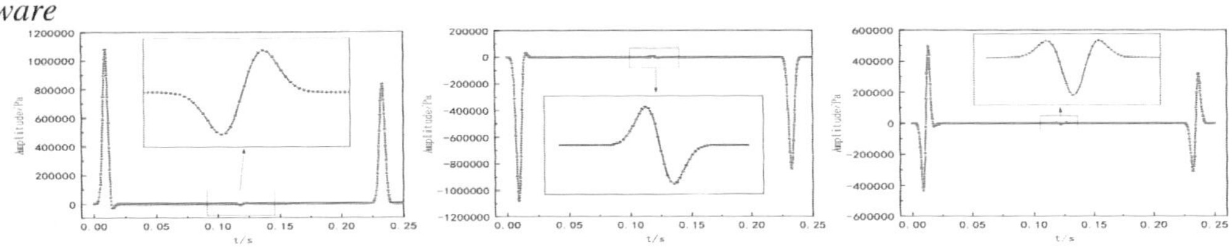

(a) (b) (c)

Figure 6. Positive wave source (a), negative signal source (b) and positive and negative superimposed signal source (c) excited perforated acoustic wave response

Figure 6 shows that the positive signal source excites and the perforated reflection signal is negative (falling) and then positive (rising) (Figure 6a); the negative signal source excites and the perforating reflection signal is positive (rising) and then negative (falling) (Figure 6b); the positive and negative superimposed signal sources are excited, and the polarities of the first peak of the perforated reflected signal are opposite to the signal source, and the reflected signals are superimposed by the responses generated by the negative and positive signal sources, respectively (Figure 6c). When the acoustic wave reaches the upper interface of the perforation, the equivalent of the cross-sectional area is from small to large, and the polarity of the reflected signal is opposite to that of the incident. When the acoustic wave leaves the lower interface of the perforation, the cross-sectional area is from large to small, and the polarity of the reflected signal is the same as that of the incident. It can be obtained that the perforation signal is obtained by superimposing the two signals.

3.3. Using COMSOL software to perform numerical simulation and characteristic analysis of the reflected wave when the internal interface of the casing is corroded or the casing collapses and sags

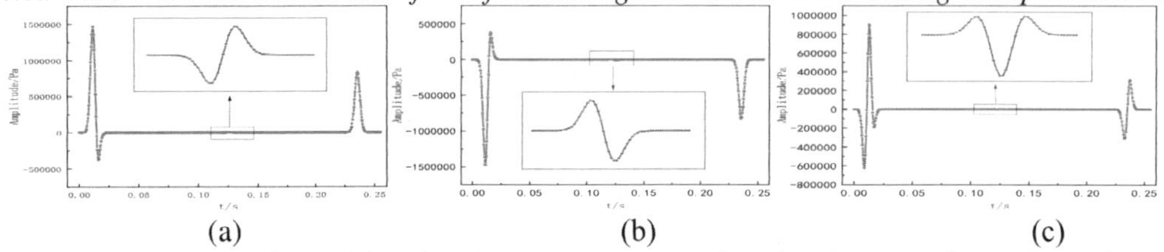

(a) (b) (c)

Figure 7. Sag response of a positive signal source (a), a negative signal source (b), and a positive and negative superimposed signal source (c)

Figure 7 shows that the positive signal source excites and the concave reflection signal is negative (fall) and then positive (rising) (Figure 7a); the negative signal source excites and the concave reflection signal is positive (rising) and then negative (falling) (Figure 7b); Positive and negative superimposed signal sources are excited. The polarity of the first peak of the concave reflection signal is opposite to that of the signal source, and its reflected signal is superimposed by the responses generated by the negative and positive signal sources respectively (Figure 7c). When the sound wave reaches the upper interface of the depression, it is equivalent to the cross-sectional area from small to large, and the polarity of the reflected signal is opposite to the incident. When the sound wave leaves the lower interface of the depression, the cross-sectional area is from large to small, and the reflected signal has the same polarity as the incident. The available sag signals are obtained by superimposing the two signals.

3.4. Using COMSOL software to carry out numerical simulation and characteristic analysis of the reflected waves in the annular space of the oil jacket

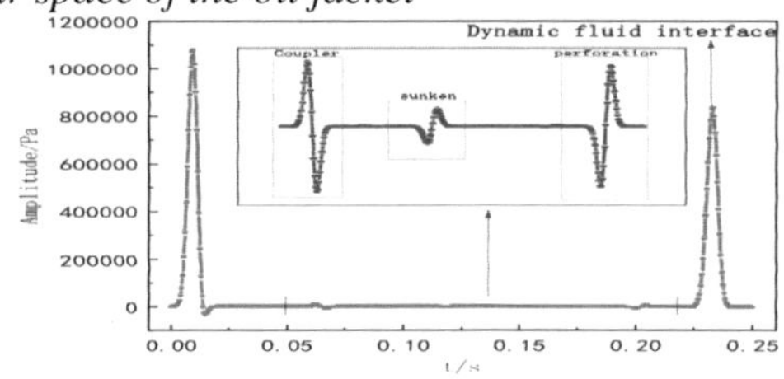

Figure 8. Acoustic response in the presence of couplings, perforations, and depressions

Figure 8 shows that: under the excitation of a positive signal source, the coupling reflection signal is positive (rising) and then negative (falling); the perforated reflection signal is completely opposite, negative and then positive; the shape of the concave reflection signal is basically the same as the perforated reflection signal; The signal of the hydrodynamic surface reflects the same polarity as the emission. When the acoustic wave reaches the coupling, it is equivalent to the cross-sectional area from large to small, and the polarity of the reflected signal is the same as that of the incident. When the acoustic wave leaves the lower interface of the coupling, the cross-sectional area is from small to large, and the polarity of the reflected signal is opposite to the incident. When a sound wave passes through a perforation or depression, the cross-sectional area is from small to large, and then from large to small, as opposed to a coupling. At the interface of the hydrodynamic surface, the sound wave is incident from the air with a small wave impedance to the oil and gas interface with a large wave impedance. According to Equation (1), the sound wave is approximately totally reflected and has the same polarity as the transmitted signal.

4. Conclusions

In this paper, under different signal source conditions, COMSOL software is used to numerically simulate the reflected waves generated by the corrosion of the inner interface of the coupling, perforation and casing or the collapse of the casing, and the simulation data is analyzed in detail. It can be known that when the polarity of the signal source is positive (negative), the coupling reflection signal is first positive (negative) and then negative (positive), and the perforation reflection signal is first negative and then positive (positive and then negative); The waveform of the reflected signal in the case of corrosion of the internal interface of the casing or the collapse of the casing is consistent with the waveform of the perforated reflection; the dynamic fluid interface reflection signal and transmitting signal source have the same polarity. When the signal source is a positive-negative superimposed source, the conclusion obtained by applying the above-mentioned positive (negative) polarized signal source is based on the polarity of the first peak of the superposed source.The numerical simulation of the reflection characteristics of the sound waves in the air of the oil jacket ring reveals the reflection mechanism of the coupling wave and the hydrodynamic surface wave in the case of corrosion of the inner interface of the coupling, perforation, and casing, or the collapse of the casing. The characteristics analysis of the signal received by the instrument, the performance verification of the echo signal processing method, and the correct identification of the position of the hydrodynamic surface provide theoretical basis and reference standards.

Acknowledgments

This research was financially supported by National Natural Science Foundation of China(Program No. 41704106,41804115),The Project Supported by Natural Science Basic Research plan in Shaanxi Province of China(Program No. 2018JQ4045， 2018JQ4008),and Scientific Research Program Funded by Shaanxi Provincial Education Department (Program No. 19JK0665).

References

[1] Li X.Y. (2016) Soft-sensor Modeling for Dynamic Fluid Level of Sucker-rod Pumping Process, Northeastern University.
[2] Nie S.T. (2014) The Research on the Soft-sensing Technique of Dynamic Liquid Level for Sucker Rod Pumping System Based on Multiple Models. Northeastern University.
[3] Wang T., Duan Z.W. (2019) Soft sensor modeling for dynamic liquid level of oil well based on fuzzy inference adaptive updating. CIESC Journal, 1-17
[4] Li X.Y., Gao X.W., Cui Y.B., et al. (2013) Dynamic liquid level modeling of sucker-rod pumping systems based on Gaussian process regression. Ninth International Conference on Natural Computation(ICNC).Shenyang, China: IEEE, 2013: 917-922.
[5] Wang T., Gao X.W., Liu W.F. (2014) Adaptive soft sensor method and application in determination of dynamic fluid levels. CIESC Journal, 65(12):4898-4904.
[6] Li X.Y., Gao X.W., Li K., et al. (2016) Ensemble soft sensor modeling for dynamic liquid level of oil well based on multi-source information feature fusion. CIESC Journal, 67(06):2469-2479.
[7] Zhang P. (2016) Downhole liquid level depth detection research based on acoustic signal blind separation,Southwest University of Science and Technology.
[8] Liu Y.X., Yang Y.C., Han B.k., et al. (2015) Liquid level detection method of oil well with acoustic waves in a low frequency. Journal of Applied Acoustics, 34(01):24-31.
[9] Zhou W., Jia W., Guo X.Y., et al.(2015) A New Method for Oil Well Dynamic Fluid Level Detection Based on the Column Sound Field Model. Journal of Southwest Petroleum University(Science & Technology Edition), 37(04):166-172.
[10] Jia W. (2014) Research on a Detection Method of Oil Well Dynamic Fluid Level Based on the Column Sound Field Model. Xi'an Shiyou University.
[11] Du G.H., Zhou Z.M., Gong X.Z. (2012) Foundamentals of Acoustics.Nanjing University Press.

IOP Conf. Series: Materials Science and Engineering **799** (2020) 012017 doi:10.1088/1757-899X/799/1/012017

Investigation into position deviation effect on micro newton force sensor

Li Xu[1*], Shuxiang Wang[2], Peiliang Zheng[1], Zhenyu Huang[1] Xudong Luo[1]

[1]Guangdong Provincial Institute of Metrology, Guangdong Provincial Key Laboratory of Modern Geometric and Mechanical Metrology Technology, Guangzhou, 510405, China

[2]Guangzhou Maritime University, Guangzhou, 510725, China

[*]Corresponding author's e-mail: li_xu_scm@163.com

Abstract. The position deviation effect on force sensor with a resolution of micro newton scale was investigated by theory model analysis and numerical simulation method. The finite element method and theoretical model were used to analyse the influence of the electrostatic micro force sensors when the electrode position of the cylindrical capacitor changed. Relationship between the sensor output micro force and the position deviation of the internal electrode in the cylindrical capacitance was obtained. It was found that different position deviation form had different degrees of influence on the sensor's output result and the regions with little effect by position deviation were existed. For translation along Z-axis, the region was [-4 mm, 14 mm]. For translation along r-axis, the region was [-0.2 mm, 0.2 mm]. And for rotation around r-axis, the region was [-1°, 1°]. The corresponding maximum errors of the micro force output were 1.49%, 2.09% and 2.20%, respectively.

1. Introduction

With the fast advances of nanotechnology, the micro force measurement sensors are very important in various fields of science and technology [1~3]. In aerospace technology, the accurate and reliable micro force measurement can provide the microsatellites for its maneuvers of station keeping, orbital corrections, and attitude control [4,5]. In biotechnology, the biomechanical research and single-cell operation require accurate measurement and control of micro forces[6,7]. In material science, the micro force measurement sensors are used for determining the elastic modulus and nano-hardness of material [8,9]. In fundamental science, AFM (atomic force microscope) and Casimir force measurement are both related to the force measurement sensor with a resolution of micro newton scale [10,11]. Researchers found it was a good method to reproduce and measure micro forces by using the micro electrostatic force between capacitor plates. NIST (National Institute of Standards and Technology) developed an electrostatic balance, using cylindrical capacitors as micro force sensors [12,13]. NPL (National Physical Laboratory) had also proposed a micro force sensor device by using the electrostatic force balance between cylindrical capacitors [14]. The AFM calibration device researched by KRISS(Korea Research Institute of Standars and Science) [15,16] and the micro force measurement and value traceability device developed in NIM (National Institute of Metrology, China) [17] were all used the cylindrical capacitors as the micro force sensor.

The position deviation of cylindrical capacitors is one of limits for micro force sensor accuracy. However, very few research was focus on the position deviation effect on force measurement sensor. In present work, theoretical and numerically simulated analysis on the position deviation effect on force measurement sensor had been carried out. Relationship between the micro force sensor output and the position deviation of internal and external electrodes of the cylindrical capacitance was investigated. The results is important for the manufacture and application of cylindrical capacitors in micro force sensors.

2. The basic theory of the cylindrical capacitance micro force measurement sensor

The structure of a cylindrical capacitor in the micro force measurement sensor is shown in Fig.1. When the sensor device is working, the external electrode is fixed, and the internal electrode can move freely along the axis direction. When the internal electrode moves, the internal and external electrode overlap length Z changes, leading to the capacitance of the cylindrical capacitor changed. If the fringe effect is ignored, the capacitance of the cylindrical capacitor can be obtained:

$$C = \frac{2\pi\varepsilon Z}{\text{In}(D_i/d_o)} \tag{1}$$

where D_i is the inner diameter of the external electrode; d_o is the outer diameter of the internal electrode; Z is the internal and external electrode overlap length; ε is the permittivity of the air between the internal and external electrode.

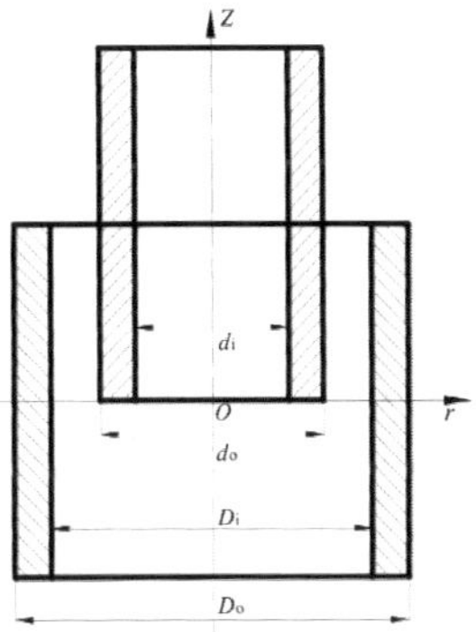

Figure 1. Cylindrical capacitor in the micro force sensor

When a constant voltage is maintained between the internal and external electrodes, the displacement dZ of the internal electrode movement needs to do work:

$$dW = F \bullet dZ = \frac{1}{2}U^2 dC \tag{2}$$

where dW is the energy change; F is the micro force acting on the internal electrode; dZ is the relative position change of the internal and external electrode of the cylindrical capacitor; and U is the potential difference between the internal and external electrode.

The force measurement sensor output value by theoretical model can be obtained:

$$F = \frac{1}{2}U^2 \frac{dC}{dZ} = \frac{\pi\varepsilon}{\text{In}(D_i/d_o)}U^2 \tag{3}$$

During the analysis of position deviation effect on micro force measurement sensor, the external electrode is fixed and the internal electrode is offset from the ideal location that is no position deviation. Figure 1 shows that when the external electrode is fixed, there are three ways about the positional deviation of the internal electrode, including translation along Z-axis (axial), along r-axis (radial) and rotation around r-axis.

3. Results and discussion

3.1. Internal electrode position deviation in Z-axis (axial)

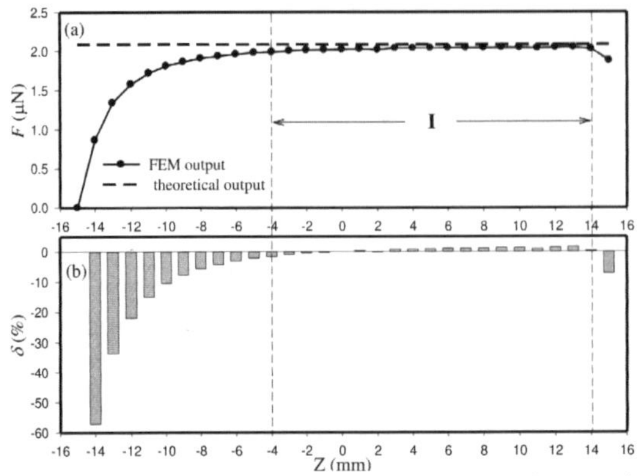

Figure 2. Force sensor outputs and errors with internal electrode translation along the Z-axis

When the internal electrode moves along *Z*-axis in the external electrode, the output of micro force sensor is shown in Fig.2(a). The theoretical model predicts a constant micro force of 2.08 μN with the fringe effect ignored, which lies just above the data predicted by finite element method (FEM) in region I [-4 mm, 14 mm]. The range of micro force predicted by FEM is from 2.00 μN to 2.05 μN in region I, the max difference of 3.8% from the theoretical model. These results show that the data trend predicted by FEM matches the theoretical model well in region I that the fringe effect can be ignored. The internal electrode is controlled in region I during the sensor is working, in order to avoid the fringe effect on the sensor output. In this region, the micro force sensor output is stable. Fig.2(b) is shown the error of the sensor output when the internal electrode position deviation in *Z*-axis by FEM. The position that the internal electrode is in the middle of the external electrode in *Z*-axis (shown in Fig.1) was set as the ideal position without deviation, and the sensor force output at this position is the standard value. Fig.2(b) shows that the maximum error does not exceed 1.49% by FEM in region I. This result indicated that the internal electrode position deviation in Z direction has little effect on the micro force sensor output at this region.

3.2. Internal electrode position deviation in r-axis (radial)

The micro force sensor with cylindrical capacitor has a symmetrical structure. When the internal electrode is in an ideal location (that is the internal and external electrodes are coaxial), the internal electrode is only acted by the micro forces in axial direction. In the radial direction, due to the symmetry of the cylindrical capacitor, the micro force acting on the internal electrode is zero.

When the internal electrode is moving along the radial direction, the positions of the internal and external electrodes are no longer symmetrical about the *Z* axis. The internal electrodes will be acted by the micro electrostatic forces in the radial direction. However, the micro force measurement sensor with cylindrical capacitor only used the micro force in *Z*-axis (axial) direction, only force in this direction is considered in the paper.

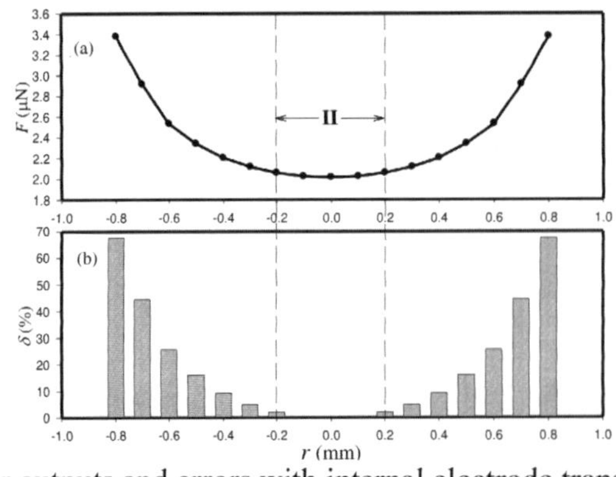

Figure 3. Force sensor outputs and errors with internal electrode translation along the r-axis

Figure 3 shows the output characteristics of the sensor when the internal electrode is translating in the *r*-axis direction. Due to the symmetrical structure of the sensor's cylindrical capacitor, when the internal electrode moves along the positive and negative directions of the r-axis, its output characteristics are symmetrically distributed with $r = 0$ mm. With the position deviation of internal electrode increased, the sensor's output micro force increased rapidly. When the position deviation is limited to the region II (-0.2 mm, +0.2 mm), the output variation is small. It can be seen in Figure 3b that when the radial position deviation is controlled in region II, the micro force error of the sensor output is within 2.09%.

3.3. Internal electrode rotation around r-axis (radial)

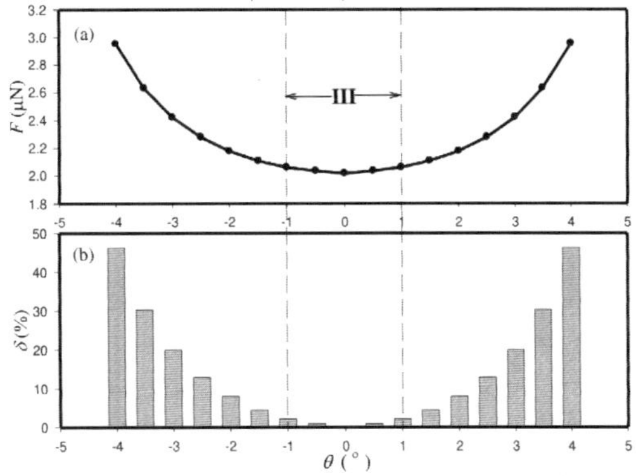

Figure 4. Force sensor outputs and errors with internal electrode rotation around the r-axis

Besides the sensor's internal electrode position deviation in *Z*-axis and in *r*-axis, the rotation around *r*-axis could occur during the sensor assembly. This position deviation would also affect the micro force output of the sensor. Fig. 4(a) shows the micro force output with the internal electrode rotating around *r*-axis. Fig. 4(b) is the output errors relative to the ideal position. The results show that the internal electrode rotation around *r*-axis has great impact on the sensor's output. The error of micro force output could exceed 2.2% even if the internal electrode rotation angle is only 1°. The output micro force error would rapidly increase to 12.9%, when the internal electrode position deviation exceeds 2.5°. Therefore, position deviation with rotation around the *r*-axis should be strictly controlled as much as possible to reduce the output error of the sensor.

Only by limiting the position deviation within region III [-1°, +1°], the accuracy of the sensor output could be effectively guaranteed.

Table 1. Regions with litter effect by the internal electrode position deviation

position deviation form	region	Max error
translation along Z-axis	[-4 mm, 14 mm]	1.49%
translation along r-axis	[-0.2 mm, 0.2 mm]	2.09%
rotation around *r*-axis	[-1°, 1°]	2.20%

Table 1 shows the regions with litter effect by the internal electrode position deviation. This table also indicates the error limitation required of the internal electrode during the micro force sensor manufacturing process.

4. Conclusions

Investigation into position deviation effect on force measurement sensor with a resolution of micro newton scale was carried out. The positional deviation of the internal electrode, including translation along *Z*-axis, along *r*-axis and rotation around *r*-axis were analyzed separately. The following conclusions were obtained:

(1) The position deviation of the internal electrode in different directions would affect the output of the micro force sensor.

(2) Different position deviation form had different degrees of influence on the sensor's output result. When the position deviation was translation along *Z* axis, there was little effect on the sensor's output. The internal electrode translate along or rotate around *r* axis had great influence on the output of the sensor.

(3) The regions with little effect of position deviation had been found in every position deviation form, and there were [-4 mm, 14 mm], [-0.2 mm, 0.2 mm] and [-1°, 1°].

(4) When position deviation could be restrict into these regions, the maximum errors of the micro force output were 1.49%, 2.09% and 2.20%, respectively.

Acknowledgments
The authors thank for the financial support by the National Natural Science Foundation of China with the contract numbers of 51806041, the Guangdong Science and Technology Project (2014A040401044, 2018A0303130013) and the State Administration for Market Regulation Science and Technology Project (2016QK028).

References

[1] Ma C Z, Du J S, Liu Y Y, et al. Overview of micro-force sensing methods[C]//Applied Mechanics and Materials. Trans Tech Publications, 2014, 462: 25-31.
[2] Wei Y, Xu Q. An overview of micro-force sensing techniques[J]. Sensors and Actuators A: Physical, 2015, 234: 359-374.
[3] Lay A, Wang D S, Wisser M D, et al. Upconverting nanoparticles as optical sensors of nano to micro Newton forces[J]. Nano letters, 2017, 17(7): 4172-4177.
[4] Kolbeck J, Porter T E, Keidar M. High Precision Thrust Balance Development at The George Washington[C]. Proceedings of the 35th International Electric Propulsion Conference, Georgia, USA, 2017.
[5] Karadag B, Cho S, Funaki I. Note: Precision balance for sub-miliNewton resolution direct thrust measurement[J]. Review of Scientific Instruments, 2018, 89(8): 086108.
[6] Metzger A, Baltisberger S, Burkhard H R, et al. Weighing cell based on the principle of electromagnetic force compensation with optoelectronic position sensor: U.S. Patent 9,086,315[P]. 2015-7-21.
[7] Panda B, Sidhu M S, Munjal P, et al. Time resolved nano Newton force spectroscopy in air and vacuum using a load cell of ultra micro balance[J]. Review of Scientific Instruments, 2019, 90(4): 043117.
[8] Molaei S. The measurement of Young's modulus of thin films using secondary laser speckle patterns[J]. Measurement, 2016, 92: 28-33.
[9] Xiang D. Capacitive micro-force sensor as a transfer standard for verification and calibration of nanoindentation instruments[C]. Micro-and Nanotechnology Sensors, Systems, and Applications XI. International Society for Optics and Photonics, 2019, 10982: 1098239.
[10] Pratt J R, Kramar J A, Newell D B, et al. Review of SI traceable force metrology for instrumented indentation and atomic force microscopy. Measurement Science and Technology. 2005, 16(11): 2129-2018.
[11] Mohideen U, Roy A. Precision measurement of the Casimir force from 0.1 to 0.9 μ m[J]. Physical Review Letters, 1998, 81(21): 4549.
[12] Shaw G A, Stirling J, Kramar J A, et al. Milligram mass metrology using an electrostatic force balance[J]. Metrologia, 2016, 53(5): A86.
[13] Shaw G A. Current state of the art in small mass and force metrology within the international system of units[J]. Measurement Science and Technology, 2018, 29(7): 072001.
[14] Leach R, Chetwynd D, Blunt L, et al. Recent advances in traceable nanoscale dimension and force metrology in the UK[J]. Measurement Science and Technology, 2006, 17(3): 467.
[15] M. S. Kim, I. M. Choi, Y. K. Park, et al. Atomic force microscope probe calibration by use of a commercial precision balance. Measurement. 2007, 40(7): 741~745.

[16] Kim M S, Pratt J R, Brand U, et al. Report on the first international comparison of small force facilities: a pilot study at the micronewton level[J]. Metrologia, 2011, 49(1): 70.
[17] Hu G., Song L, Meng F, et al. Research and Development of Small Force Standards at NIM [C]. International Journal of Modern Physics: Conference Series. World Scientific Publishing Company, 2013, 24: 1360020.

IOP Conf. Series: Materials Science and Engineering **799** (2020) 012018 doi:10.1088/1757-899X/799/1/012018

Observing the Polarization Dependent Light Shifts Using an Atom Fountain

Qingqing Hu[1,2], Qian Wang[3], Yukun Luo[1,2], Fufang Xu[1,2], Wei Jiang[4]*, Rong Wei[3]

[1]Advanced Interdisciplinary Technology Research Center, National Innovation Institute of Defense Technology, Beijing 100010, China

[2]Beijing Academy of Quantum Information Sciences, Beijing, 100010, China

[3]Key Laboratory for Quantum Optics, Shanghai Institute of Optics and Fine Mechanics, Chinese Academy of Sciences, Shanghai 201800, China

[4]College of Artificial Intelligence, National University of Defense Technology, Changsha 410073, China

*Corresponding author's e-mail: weijiang@nudt.edu.cn

Abstract. We theoretically predicted and experimentally demonstrated the sinusoidal shifts of the ^{87}Rb atom Raman transition frequencies as the polarization vectors of the Raman lasers rotate. We also presented the vanish of the differential light shifts for the magnetic sensitive Raman transitions at some elliptically polarized configurations while no vanish for the magnetic insensitive Raman transition. These results are helpful for the evaluation and elimination of the light shifts induced systemic errors in the atomic magnetometer, atomic interferometer, atomic clock and quantum computing.

1. Introduction

Light shifts (LSs) were first discovered in 1960s [1], which refers to the alteration of the atomic energy levels due to the external oscillatory electric fields and plays a fundamental role in a vast variety of physical applications ranging from basic atomic physics [2] to applied quantum registers [3], quantum computing [4], atomic magnetometers [5], atomic clocks [6] and atomic interferometers [7]. In particular, for atom interferometry in a warm vapor [8] and quantum memory with a photonic polarization qubit [9], the LSs are significant because either the Raman laser Rabi frequency is up to 1.2 MHz or the Zeeman sublevels other than the clock states are employed. To date, the research on LSs mainly focuses on two aspects. On one hand, it's about methods to calculate [10, 11], measure [12-14] and eliminate [15, 16] the adverse effects of the LSs that reduce the fidelity and accuracy of the experiments. These methods include hyper-Ramsey schemes [17], magic wavelength lattices [18], and intensity ratio of the lasers [19]. On the other hand, it's about making use of the LSs to achieve sensitive measurement such as absolute polarization [20] and optical potential [21], to cancel some effects such as Zeeman effects [22], and to achieve new physical such as photon-echo [23].

This Letter aims to study the effect of laser polarization on the differential light shifts (DLSs) of the ^{87}Rb atom Raman transitions in near-resonant case, and demonstrate the vanish of the DLSs in elliptically polarized laser configurations. These results are helpful for the evaluation and elimination of the light shifts related systemic errors in the atomic magnetometer, atomic interferometer, atomic clock and quantum computing.

2. Theoretical model

According to the electric-dipole selection rules, the π transitions are forbidden and only the σ transitions are allowed when a quantization magnetic field parallel to the propagation direction of the Raman beams is applied. Assuming the atoms are distributed evenly among the magnetic sublevels of the F = 2 hyperfine states, only seven transitions with $\Delta F = 1$ and $\Delta m_F = 0, \pm 2$ will happen considering arbitrarily polarized lasers are composed of the left and right circularly polarized lasers. Figure 1(a) depicts these transitions in the presence of Zeeman shifts and light shifts. The frequency differences of the two lasers driving the Raman transitions satisfy

$$\mathrm{h}\left(\omega_1 - \omega_2\right) = \mathrm{h}\omega_{\mathrm{hfs}} + \Delta E_{\mathrm{Zeeman}} + \mathrm{h}\delta_D + \Delta E^{ac}. \tag{1}$$

In general, the LSs are different for any two internal states of an atom and depend on the frequency and local intensity of the applied laser. For a particular hyperfine state $|F, m_F\rangle$ interacting with a laser field $\vec{E}(t) = \varepsilon\hat{\zeta}e^{-i(\omega t + \hat{k}\cdot\hat{r})} + c.c.$, where ε, ω and $\hat{k}$ are the amplitude, frequency and wave vector of the laser, the LS is quantified using the dynamic polarizability by [24]

$$\Delta E^{ac}_{F,m_F} = -\alpha_{F,m_F}(\omega)\left(\frac{\varepsilon}{2}\right)^2, \tag{2}$$

and the dynamic polarizability $\alpha_{F,m_F}(\omega)$ is composed of

$$\alpha_{F,m_F}(\omega) = \alpha_F^S(\omega) + (\hat{k}\cdot\hat{B})A\frac{m_F}{2F}\alpha_F^V(\omega) + \left(3\left|\hat{\zeta}\cdot\hat{B}\right|^2 - 1\right)\frac{3m_F^2 - F(F+1)}{2F(2F-1)}\alpha_F^T(\omega) \quad , \tag{3}$$

where the m_F independent factors $\alpha_F^S(\omega)$, $\alpha_F^V(\omega)$ and $\alpha_F^T(\omega)$ are named as scalar, vector and tensor polarizabilities, respectively.

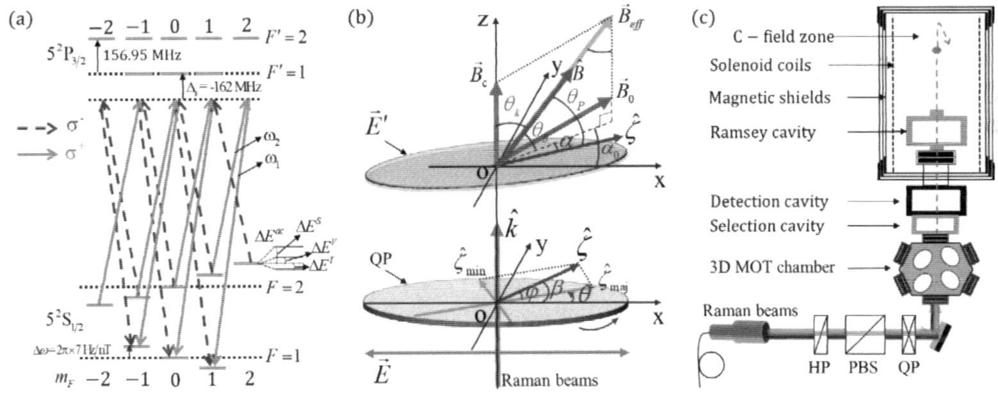

Figure 1. Schematic diagram of the theoretical model and setup. (a) The selection rules allowed [87]Rb atom D2 line Raman transitions. (b) The resultant magnetic field vector $\vec{B}_{eff}$ and the Raman laser complex polarization vector $\hat{\zeta}$ in a Cartesian coordinate. (c) The atomic fountain experimental setup.

As shown in Figure 1(b), $\hat{B}$ is the unit vector along the effective magnetic field $\vec{B}_{eff}$, which is composed of a residual magnetic field $\vec{B}_0$ and a quantization magnetic field $\vec{B}_c$ provided by a solenoid coils inside the innermost magnetic shield of C-field zone. $\vec{B}_c$ parallels to the wave vector $\hat{k}$ (z-axis) of the Raman beams, $\vec{B}_{eff}$ makes an angle θ_k with the z-axis, an angle $\theta_B(\pi/2-\theta_k)$ with the xoy plane, an angle θ_P with the complex polarization unit vector $\hat{\zeta}$, and an angle α (α_0) between its projection in the xoy plane and $\hat{\zeta}$ (the x-axis), respectively. The complex polarization unit vector $\hat{\zeta}$ can be expressed as $\hat{\zeta}=e^{i\gamma}\left(\cos\varphi\hat{\zeta}_{\rm maj}+i\sin\varphi\hat{\zeta}_{\rm min}\right)$, in which γ is a real number, the unit vectors $\hat{\zeta}_{\rm maj}$ and $\hat{\zeta}_{\rm min}$ ($\hat{\zeta}_{\rm maj}\times\hat{\zeta}_{\rm min}=\hat{k}$) align with the semi-major-axis and semi-minor-axis of the ellipse which swept out by the tip of the electric field vector of the laser in one period, respectively [10]. The parameter φ, represented in the $\hat{\zeta}_{\rm maj}$ - $\hat{\zeta}_{\rm min}$ plane, is related to the degree of circular polarization with $A=\sin 2\varphi$.

Figure 1(c) depicts the experimental setup and a detailed description of the whole system can be found in Refs. [25, 26]. The intensity ratio and relative frequency of the two Raman beams are determined by the 0th to +1th sideband of a fiber electro-optical modulator (FEOM). The frequencies of the two Raman lasers are red shifted 162 MHz to the $F=2\rightarrow F'=1$ transition frequency by a double-pass optical path using an acoustic optical modulator (AOM), which also serves as the switch and intensity controller of the Raman beams. The Raman beams coming from a polarization maintaining (PM) fiber are purified to linear polarization with a half-wave plate (HP) and a polarization beam splitter (PBS) firstly, and are adjusted to arbitrary polarization with a quarter-wave plate (QP) afterwards. According to the principles of optics [27, 28], as the fast-axis of the QP rotates an angle θ with the x-axis, the semi-major-axis $\hat{\zeta}_{\rm maj}$ of the polarization ellipse rotates along with θ, the axis angle β of the complex polarization vector rotates 2 times of θ, and the ellipticity angle φ is related to θ by $\sin 2\varphi=-\sin 2\theta$.

With above knowledge, the dynamic polarizability Eq. (3) applied to the Zeeman sublevels of the two ground states are simplified to

$$\alpha_{1,m_{F1}}(\omega)=\alpha_1^S(\omega)-\cos\theta_k\sin 2\theta\frac{m_{F1}}{2}\alpha_1^V(\omega)+\left[3\cos^2\theta_{\rm P}-1\right]\frac{3m_{F1}^2-2}{2}\alpha_1^T(\omega), \tag{4}$$

$$\alpha_{2,m_{F2}}(\omega)=\alpha_2^S(\omega)-\cos\theta_k\sin 2\theta\frac{m_{F2}}{4}\alpha_2^V(\omega)+\left[3\cos^2\theta_{\rm P}-1\right]\frac{3m_{F2}^2-6}{12}\alpha_2^T(\omega) \quad , \tag{5}$$

where $m_{F1}=0,\pm1$, $m_{F2}=0,\pm1,\pm2$. Noting the amplitude of the laser field $\vec{E}_1(\omega_1)$ as ε_1, the intensity ratio between the two Raman lasers as R $(I_2/I_1=3:1$ here$)$, and taking the polarizabilities $\alpha_F^l(\omega_2)$ calculated in our previous work [24], in which $l=S,V,T$ represent the scalar, vector and tensor components, $F=1,2$ represent the two ground state hyperfine levels, $j=1,2$ represent the two Raman beams, the DLSs of the $|F=2,m_{F2}\rangle\rightarrow|F=1,m_{F1}\rangle$ transitions are composed of the differential scalar light shifts (DSLSs), differential vector light shifts (DVLSs) and differential tensor light shifts (DTLSs) by

$$\begin{aligned}\Delta E^{ac}&=\Delta E^S+\Delta E^V+\Delta E^T=\\ &-\left(\frac{\varepsilon_1}{2}\right)^2\left\{\left[\alpha_2^S(\omega_1)+q\alpha_2^S(\omega_2)\right]-\left[\alpha_1^S(\omega_1)+q\alpha_1^S(\omega_2)\right]\right\}\\ &+\left(\frac{\varepsilon_1}{2}\right)^2\cos\theta_k\sin 2\theta\left\{\frac{m_{F2}}{4}\left[\alpha_2^V(\omega_1)+q\alpha_2^V(\omega_2)\right]-\frac{m_{F1}}{2}\left[\alpha_1^V(\omega_1)+q\alpha_1^V(\omega_2)\right]\right\}\\ &-\left(\frac{\varepsilon_1}{2}\right)^2\left[3\cos^2\theta_{\rm P}-1\right]\left\{\frac{3m_{F2}^2-6}{12}\left[\alpha_2^T(\omega_1)+q\alpha_2^T(\omega_2)\right]-\frac{3m_{F1}^2-2}{2}\left[\alpha_1^T(\omega_1)+q\alpha_1^T(\omega_2)\right]\right\}\end{aligned} \tag{6}$$

3. Experimental setup and procedure

The sensor used for measuring the DLSs is developed based on the atomic fountain clock in Shanghai Institute of Optics and Fine Mechanics (SIOM), and its schematic diagram is shown in Figure 1 (c). Here we only describe the major parameters related to this experiment. The intensity ratio and frequency difference of the two Raman beams are determined by the 0th to +1th sideband of a fiber electro-optical modulator (FEOM). The frequencies of the two Raman lasers are red shifted 162 MHz to the $F=2\rightarrow F'=1$ resonant frequency by a double-pass optical path using an acoustic optical modulator (AOM), which also serves as the switch and intensity controller of the Raman beams. The Raman beams coming from a polarization maintaining (PM) fiber are purified to linear polarizations with a half-wave

IOP Conf. Series: Materials Science and Engineering **799** (2020) 012018 doi:10.1088/1757-899X/799/1/012018

plate (HP) and a polarization beam splitter (PBS) firstly, and then are adjusted to arbitrary polarizations with a quarter-wave plate (QP).

In this experiment, about 10^8 ^{87}Rb atoms are trapped and cooled to 2.5 μK in the magneto-optical trap (MOT) chamber using a fold optical path [29] in 1 s, and then are launched vertically with a velocity of 4.188 m/s in 2 ms. The repumping laser is switched off 1 ms later than the cooling laser to ensure all atoms are in the $F = 2$ ground state. At 1.4 s, when the atoms reaching a launch height of 891 mm, a pair of co-propagating Raman beams with a Gaussian diameter of 16 mm and a duration time of 8 ms are injected from the bottom window in order to interact with atoms to create two photon Raman transition. When the atoms fall back through the detection zone, the atom population in the $F = 1$ and $F = 2$ levels are detected to give a transition probability. The relative frequency of the Raman beams is increased progressively at each launch-detection cycle by modulating the FEOM to generate a Raman transition spectrum. The above processes are repeated with four laser intensities (9.54, 19.18, 31.95, 47.31 μW/cm^2) firstly and also repeated with 25 sets of QP angles (0°, 15°, 30°, ..., 345°, 360°) afterwards in order to examine the effects of the laser intensity and polarization on the Raman transition frequencies.

4. Results

The dependences of $\omega_{-1,-1}$, $\omega_{0,0}$ and $\omega_{1,1}$ on the rotation angle θ of the QP are depicted in Figure 2 (a) as Blue diamonds, Red circles and Magenta triangles connected by dashed lines, respectively. The green squares (noted as $\omega_{\pm1,\mp1}$) represent the transition frequencies of $|2,1\rangle \rightarrow |1,-1\rangle$ and $|2,-1\rangle \rightarrow |1,1\rangle$ in the Lin-Lin configuration ($\theta = k * 90°$, k=0, 1, 2, 3, 4), in which situation the $|2,0\rangle \rightarrow |1,0\rangle$ transition is forbidden. The solid lines are the theoretically calculated Raman transition frequencies using Eq. (1)-(6). In this calculation, the first-order Zeeman shifts on $\omega_{-1,-1}$ and $\omega_{1,1}$ are calculated as $\mp$1103.75 Hz (the absolute magnetic field is determined using the method proposed in Ref. [30]), the Doppler shifts of the three resonant frequencies are calculated as 6.15 Hz at the Raman laser irradiation time of 1.4 s, the angles of θ_k and α_0 are measured as 3.4° and 17.72° by probing the line strengths of the stimulated Raman transitions [26], and the Raman laser intensity is measured as 31.95 μW/cm^2 by detecting at the output port of the polarization maintaining fiber using a photon diode. The experimental results are in good agreement with the theoretical results, and both show the sinusoidal shift periods of $\omega_{-1,-1}$ and $\omega_{1,1}$ are approximately 180° while the shift period of $\omega_{0,0}$ is approximately 90°.

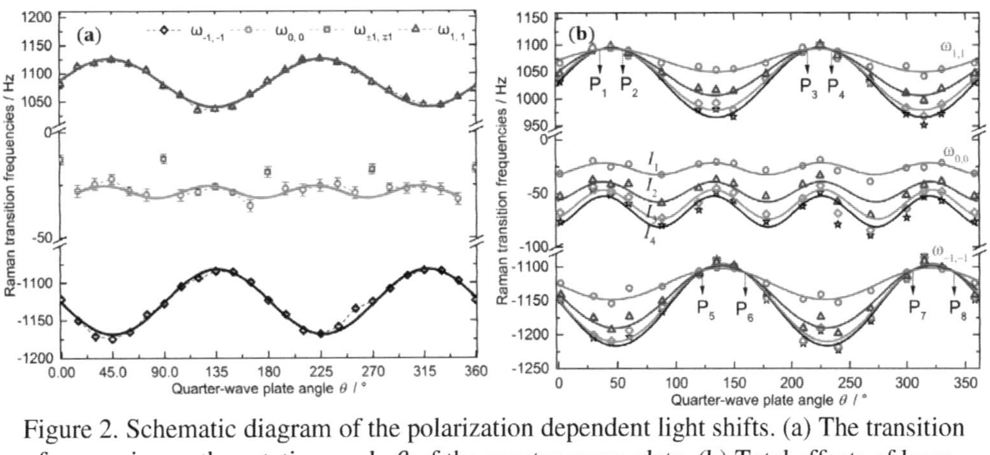

Figure 2. Schematic diagram of the polarization dependent light shifts. (a) The transition frequencies on the rotation angle θ of the quarter-wave plate. (b) Total effects of laser polarization and intensity on the differential light shifts of the Raman transition frequencies.

Figure 2(b) depicts the total effects of the laser polarization and intensity on the DLSs of the Raman transitions. Besides sinusoidal shifts with Raman laser polarization, the DLSs increase with laser intensity. Furthermore, the vanish of the DLSs for $\omega_{-1,-1}$ and $\omega_{1,1}$ at some elliptically polarized configurations (marked as P_1 to P_8) are observed, which means the DVLSs and DTLSs cancel the DSLSs exactly. However, due to the absence of the DVLSs, the DLS for $\omega_{0,0}$ doesn't vanish at any Raman laser polarization configuration.

5. Conclusion

In conclusion, we have presented the polarization dependent DLSs and the vanish of the DLSs in the near-resonant ^{87}Rb atom Raman transitions using a cold atomic fountain. Analytical theory has predicted that the transition frequencies of the magnetic sensitive and insensitive transitions would shift sinusoidally with periods of approximately 180° and 90° respectively, and the results have been confirmed experimentally. Furthermore, we observed the vanish of the DLSs for the magnetic sensitive Raman transitions at some elliptically polarized laser configurations while no vanish for the magnetic insensitive transition with our experimental parameters. These results are helpful for measuring and eliminating the Light shift related systemic errors in the atomic magnetometer, atomic clock, atomic interferometer and quantum computing. Furthermore, the scalar, vector and tensor components of the DLSs and the effective magnetic field vector can be extracted from the sinusoidal shifted transition frequencies.

Acknowledgment

This paper was financially supported by National Science Foundation (NSF) (11904408), and Strategic Priority Research Program of the Chinese Academy of Sciences (XDB21030200).

References

[1] Mathur, B. S.; Tang, H.; Happer, W. (1968) Phys. Rev., 171: 11-19.
[2] Brown, J. M.; Smullin, S. J.; Kornack, T. W.; Romalis, M. V. (2010) Phys. Rev. Lett., 105: 151604.
[3] Xu, Z.; Wu, Y.; Tian, L.; Chen, L.; Zhang, Z.; Yan, Z.; Li, S.; Wang, H.; Xie, C.; Peng, K. (2013) Phys. Rev. Lett., 111: 240503.
[4] Brion, E.; Molmer, K.; Saffman, M. (2007) Phys. Rev. Lett., 99: 260501.
[5] Kominis, I. K.; Kornack, T. W.; Allred, J. C.; Romalis, M. V. (2003) Nature, 422: 596-9.
[6] Derevianko, A.; Katori, H. (2011) Rev. Mod. Phys., 83: 2015-2019.
[7] Biedermann, G. W.; Wu, X.; Deslauriers, L.; Roy, S.; Mahadeswaraswamy, C.; Kasevich, M. A. (2015) Phys. Rev. A, 91: 033629.
[8] Biedermann, G. W.; McGuinness, H. J.; Rakholia, A. V.; Jau, Y. Y.; Wheeler, D. R.; Sterk, J. D.; Burns, G. R. (2017) Phys. Rev. Lett., 118: 163601.
[9] Lettner, M.; Mücke, M.; Riedl, S.; Vo, C.; Hahn, C.; Baur, S.; Bochmann, J.; Ritter, S.; Dürr, S.; Rempe, G. (2011) Phys. Rev. Lett., 106: 210503.
[10] Beloy, K. Theory of the ac Stark effect on the atomic hyperfine structure and applications to microwave atomic clocks. University of Nevada, 2009.
[11] Le Kien, F.; Schneeweiss, P.; Rauschenbeutel, A. (2013) Eur. Phys. J. D, 67: 92.
[12] Simon, E.; Laurent, P.; Clairon, A. (1998) Phys. Rev. A, 57: 491-3.
[13] Gregoire, M. D.; Hromada, I.; Holmgren, W. F.; Trubko, R.; Cronin, A. D. (2015) Phys. Rev. A, 92: 052513.
[14] Costanzo, G. A.; Micalizio, S.; Godone, A.; Camparo, J. C.; Levi, F. (2016) Phys. Rev. A, 93: 063404.
[15] Wood, A. A.; Turner, L. D.; Anderson, R. P. (2016) Phys. Rev. A, 94: 052503.
[16] Novikova, I.; Phillips, D. F.; Zibrov, A. S.; Walsworth, R. L.; Taichenachev, A. V.; Yudin, V. I. (2006) Opt. Lett., 31: 622.
[17] Zanon Willette, T.; de Clercq, E.; Arimondo, E. (2016) Phys. Rev. A, 93: 042506.
[18] Derevianko, A. (2010) Phys. Rev. Lett., 105: 033002.
[19] Gerginov, V.; Beloy, K. (2018) Phys. Rev. Applied, 10: 014031.
[20] Zhu, K.; Solmeyer, N.; Tang, C.; Weiss, D. S. (2013) Phys. Rev. Lett., 111: 243006.

IOP Conf. Series: Materials Science and Engineering **799** (2020) 012018 doi:10.1088/1757-899X/799/1/012018

[21] Bertoldi, A.; Bernon, S.; Vanderbruggen, T.; Landragin, A.; Bouyer, P. (2010) Opt. Lett., 35: 3769-71.
[22] Jensen, K.; Acosta, V. M.; Higbie, J. M.; Ledbetter, M. P.; Rochester, S. M.; Budker, D. (2009) Phys. Rev. A, 79: 023406.
[23] Chaneliere, T.; Hetet, G. (2015) Opt. Lett., 40: 1294-7.
[24] Hu, Q. Q.; Freier, C.; Sun, Y.; Leykauf, B.; Schkolnik, V.; Yang, J.; Krutzik, M.; Peters, A. (2018) Phys. Rev. A, 97: 013424.
[25] Dong, R.; Wei, R.; Du, Y.; Zou, F.; Lin, J.; Wang, Y. (2015) Appl. Phys. Lett., 106: 152402.
[26] Wang, W.; Dong, R.; Wei, R.; Lin, J.; Zou, F.; Chen, T.; Wang, Y. (2016) Appl. Phys. Lett., 108: 122401.
[27] Born, M.; Wole, E.; Born; Wole, *Principles of Optics*. Pergamon Press: 2001; p 986.
[28] Li, R. B.; Wang, P.; Yan, H.; Wang, J.; Zhan, M. S. (2008) Phys. Rev. A, 77: 033425.
[29] Wei, R.; Zhou, Z. C.; Shi, C. Y.; Zhao, J. B.; Li, T.; Wang, Y. Z. (2011) Chin. Phys. B, 20: 089501.
[30] Hu, Q. Q.; Freier, C.; Leykauf, B.; Schkolnik, V.; Yang, J.; Krutzik, M.; Peters, A. (2017) Phys. Rev. A, 96: 033414.

IOP Conf. Series: Materials Science and Engineering **799** (2020) 012019 doi:10.1088/1757-899X/799/1/012019

Research on Analog Signal Source of Gamma Ray Spectrum Logging Technology

Jun Tang, Yiying Li, Jia He, Simeng Wang, Hui Wang, Chaoyang Zhang,

College of Electronic Engineering, Xi'an Shiyou University, Xi'an, Shaanxi Province, ZIP code 710065, China

Corresponding author's e-mail: 657111481@qq.com

Abstract. Gamma ray spectrum logging tool acquires reservoir information by detecting gamma ray from formation. It has strict requirements for preservation and use, which is not convenient for college teaching. This paper introduces the design of an analog signal source for gamma ray spectrum logging tool, which generates random number based acceptance-rejection method. These random numbers are converted into electronic pulse signal by data acquisition system. It can simulate the electric pulse signal obtained from gamma ray through stratum. The statistical law of energy spectrum from simulation signal source accords with the probability distribution law of pulse signal generated gamma ray from formation. The analog signal source of gamma ray spectrum logging tool can generate simulation gamma ray logging signal without formation model and radiation source, which is of great value for efficient teaching.

1. Introduction

In the process of oil and gas exploration, logging technology is an engineering technology that solves geological problems with the principles of physics. Its measuring principles include electrical logging, acoustic logging and nuclear logging [1][2]. Gamma spectrum logging is an important nuclear logging method. In the borehole, the energy spectrum information of the gamma ray is recorded and analyzed, which evaluates the oil and gas reservoir information of the reservoir and solves the problems in geology and oilfield [3-6]. In the process of logging, gamma ray spectrum logging tool detects the gamma ray from the stratum. Therefore, when teaching gamma ray logging technology in colleges and universities, a realistic stratum model should be built for the calibration of logging tool. Furthermore, the radiation protection level of the laboratory must meet the basic standards of national ionizing radiation protection and radiation source safety, and must have the relevant qualification recognized by the state. It raises the threshold of nuclear logging teaching activities for logging instrument majors in colleges and universities. In order to reduce the threshold of nuclear logging teaching in Colleges and universities, this paper studies the analog signal source technology of gamma energy spectrum logging, analyzes the energy spectrum of scattering gamma ray from the formation, discusses the algorithm of generating signal pulse in accordance with the energy spectrum distribution, and puts forward the working program of analog signal source, which simulates the photoelectric pulse signal generated by gamma ray.

2. Generation principle of gamma ray energy spectrum

Gamma spectrum logging analyzes the spontaneous or scattered gamma spectrum distribution information from the formation to estimate the formation information [7-9]. Fig. 1 is the gamma ray energy spectrum curve generated by the stratum. The horizontal axis is the energy and the vertical axis is the counting rate of unit energy area. The counting rate of gamma ray is the statistics of the total number of electric pulses received in the unit energy area. In measurement, even if the measurement conditions are unchanged, the counting rate of each measurement is not constant, but the fluctuation around a certain value, that is, the statistical fluctuation. The counting rate changes with the change of gamma ray energy, and the change rule of energy spectrum is affected by the lithology, density and other parameters of surrounding strata.

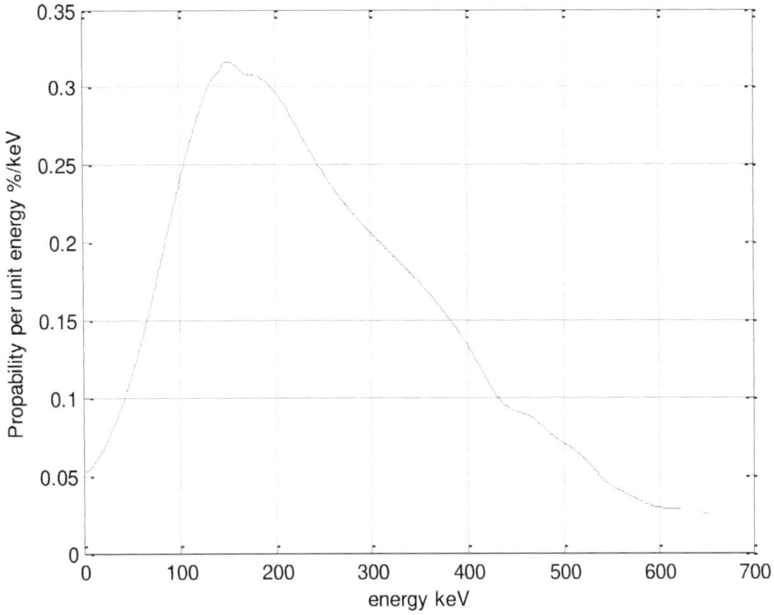

Figure 1. gamma ray energy spectrum curve of strata

According to the design purpose, the simulation signal source of gamma ray energy spectrum logging generates electronic pulse signal, whose pulse amplitude is directly proportional to the energy of gamma ray pulse signal, and the magnitude is random. The probability density distribution of pulse amplitude statistics should be consistent with the energy spectrum curve law shown in Figure 1.

3. Algorithm of random signal generation

In the design of analog signal source of gamma spectrum logging, it is necessary to generate random number with specific distribution. The main methods of generating random numbers according to the specified probability distribution are inverse function method, transformation method and acceptance-rejection method.

3.1. Inverse function method

The inverse function method is one of the methods to generate random numbers with specific distribution. Its theoretical principle is: Let the random number y obey the uniform distribution on the [0,1] interval, the random number function of the specific probability distribution is f (x), the inverse function of F (x) is F-1, let x = F-1 (y), then x is the random number obeying the probability distribution function f (x). Therefore, as long as we obtain the inverse function of the specific probability distribution function, we can generate the random number which obeys the specific distribution from the [0,1] uniformly distributed random number.

3.2. Transformation method

A random number of the specific distribution can be transformed into a random number of different distributions by a transformation. For example, linear transformation can transform the distribution on a finite interval [a, b] to any real number interval [c, d]. For each x, y can be given according to formula (1):

$$y=x\ (c-d)/(b-a)+c \tag{1}$$

And the box-Muller transformation can transform two random variables uniformly distributed on the interval [0,1] into two random variables with normal distribution.

3.3. Acceptance-rejection method

The acceptance-rejection method applied in this paper was first proposed by Von Norman. Its basic idea is: Assuming that the probability density distribution function is f(x). The random number sequence {r1,r2,r3,ri} is processed with acceptance-rejection method. The rule to acceptance-rejection is: when the density function f(x) is large, more random numbers ri are reserved; when the density function f(x) is small, less random numbers ri are reserved, so the distribution law of ri in the sub samples will meet the requirements of probability density distribution function f(x).

The algorithm of acceptance-rejection method is described as follows: [10-11]

(1)If the probability density function is f(x), a uniformly distributed random number X~Uni(xmin,xmax) is generated;

(2) Generate another uniformly distributed random number Y~Uni(ymin,ymax) independently;

(3) If $Y\leq f(X)$, return to x, otherwise return to step (1).

Acceptance-rejection method is essentially a simulation method, not a direct mathematical method. Each time a new random number is generated, another random number is used to ensure that the probability of the new random number being accepted obeys the specified probability density function. In the design of analog signal source of gamma spectrum logging, the probability distribution of pulse signal is determined, and the random data of corresponding distribution can be generated by acceptance-rejection method.

4. Program design

The analog signal source of gamma spectrum logging uses data acquisition card to realize the output of signal pulse. The acquisition card program first generates random numbers that obey the uniform distribution, processes the random numbers according to acceptance-rejection method, obtains the new random numbers that obey the same probability distribution as the gamma spectrum logging signals, and converts them into pulse simulation signals and outputs them. The data acquisition card program design flow is shown in Figure 2.

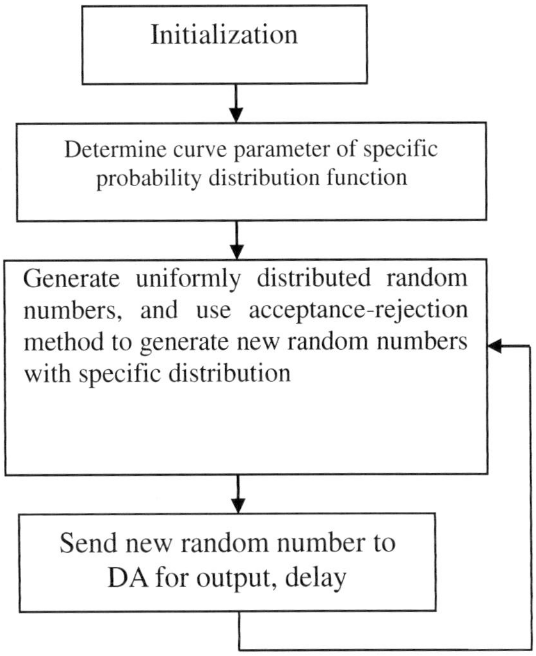

Figure 2. program flow chart of acquisition card

It can be seen from the flow chart that after the acquisition card generates the random data subject to the distribution of gamma spectrum logging signal, it is converted into pulse signal through DA. According to the sampling theorem, in order to distinguish the pulse signal in the subsequent signal receiving circuit, the delay time in each pulse generation cycle is about 4 times of the acquisition time interval.

5. Result analysis

According to the design requirements, the gamma spectrum logging analog signal source is built, which can generate analog pulse signals that meet the requirements of probability distribution. The pulse signals are processed statistically, and the statistical results are shown in Figure 3.

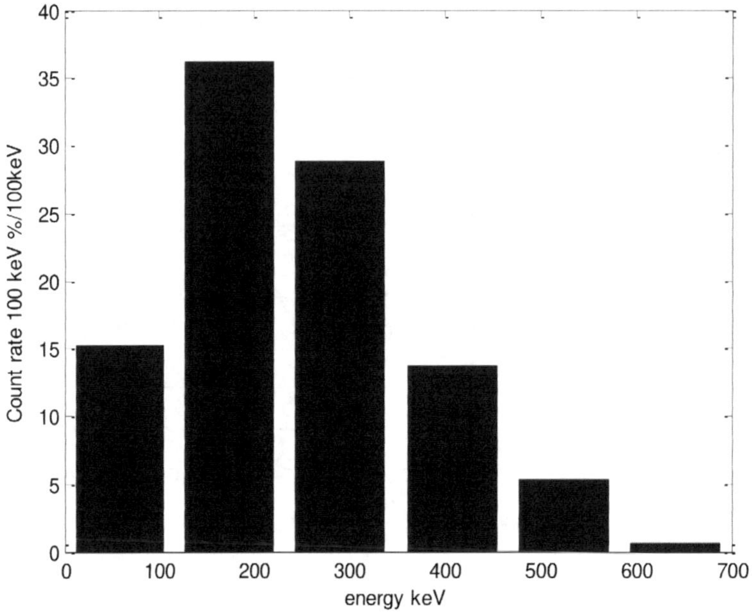

Figure 3. statistical results of output signals of analog signal source

From the statistical results in Fig. 3, it can be seen that the probability distribution of the pulse signal is close to that of the gamma ray energy spectrum curve shown in Fig. 1, indicating that the analog signal source of the gamma ray energy spectrum logging can realize the simulation of the gamma ray pulse signal.

6. Conclusion

The simulated signal source of gamma ray logging discussed in this paper can generate the simulated electronic pulse signal from the gamma ray. The probability distribution of the pulse signal conforms to the parameters of the gamma ray spectrum curve generated by the actual formation. During the use process, there is no need for the actual formation model, no risk of nuclear radiation and the limitation of the use of radioactive equipment, which has practical value in teaching.

References

[1] Wu Y., Mao J.L., Tang J., Yang J.F. (2014) Effects of casing on cross-hole electromagnetic logging , Journal of Northwest University (Natural Science Edition), 44:17-22.

[2] Tang J. (2009)Design of remote inversion processing system for array induction logging. Journal of Xi'an Shiyou University (Natural Science Edition), 24:96-98+102+114.

[3] Luo T.,Yan S. L., Wang J.B., Han Z.K., Kou Y.(2012)Interface circuit design of EILOG-06 litho-density logging tool. Petroleum Instruments, 26:11-12+15+101.

[4] Mai T. M., Vu N. B., Dang T. T. M., Dang N. P., Truong T. H. L., Ngo Q. H. (2019) Full spectrum analysis with the non-negative least square constraint for laboratory HPGe gamma spectrometry. Journal of Radioanalytical and Nuclear Chemistry, 322:961-967.

[5] Zhang F.,Zhao L.,Zhang Q.Y.,Li X.H.,Di Y.Z. (2019)Numerical simulation on gas reservoir identification using element gamma spectrum well logging . Journal of China University of Petroleum, 43:60-68.

[6] Tang J. (2008) Design of borehole correction network system based on MATLAB. Journal of Southwest Petroleum University (Science & Technology Edition), 5:152-154+1.

[7] Jia Q.G., Zhang T.K., Xu H.B. (2017) Optimization design of a Gamma-to-electron spectrometer for high energy gammas induced by fusion. Acta Phys. Sin., 66:55-62.

[8] Dong H.C. (2017) Application of natural gamma spectrum logging. Petrochemical Industry Technology, 24:260.
[9] Zhangjin Q.N.,Jia H.T.,Peng H.,Ai W.P.,Gao W.K.,Guan K.,Fan J.H. (2017)Neutron Gamma spectrometric system working with drill bit . Modern Electronics Technique, 40:6-11.
[10] Wolf-Dieter R., Kay S. (2017) Simulation of polyhedral convex contoured distributions. Journal of Statistical Distributions and Applications, 4:1-19.
[11] Nguyet N., Giray Ö. (2016) The acceptance-rejection method for low-discrepancy sequences.Monte Carlo Methods and Applications, 6:133-148.

IOP Conf. Series: Materials Science and Engineering **799** (2020) 012020 doi:10.1088/1757-899X/799/1/012020

A Reconstructing Minimization Approach for Total Variation Based Transient Electromagnetic Inversion

Liting Rao*

School of Electronic Engineering, Xi'an Shiyou University, Xi'an, Shaanxi, 710065, China

*Corresponding author's e-mail: ltrao@xsyu.edu.cn

Abstract. Transient electromagnetic (TEM) data are conveniently inverted to visualized underground structure with the widely used Occam's inversion. However, in sedimentary environments, Occam's inversion performs poorly in reproducing the sharp boundaries. This issue can be overcome by adopting total variation (TV) regularization. The existing computational methods for TV-based regularization are more complex than Occam's inversion, which limits the application of TV-based regularization in TEM inversion. Here, we develop a reconstructing minimization (RM) approach for TV-based transient electromagnetic inversion. The TV stabilizer is reconstructed as the weighted squared L2-norm of a kernel function. The obtained iterative model update expression of TV-based inversion is extremely close to that of Occam's inversion, which is rather simple to implement. Synthetic examples validate the efficiency and accuracy of the proposed RM approach.

1. Introduction

Transient electromagnetic (TEM) method is a well-established geophysical tool with applications in mineral exploration, geological mapping, unexploded ordnance detection, etc. [1]. In the TEM survey, a primary electromagnetic (EM) impulse is transmitted to the underground and then a secondary EM field induced by the eddy current can be measured. These secondary field data contain abundant geoelectrical information and we can transform these data into visualized underground structure by means of inversion method [2-3].

As with most geophysical inverse problems, the TEM inverse problem is always ill-posed. In order to reduce the non-uniqueness and stabilize the solution, a priori geologic information can be included in the inversion process and formalized by means of the stabilizer, which is classified as regularization method [4]. Currently, one of the most widely used stabilizers in TEM inversion is the maximum smoothness stabilizer and the corresponding inversion method is what we refer to as Occam's inversion [5], which is rather simple to implement. However, Occam's inversion produces smoothed results of subsurface geophysical properties [6], and therefore sometimes it looks geologically unrealistic, especially for sedimentary areas. This issue can be overcome by adopting another stabilizer, called total variation (TV) stabilizer. The TV-based regularization was originally developed to reconstruct noisy images [7], which can produce better quality images for blocky structures than the maximum smoothness stabilizer. The success of TV-based regularization relies on its ability of favouring the bounded variation, without penalizing discontinuities. With the advantage over Occam's inversion, TV-based regularization should also be widely used in TEM inversion, while the fact is not quite so. Due to the highly nonlinearity of TV stabilizer, the optimization problem of TV-based regularization is theoretically more difficult to solve. So far, compared with Occam's inversion, widely recognized computation methods, the fixed-point iteration scheme [7], and the Split-Bregman iterative method [8-9], are still quite complex so that the application of TV-based regularization is limited in TEM data inversion, and, more broadly, in geophysical data inversion.

Here, we proposed a reconstructing minimization (RM) approach for TV-based TEM inversion, which has a computational complexity comparable to Occam's inversion. We reconstruct the TV stabilizer as the weighted squared L2-norm of a kernel function and linearize the TV stabilizer by fixing its weight value at each iteration. The obtained iterative model update expression of TV-based inversion is extremely close to that of Occam's inversion. In this sense, the computational complexity of TV-based inversion is comparable to that of Occam's inversion. Moreover, the reconstruction idea of TV stabilizer also provides a new perspective of better understanding how the stabilizers differ from each other.

In the following text, we first explain the forward modeling and inversion procedure in details. Then we validate the proposed RM approach and compare the capabilities of TV-based inversion against Occam's inversion by means of synthetic TEM data sets.

2. Formulation

2.1. TEM Forward Modeling

In TEM method, the transmitter first creates a primary time varying EM field, which causes eddy currents in the subsurface conductors, then a secondary EM field is generated by these eddy currents. One can infer the underground electrical distribution by inverting the measured secondary field data. Up to date, basing forward modeling and inversion on the stratified model is still a routine practice. In this letter, we implement our modeling and inversion on layered earth model. The circular loop source is adopted for the transmitter. The system parameters, including the radius of transmitting loop a, the transmitting current I, the height of transmitter h_0, the height of the receiver z, and the center offset between the transmitter loop and receiver coil ρ are known in general. The received vertical magnetic field can be calculated by [10]

$$\boldsymbol{H}_z = \frac{Ia}{2}\int_0^{\infty}\frac{\lambda^2}{u_0}\left[e^{-u_0(z+h_0)} + r_{\mathrm{TE}}e^{u_0(z-h_0)}\right]J_0(\lambda\rho)J_1(\lambda a)d\lambda \quad (1)$$

where J_0 and J_1 are Bessel functions of the first kind, λ is the beam of electromagnetic wave, and r_{TE} is the reflection coefficient which can be iteratively calculated as

$$r_{TE} = \frac{Y_0 - \hat{Y}_1}{Y_0 + \hat{Y}_1} \quad (2)$$

where

$$\hat{Y}_i = Y_i\frac{\hat{Y}_{i+1} + Y_i\tanh(u_i h_i)}{Y_i + \hat{Y}_{i+1}\tanh(u_i h_i)}, i = N-1, N-2, \cdots, 1 \quad (3)$$

starts from $\hat{Y}_N = Y_N$. $Y_i = u_i/(i\mu_i\omega)$, $u_i = (\lambda^2 + i\omega\mu_i\sigma_i)^{1/2}$.

Fast Hankel and Fourier transformation methods are widely used to calculate the received magnetic field in the frequency domain and transform it to time domain. These methods are not described in detail here and can be referred to [2] and [11].

2.2. Inversion

Introducing specially selected stabilizing functional to narrow the class of models within the correctness set is a simple and efficient way to overcome the ill-posedness of inverse problem. For TEM inversion, the objective functional can be constructed as

$$P^{\alpha}(\boldsymbol{m}) = \phi(\boldsymbol{m}) + \alpha s(\boldsymbol{m}) \quad (4)$$

where $\boldsymbol{m}$ represents the model parameters, $\phi(\boldsymbol{m})$ is the misfit functional that determines the difference between observed data and theoretical calculation, $s(\boldsymbol{m})$ is a stabilizing functional (or a stabilizer) and α is a regularization parameter.

In this paper the misfit functional is calculated based on least-squares norm, which can be expressed as

$$\phi(\boldsymbol{m})=\left\|\boldsymbol{W}_d\left(F(\boldsymbol{m})-\boldsymbol{d}\right)\right\|_{L_2}^2 \tag{5}$$

where $F(\cdot)$ represents the forward modeling operator, $\boldsymbol{W}_d$ is the data weight matrix, $\boldsymbol{d}$ is the observed TEM data set. Here $F(\cdot)$ represents the forward modeling procedure described in last section and the model parameters consist of each layer's resistivity and thickness. Note that we usually apply logarithmic data and logarithmic model parameters in TEM inversion, to minimize nonlinearity and to impose positivity [12].

For TV-based inversion, we select the TV stabilizer to constrain the model parameters, which is essentially the L_1-norm of the gradient of the model parameters:

$$s_{TV}(\boldsymbol{m})=\left\|\nabla\boldsymbol{m}\right\|_{L_1} \tag{6}$$

The minimization of TV stabilizer is the key part in solving the optimization problem of objective functional. To deal with the highly nonlinearity of TV term, a lot of research on efficient algorithms for computing optimal of nearly optimal solutions have been conducted. However, compared with widely used Occam's inversion for TEM data, these recognized methods for TV stabilizer are still quite complex. In order to reduce the computational complexity and make the inversion easy to implement, we propose a reconstructing minimization (RM) approach for TV-based inversion to solve the optimization problem.

First, reconstruct TV stabilizer as the weighted squared L2-norm of a kernel function:

$$s_{TV}(\boldsymbol{m})=\left\|w_{TV}(\boldsymbol{m})k(\boldsymbol{m})\right\|_{L_2}^2 \tag{7}$$

where $w_{TV}(\boldsymbol{m})$ is weight function, $k(\boldsymbol{m})$ is kernel function, and their specific expressions are as follows:

$$w_{TV}(\boldsymbol{m})=\frac{\sqrt{|\nabla\boldsymbol{m}|}}{\sqrt{|\nabla\boldsymbol{m}|^2+\varepsilon^2}} \tag{8}$$

$$k(\boldsymbol{m})=\nabla\boldsymbol{m} \tag{9}$$

where ε is a very small number for reconstruction.

Now the objective functional with reconstructed TV stabilizer can be expressed as:

$$P^\alpha(\boldsymbol{m})=\left\|\boldsymbol{W}_d\left(F(\boldsymbol{m})-\boldsymbol{d}\right)\right\|_{L2}^2+\alpha\left\|w_{TV}(\boldsymbol{m})\nabla\boldsymbol{m}\right\|_{L2}^2 \tag{10}$$

Suppose that $\boldsymbol{m}$ is a vector with length N and $\boldsymbol{d}$ is a vector with length M. Rewrite the objective functional using matrix notations, we have

$$P^\alpha(\boldsymbol{m})=\left[\boldsymbol{W}_d\left(F(\boldsymbol{m})-\boldsymbol{d}\right)\right]^T\left[\boldsymbol{W}_d\left(F(\boldsymbol{m})-\boldsymbol{d}\right)\right]+\alpha\left(\boldsymbol{W}_{TV}\boldsymbol{G}\boldsymbol{m}\right)^T\boldsymbol{W}_{TV}\boldsymbol{G}\boldsymbol{m} \tag{11}$$

where $\boldsymbol{W}_{TV}$ is the diagonal matrix representation of the weight function, the symbol T denotes a transposed complex conjugate matrix, and $\boldsymbol{G}$ is the matrix representation of gradient operator ∇ which is written as

$$G=\begin{bmatrix} 0 & & & & 0 \\ -1 & 1 & & & \\ & -1 & 1 & & 0 \\ & & \cdots & & \\ & & 0 & -1 & 1 \end{bmatrix}_{N\times N} \tag{12}$$

Suppose an initial model $\boldsymbol{m}_0$ and assume the forward modeling operator is differentiable at $\boldsymbol{m}_1=\boldsymbol{m}_0+\delta\boldsymbol{m}$ for sufficiently small vector $\delta\boldsymbol{m}$, we have

$$F(\boldsymbol{m}_1)=F(\boldsymbol{m}_0)+\boldsymbol{J}_0(\boldsymbol{m}_1-\boldsymbol{m}_0)+o\left\|(\delta\boldsymbol{m})^2\right\| \tag{13}$$

Where $\boldsymbol{J}_0$ is the Jacobian matrix evaluated at the vector $\boldsymbol{m}_0$, $o\left\|(\delta\boldsymbol{m})^2\right\|$ is the higher order remainder of Taylor expansion.

Then approximate the $F(\boldsymbol{m}_1)$ by dropping the remainder term and substitute its approximate expression into equation (12). Meanwhile, one can linearize the stabilizer by substituting $\boldsymbol{m}_0$ into the weight function to obtain the weight matrix $\boldsymbol{W}_{TV0}$. Under this approximation and linearization, we have returned to a linear problem of $\boldsymbol{m}_1$. Assume $\boldsymbol{m}_1$ as the model that minimizes the objective functional, we have

$$\nabla_{\boldsymbol{m}_1} P^{\alpha}(\boldsymbol{m}_1)=0 \tag{14}$$

Then the model vector $\boldsymbol{m}_1$ satisfies the equation:

$$(\boldsymbol{W}_d\boldsymbol{J}_0)^T\boldsymbol{W}_d\left[F(\boldsymbol{m}_0)+\boldsymbol{J}_0(\boldsymbol{m}_1-\boldsymbol{m}_0)-\boldsymbol{d}\right]+\alpha(\boldsymbol{W}_{TV0}\boldsymbol{G})^T(\boldsymbol{W}_{TV0}\boldsymbol{G})\boldsymbol{m}_1=0 \tag{15}$$

After some algebraic operations, we get

$$\boldsymbol{m}_1=\left[(\boldsymbol{W}_d\boldsymbol{J}_0)^T\boldsymbol{W}_d\boldsymbol{J}_0+\alpha(\boldsymbol{W}_{TV0}\boldsymbol{G})^T\boldsymbol{W}_{TV0}\boldsymbol{G}\right]^{-1}(\boldsymbol{W}_d\boldsymbol{J}_0)^T\boldsymbol{W}_d\hat{\boldsymbol{d}}_0 \tag{16}$$

$$\hat{\boldsymbol{d}}_0=\boldsymbol{d}+\boldsymbol{J}_0\boldsymbol{m}_0-F(\boldsymbol{m}_0) \tag{17}$$

Through n steps of iteration,

$$\boldsymbol{m}_{n+1}=\left[(\boldsymbol{W}_d\boldsymbol{J}_n)^T\boldsymbol{W}_d\boldsymbol{J}_n+\alpha(\boldsymbol{W}_{TVn}\boldsymbol{G})^T\boldsymbol{W}_{TVn}\boldsymbol{G}\right]^{-1}(\boldsymbol{W}_d\boldsymbol{J}_n)^T\boldsymbol{W}_d\hat{\boldsymbol{d}}_n \tag{18}$$

$$\hat{\boldsymbol{d}}_n=\boldsymbol{d}+\boldsymbol{J}_n\boldsymbol{m}_n-F(\boldsymbol{m}_n) \tag{19}$$

Although the reconstruction seems to make the expression for TV stabilizer more complicated in the first place, it turns out that one can linearize the TV stabilizer by fixing its weight value at each iteration and the iterative model update expression is extremely close to that of Occam's inversion. The difference between their expressions can be considered as replacing the identity matrix in Occam's inversion with the diagonal matrix $\boldsymbol{W}_{TV}$ determined by weight function. In this sense, the computational complexity of TV-based inversion is comparable to that of Occam's inversion. Moreover, it should be noted that the weight function depends on the gradient of model parameters rather than the model parameters themselves since the core of the minimization of TV stabilizer is the gradient of model parameters, and a false reconstruction of weight function directly using the model parameters leads to wrong results.

To examine whether the updated $\boldsymbol{m}_n$ meets the requirement, the rms misfit is defined as follow:

$$RMS=\sqrt{\frac{1}{M}\sum_{n=1}^{M}\left(\frac{F(\boldsymbol{m}_n)-\boldsymbol{d}}{\boldsymbol{d}}\right)^2} \tag{20}$$

The iteration procedure continues until the rms misfit reaches an acceptable level.

3. Numerical examples

We demonstrate the application of TV-based inversion using the RM approach on the synthetic TEM data sets. In particular, we compare the capabilities of TV-based inversion against Occam's inversion which is based on maximum smoothness model. The goals are to validate the RM approach and understand the differences between TV-based inversion and traditional Occam's inversion.

In the following cases, the synthetic data are simulated by means of (1) for a ground-based TEM instrument. We set the transmitting loop with radius 100 m and current strength 20 A. The receiver coil is located at the center of transmitting loop. Noise with Gaussian distribution $(0, 0.01)\text{nV/m}^2$ is added into the synthetic data. For all the inversions, a $100\,\Omega\text{m}$ uniform half-space is used as a starting model, and the ground in vertical direction is discretized with 50 layers, with the first layer thickness 5m and increasing ratio 1.05. The choice of regularization parameter is significant for results of the inversion procedure. Our selection strategy of regularization parameter is similar with that of Occam's inversion, which emphasizes more on misfit functional. At each iteration, the ratio of the misfit functional to the stabilizing functional is calculated as the initial value of regularization parameter. The optimal regularization parameter is chosen based on the minimization of rms misfit with a 1-D linear search and the zoom range for regularization parameter is set linearly partitioned in the log10 domain from 10^{-3} to 10^{0}. The target rms misfit is set to 1%.

3.1. 1D inversion

Four synthetic models describing different underground structures are considered. Table 1 shows the model parameters of these models. Their forward responses with noise perturbation are presented in figure 1, which will then be used for inversions. Comparing their responses, we find that the response of model A coincides with that of model C in the early time, while their response curves differ from each other in the late time. This phenomenon also occurs on the response curves of model B and model D. The reason for this phenomenon is that the resistivity values of their shallow layers are set the same and the values of deep layers become different. According to the physical principle of TEM method, the early stage response mainly reflects the property of shallow layer while the late time response is determined by the deep structure.

Table 1. Model parameters

Models	Resistivity (Ωm)	Thickness (m)
Model A	[100, 20, 300]	[50, 50, ∞]
Model B	[100, 300, 20, 100]	[50, 100, 100, ∞]
Model C	[100, 20, 300, 10, 100]	[50, 50, 100, 100, ∞]
Model D	[100, 300, 20, 300, 10, 100]	[50, 100, 100, 200, 200, ∞]

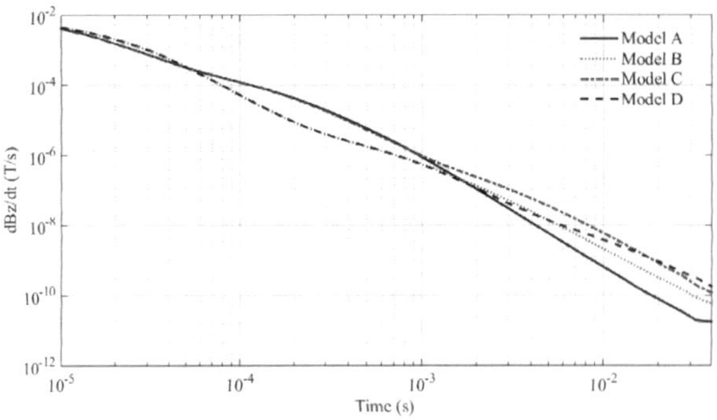

Figure 1. Responses of four synthetic models, with noises added.

The results of TV-based inversion and Occam's inversion are shown in figure 2. Iteration times of Occam's inversion for these four models are 5, 4, 8, and 8, respectively. Iteration times of TV-based inversion are 5, 4, 8, and 9, respectively. Average running time of each iteration for both methods is basically the same. Comparing the inversion results, we find that both inversions retrieve the main features of the true models reasonably well. However, from the specific details of the recovered structures, TV-based inversion, represented by black solid lines, has better quality than Occam's inversion. The differences between the two inversions are discussed in terms of the boundary location and resistivity value. In the case of Occam's inversion, the recovered boundaries appear less steep than they actually are due to the slow transition between the high resistivity interface and the low resistivity interface, and the retrieved resistivity presents spurious oscillation around the true resistivity value especially in homogenous top layer and bottom layer. As for TV-based inversion, the boundaries in TV-based inversion are more precisely retrieved which presents the capability to reproduce sharp boundary, and the recovered resistivity values of each layer are better determined and also more homogenously reproduced. As can be seen in all the examples in figure. 2, the spurious oscillation in Occam's inversion has been almost completely removed by the TV-based inversion. Clearly, the TV-based inversion enhances the sharp features and also eliminates spurious oscillations. Combining the above statements, in terms of efficiency and quality, we have validated the proposed RM approach in solving the optimization problem of TV-based inversion.

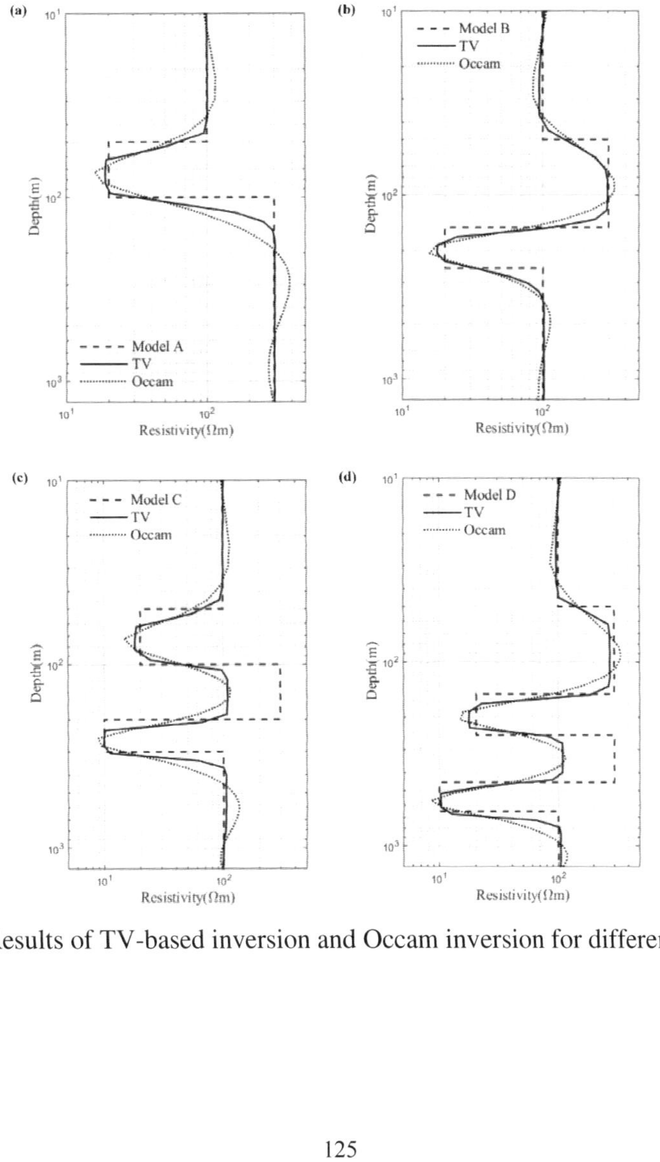

Figure 2. Results of TV-based inversion and Occam inversion for different structures.

Besides, in figure 2c and 2d, we notice that the retrieved resistivity level of the middle high resistivity layer is underestimated in both inversion methods. This is because the TEM method is more sensitive to conductive abnormal formation than high resistance formation, meanwhile the middle high resistivity layer locates at relatively deeper area which further aggravates the shielding effect of the conductive layer. In other words, the local underestimation of high resistivity layer results from the insufficient data sensitivity.

3.2. Quasi-2D inversion

Here the considered synthetic model (see figure 3a) is generated by using 35 five-layered models to form a 2D section. As shown in figure 3a, the true model represents a five-layered sedimentary environment. The first layer is characterized by the resistivity of 90 Ωm to 95 Ωm and the thickness about 40 m to 50m. The second layer represents the conductive structure with the resistivity of 18 Ωm to 22 Ωm and the thickness ranging from 30 m to 60 m. The third layer has the resistivity of 80 Ωm to 85 Ωm and the thickness about 100 m. The fourth layer is another conductive structure with the resistivity of 8 Ωm to 9 Ωm and the thickness ranging from 90 m to 140 m. The bottom layer has a relatively high resistivity of 100 Ωm to 105 Ωm.

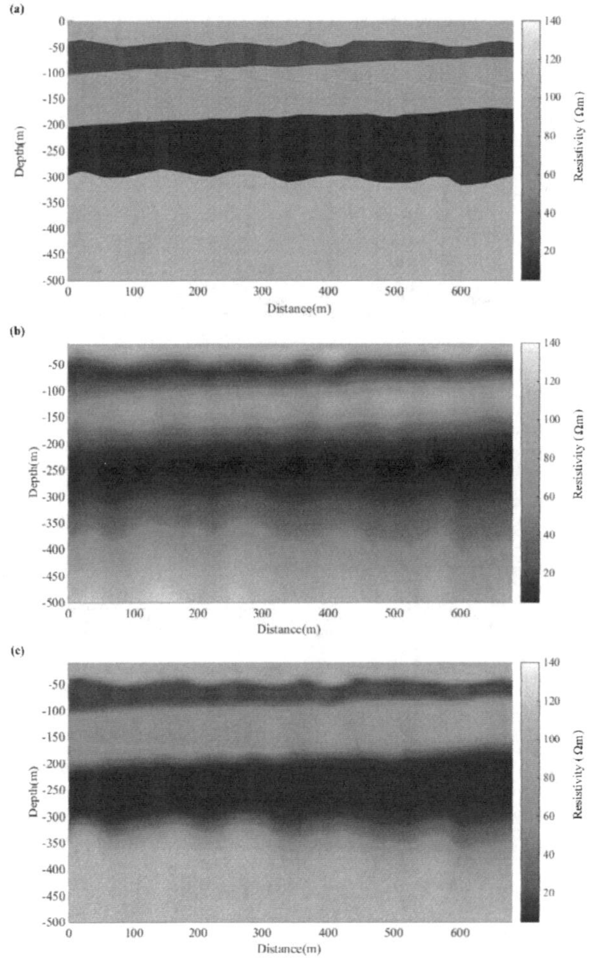

Figure 3. True model and results of TV-based inversion and Occam inversion. (a) The true model, (b) Occam's inversion, (c) TV-based inversion.

IOP Conf. Series: Materials Science and Engineering **799** (2020) 012020 doi:10.1088/1757-899X/799/1/012020

The inversion results for both methods are shown in figure 3b and 3c. The average iteration times for Occam's inversion and TV-based inversion are 6.94 and 6.87, respectively, which are basically the same on efficiency. From figure 3b, we find that Occam's inversion is fairly successful in recovering the overall section: the outline of the anomalous area can be seen, the mean resistivity value of the conductive structure is well reconstructed, the overall resistivity level of the background is slightly overshooting and not homogeneous enough, and the lower boundary is quite blurred especially in the deeper area. Similar to that in figure 2, the Occam's inversion is prone to have spurious oscillation around the true resistivity value, which results in inhomogeneous reproduced background. In the case of TV-based inversion (see figure 3c), we again find that the recovered section is significantly improved: the location of the boundary is more accurate, and the resistivity values of the conductive structure and the background are better recovered and more homogenously reproduced. Along the entire section, we can notice the bottom boundary is less clear than the upper boundary and this fuzziness is aggravating in deeper area for both inversion methods. This is normal because the resolution of TEM method will decline along with the increasing of depth essentially. Even if you increase the number of layers for inversion, this phenomenon still exists.

4. Conclusion

In this paper, we have developed a reconstructing minimization (RM) approach for TV-based transient electromagnetic inversion. The computational cost of the RM approach is comparable to that of classical Occam's inversion, which is simpler to be implemented than previous computational methods for TV-based inversion. Numerical examples have verified the effectiveness of the RM approach and demonstrated the superior capability of TV-based inversion in resolving sharp resistivity changes. Concerning future applications, the proposed RM approach will broaden the application of TV-based regularization in the processing of TEM data or other geophysical data.

Acknowledgments

The author would like to acknowledge the support by the Natural Science Basic Research Plan in Shaanxi Province of China (No. 2018JQ4007) and the Natural Science Foundation of Shaanxi Provincial Department of Education (No. 19JK0669).

References

[1] Wu, X., Xue, G., Xiao, P., Li, J., Liu, L., Fang, G. (2019) The removal of the high-frequency motion-induced noise in helicopter-borne transient electromagnetic data based on wavelet neural network. Geophysics, 84: K1-K9.

[2] Qi, Y., Huang, L., Wang, X., Fang, G., Yu, G. (2017) Airborne transient electromagnetic modeling and inversion under full attitude change. IEEE Geoscience and Sensing Letters, 14: 1575–1579.

[3] Rao, L., Wu, X., Wu, C., Zhang, X., Fang, G. (2016) A rapid 3D inversion based on TEM moment transformation. Chinese Journal of Geophysics, 59: 4338-4348.

[4] Zhdanov, M.S. (2015) Inverse Theory and Applications in Geophysics. The Netherlands: Elsevier, Amsterdam.

[5] Constable, S.C., Parker, R.L., Constable, C.G. (1987) Occam's Inversion: a practical algorithm for generating smooth models from electromagnetic sounding data. Geophysics, 52: 289-300.

[6] Chen, W., Li, H., Xue, G., Chen, K., Zhong, H. (2017) 1D OCCAM inversion of SOTEM data and its application to 3D models. Chinese Journal of Geophysics, 60: 3667-3676.

[7] Vogel, C.R., Oman, M. E. (1998) Fast total variation based reconstruction of noisy, blurred images. IEEE Transactions on Image Processing, 7: 813-824.

[8] Goldstein, T., Osher, S. (2009) The split Bregman method for l1 regularized problems. Siam Journal on Imaging Sciences, 2: 323-343.

[9] Lin, Y., Huang, L. (2015) Acoustic- and elastic-waveform inversion using a modified total-variation regularization scheme. Geophysical Journal International, 200: 489-502.

[10] Nabighian M.N. (1988) Electromagnetic methods in in applied geophysics (Volume 1, Theory). Society of Exploration Geophysicists.

[11] Anderson, W.L. (1989) A hybrid fast Hankel transform algorithm for electromagnetic modelling. Geophysics, 54: 263-266.

[12] Auken, E., Christiansen A.V. (2004) Layered and laterally constrained 2D inversion of resistivity data. Geophysics, 69: 752-761.

A room-temperature hydrogen gas sensor based on TiO_2/MoS_2 nanocomposite

Cheng Zhang[1], Yixue Li[2], Qiao Sun[1], Weitao Lei[1], Yankun Tang[1], Jiuhong Wang[1*] and Hairong Wang[1]

[1]State Key Laboratory for Manufacturing Systems Engineering, Mechanical Engineering School, Xi'an Jiaotong University, Xi'an, 710049, China

[2]State Key Laboratory for Mechanical Behavior of Materials, Xi'an Jiaotong University, Xi'an 710049, China

[*]Corresponding author's e-mail: jhw@xjtu.edu.cn

Abstract. A room-temperature hydrogen gas sensor based on TiO_2/MoS_2 nanocomposite which synthesized by the hydrothermal process was investigated. The nanocomposite was the flowerlike MoS_2 nanosheets adhering on end surfaces of the porous TiO_2 nanorods. The sensing properties of the TiO_2/MoS_2 based sensor to hydrogen were investigated at room temperature and compared with the sensing properties of the sensor based on pure TiO_2 nanorods. The nanocomposite-based sensor has the sensing response(S) of 4.25% to 200 ppm hydrogen at 25 °C, which was about 80 times higher than the sensing response(S) of the pure TiO_2 nanorod-based sensor. The response time of the nanocomposite-based sensor was about 96 s, and the recovery time was about 54 s to 200 ppm hydrogen at 100 °C. The enhanced sensing response(S), rapid response and recovery characteristics were attributed to the structural characteristics of the nanocomposite and the formed p-n junction between TiO_2 and MoS_2.

1. Introduction

With the exhaustion of fossil fuels and increasingly serious environmental problems, Hydrogen, which has high combustion heat (142 kJ/g), low minimum ignition energy (0.017 mJ), wide flammable range (4-75%vol), and high burning velocity, has drawn more and more attentions and has become one of the best clean energy sources because of its regenerated and clean final combustion product [1]. However, hydrogen is not a really safe energy source because it will explode violently when its concentration ranges from 4% to 75% (volume ratio) in the air [2]. Therefore, for safe production, storage, and application, it is extremely necessary to detect the hydrogen leakage rapidly and accurately.

So far, there are various types of hydrogen gas sensors [3-5], and among of them, those based on the different TiO_2 [6-8] have been widely studied and applied, due to their good stability, simple structure, low cost and simplicity of synthetic process. The nanostructured material can provide large specific surface area, e.g., a kind of uniformly distributed TiO_2 nanoparticle fabricated by Wang, LQ et al., whose specific surface area is 203.245 m^2/g [9]. The surface and size effect of the nanostructured metal oxides allow their higher sensing response. To realize low-temperature or even room-temperature gas sensors and to reduce power consumption, the TiO_2 materials are often prepared intentionally in the form of nanostructures and then doped. B. Lyson-Sypien et al. investigated that the mixture of TiO_2 and Cr/TiO_2 nanopowders has good response of to H_2 at 350 °C [10]. However, at a lower working temperature, the gas sensors usually exhibit some disadvantages, for example, the low sensing response and long response/recovery time [11].

In order to reduce the power consumption of the hydrogen gas sensor as well as to overcome those disadvantages above, researchers attempt to find some ways of lowering the working temperature in the precondition of keeping good sensing properties. One of the most effective way is doping or coating noble metal catalysts like Pd, Pt or others on metal oxide semiconductor materials [12, 13]. Hayawa et al. reported that the TiO_2 doped with Pt has good gas sensing performance at 200°C [14]. Some researchers also doped the different semiconductors for synergistic effects or forming heterojunctions to enhance the sensing properties [15-17].

Graphene has attracted much attention due to its unique thermal, electrical, biological, mechanical and sensing properties [18-20]. However, it is not beneficial to improve the gas sensing response and selectivity because of its zero band gap, limiting its application in the gas sensors. Zero band gap of graphene has inspired researchers to turn to other graphene-like materials with appropriate band gaps, such as transition-metal oxides (TMOs) or transition-metal chalcogenides (TMCs), like MoS_2 [21-26], GaS, GaSe [27, 28] and WSe_2 and so on [29, 30]. Compared with other emerging 2-D layered materials, the MoS_2 has been widely studied as the gas sensing material because of its high sensing response [31]. The nanosheet MoS_2 has better hydrogen adsorption and desorption properties, and it can be developed into a monolayer hydrogen sensor [32].

In this paper, we proposed a room-temperature hydrogen sensor based on the TiO_2 / MoS_2 nanocomposite. The nanocomposite was prepared by a two-steps hydrothermal process and characterized by a series of methods. The gas sensor was fabricated by integrating the nanocomposite with the micro electrodes on substrate, and its sensing properties to H_2 were also tested. The nanocomposite-based hydrogen sensor demonstrates a higher sensing response and faster response/recovery properties compared with the sensor based on the pure TiO_2 nanorods.

2. Experiment

All chemicals are analytical reagent grade and used without further purification. The precursor solution was prepared by mixing 10 ml HCl with 10 ml deionized water, under vigorous stirring at room temperature to form a homogeneous solution. Then appropriate amount of cetyltrimethyl ammonium bromide (CTAB) was added into the solution and stirred continuously until the CTAB was dissolved completely. Titanium butoxide of 0.4 ml was added drop by drop. The solution was kept stirring 25 min at room temperature until it was turned into totally homogeneous, then transferred into a 25 ml Teflon-lined stainless-steel autoclave and maintained at 150 °C in an electric thermostatic drying oven for 6 h. After the hydrothermal procedure, the autoclave was cooled down to the room temperature naturally. After the above hydrothermal process, some white powder, TiO_2 nanorods, grew on the inside wall of the autoclave. The powder was carefully collected and washed thoroughly by centrifugation alternately with deionized water and ethanol for 3 times, of which each one lasts for 10 min. Then the solid powder was placed in an electric thermostatic oven to dry up at 80 °C for 12 h, and subsequently annealed at 600 °C in air for 2 h. Finally, the nanostructured TiO_2 nanorods were obtained.

The hydrothermal process was used again to prepare the TiO_2/MoS_2 nanocomposite. In the synthesis process, TiO_2 powder of 20 mg prepared by the above process was dispersed into 15 mL of deionized water under vigorous stirring to be a clear solution. After that, 60.5 mg sodium molybdate ($Na_2MoO_4 \cdot 2H_2O$) and 4 mg oxalic acid($H_2C_2O_4$) were added and stirred for over 10 min until all the reagent were dissolved. Then 91.125mg L- cysteine was added and then stirred for 20 min to prepare the second precursor solution.

The solution was transferred into a 25 ml Teflon-lined stainless-steel autoclave and kept at 200°C for 12 h. After this, the autoclave was cooled down to the room temperature, and black solution was obtained. It was centrifuged, washed with deionized water and ethanol for 3 times, and dried up at

IOP Conf. Series: Materials Science and Engineering **799** (2020) 012021 doi:10.1088/1757-899X/799/1/012021

60 °C in the vacuum drying oven. After keeping it at 500 °C for 1 h in the nitrogen ambient atmosphere, the TiO_2/MoS_2 nanocomposite was obtained in the end.

The prepared pure TiO_2 and TiO_2/MoS_2 composite were characterized by X-ray diffraction (XRD) using Cu Kα radiation at 40 kV and 40 mA at the scanning rate of 0.02° in degree, 0.1 s/step, field-emission scanning electron microscopy (SEM), and transmission electron microscopy (TEM).

The gas sensor was fabricated by integrating sensing nanomaterial with micro interdigited electrodes on the silicon substrate with drop-coating process. The TiO_2/MoS_2 nanocomposite of 10 mg was put into a sampling bottle, 10 ml of absolute ethyl alcohol was injected, and 20 mg of ethocel was added to improve the adhesive property. The bottle was vibrated for 10 min in an ultrasonic cleaner to mix the solution well. A syringe was used to pipet 5 μl solution, and then the solution was dropped on the surface of the interdigitated electrodes. The sample was dried up on the heating plate at the 90 °C to make the solution volatize, and a layer of sensing materials appeared on surface of the electrodes. After aging 12 h at 120 °C, the fabrication of the nanocompsite sensor was completed, as shown in Fig. 1, in which the two pads are used for connecting the wires. By taking appropriate TiO_2 nanorods and using the same process, the sensor based on the pure TiO_2 nanorod were prepared.

The sensor chip was attached to a small PCB board and put into an airtight chamber for testing. Response properties of the sensor to different concentration of H_2 were tested at 25 °C-150 °C. The total flow rate was 200 sccm, which was controlled by a gas delivered system and a calibrated mass flow controller, as shown in Fig. 2. The airtight chamber was full of air before measurement, and the gas concentration was changed in an interval of 5 min. A DC power supplier was used to drive the sensor at the fixed value of 5 V, and the change in the resistance of the sensor was recorded by a High Resistance Meter. The gas response of the sensor was defined as |Rg-Ra|/Ra, where Ra is the electrical resistance of the gas sensor in air and Rg the resistance in the presence of target gas.

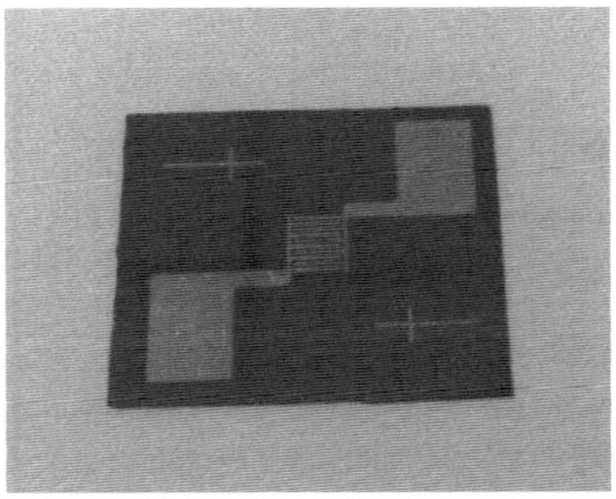

Fig. 1 Image of the hydrogen sensor.

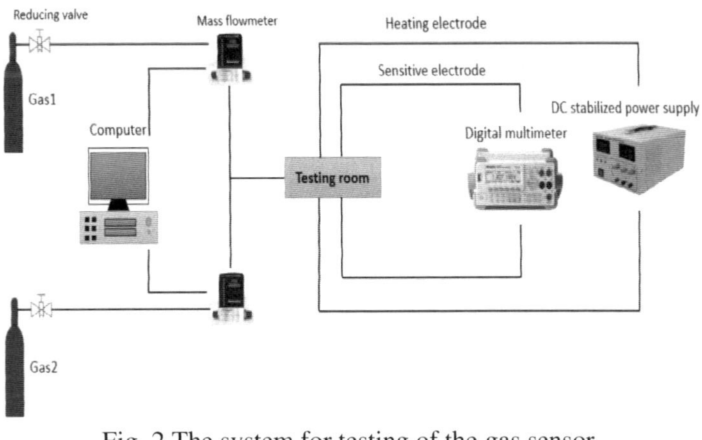

Fig. 2 The system for testing of the gas sensor.

3. Results And Discussion

Fig. 3 shows the XRD patterns of the pure TiO_2 nanorods and the TiO_2/MoS_2 composites, respectively. It indicates that both samples consist of high intensity fraction of the rutile phase. In terms of TiO_2 / MoS_2 composites, the peak at 13.396°and 32.686°are indexed with (002) and (100) of MoS_2 (JCPDS No.2-1133). Besides this, the peak at 27.499°, 36.081°and 54.372°are corresponding with (110), (101) and (211) of TiO_2 (JCPDS No.4-551). No other diffraction peak of impurities may be observed, which indicates that the final products are the mixture of TiO_2 and MoS_2 with high purity.

As shown in Fig. 4a, the SEM image indicates that the prepared TiO_2 nanorods are good dispersity, and only a few short nanorods perhaps are the undeveloped nanorods. It can be seen that the nanorods are uniform and have smooth surfaces. The diameters of the TiO_2 nanorods are ranging from 80-100 nm and lengths are about 1.5 μm. Figs. 4b and c are the TEM and high-resolution TEM (HRTEM) images of the TiO_2 nanorods. As shown in Fig. 4b, the nanorods are straight and uniform, and some short nanorods also can be found, which are as the same as SEM image. Some unshaped nanorods can be found around the periphery of the nanorods, which is in accordance with the SEM image. Fig. 4c is a magnified TEM image of the area within the red rectangle in Fig. 4b. According to the image, the lattice distances are d110 = 3.25 Å and d101 = 2.49 Å, which are in accordance with the interplanar spacings of rutile (110) and (101), respectively. And the fast Fourier transform (FFT) image of the Fig. 4b also shows regular speckles, which indicated that the TiO_2 nanorods are pure rutile crystal. These imply that the TiO_2 nanorods grow along [001] direction.

After the second hydrothermal reaction, flowerlike MoS_2 nanosheets grows outward and attaches themselves onto TiO_2 nanorods to form the TiO_2 / MoS_2 hierarchical nanocomposites. It can be found in Fig. 4d that the flowerlike MoS_2 microspheres appear after the second hydrothermal process. Moreover, it shows that the MoS_2 nanosheets wrap on the end of the TiO_2 nanorods. The same phenomenon can be found in the low-resolution TEM image (Fig. 4e). Besides, the end surfaces of TiO_2 nanorods appear much coarser, which may be a useful for MoS_2 nanosheet to form core and grow up. Fig. 4f is the high-resolution TEM image of selected region in Fig. 4e. The lattice fringes spacing of TiO_2 and MoS_2 are 2.5 Å and 6.6 Å, respectively, which are consist with the interplanar spacing of (101) planes of rutile TiO_2 and (002) planes of MoS_2.

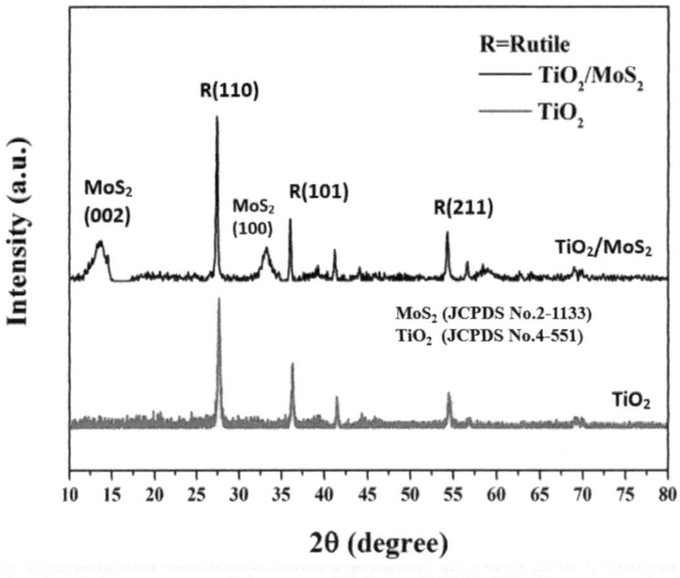

Fig. 3 XRD patterns of the pure TiO_2 and TiO_2/MoS_2 composites

IOP Conf. Series: Materials Science and Engineering **799** (2020) 012021 doi:10.1088/1757-899X/799/1/012021

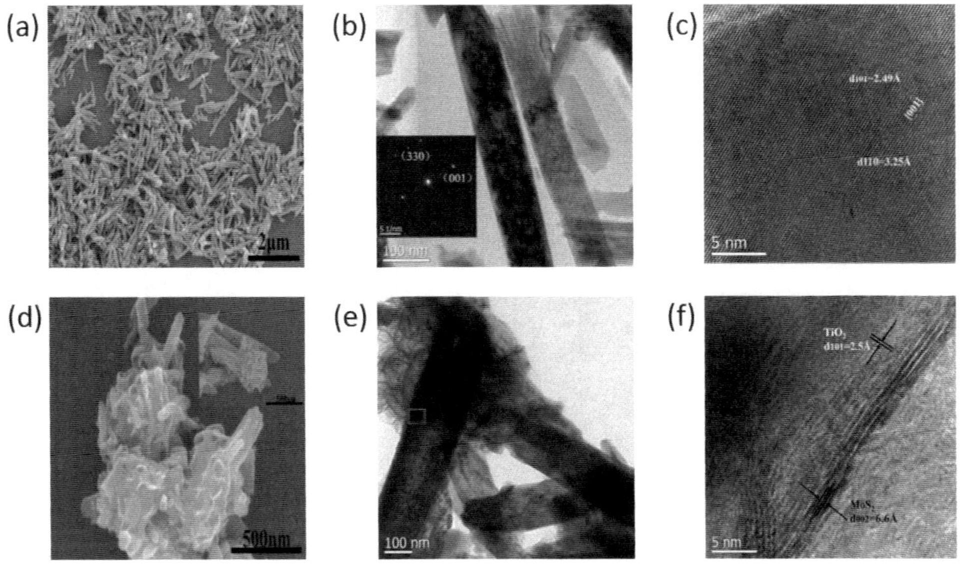

Fig. 4 SEM images and TEM images of the TiO_2 and TiO_2 / MoS_2

The optimal working temperature of gas sensors was investigated by studying their sensing response to hydrogen at 200 ppm as the temperature varying from 25 to 150 °C. The results indicate that the response of the sensor based on TiO_2 / MoS_2 nanocomposites reaches its maximum of 19.48% at 100 °C, as shown in Fig. 5a. However, the sensor based on the pure TiO_2 nanorods has the highest sensing response at 50 °C, which is only 0.56%. It can be observed that the response of the sensor based on the TiO_2 / MoS_2 nanocomposites to 200 ppm hydrogen is 40 times higher than the sensor based on the pure TiO_2 nanorods. This indicates that the gas sensing response to hydrogen can be enhanced remarkably by the TiO_2 / MoS_2 nanocomposites. Since it has a good sensing response at room temperature, the response and recovery properties of the sensors at 25°C will be investigated further in the following.

The sensor can work at room temperature, as the responses at ambient 25 °C, as shown in Fig. 5b, indicating that the sensor. Due to heterogeneous structure of the nanocomposites, it is impossible to keep porosity and uniform adsorption on the throughout surface sensing material. To simplify the question, assuming that the concentration of the surface species is uniform and then the sensing response of the sensors can be explained by the Freund lich isotherm processes [33]. As shown in Fig. 5c, higher the gas concentration, longer the response and recovery time at 25°C to different concentrations. The response and recovery time are the important factors of gas sensitivity. These phenomena can be attributed to the adsorption/desorption, and diffusion process. For higher concentration of H_2, gas molecules require longer time to diffuse into the nanostructures and adsorb/desorb on the surfaces.

The responses of the sensors based on the TiO_2/MoS_2 nanocomposites and pure TiO_2 nanosheets to various gases such as H2, NH_3 and CH_4 are shown in Fig. 5d. It can be seen that the sensors based on the TiO_2/MoS_2 nanocomposites have enhanced responses to all gases compared with the sensor based on the pure TiO_2 nanorods. And the nanocomposites sensor has the strongest response to H_2 among the three testing GaSe, which indicate that the nanocomposites based sensor has good selectivity to H_2.

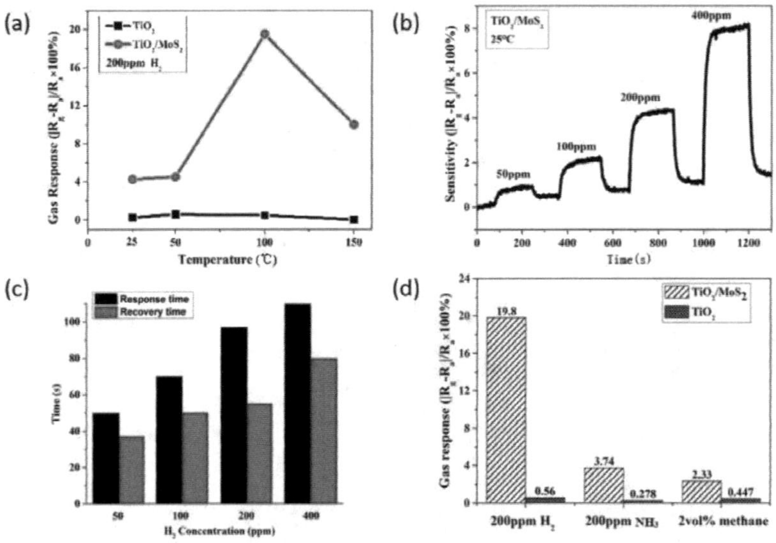

Fig. 5 Response properties of the sensors.

The sensing mechanism is resistance change of a semiconductor sensor due to the adsorption and desorption processes of oxygen molecules on the surface of semiconductor materials. When the n-type TiO_2 materials are exposed to atmosphere, oxygen molecules will be adsorbed onto the surfaces of the sensing materials and be ionized into species such as O^{2-}, O^-, and O^{2-} by transferring electrons from conduction band to valence band of the n-type semiconductor. When the materials are exposed to H_2. H2 reaction with oxygen ion on the surface produces resultants of H_2O and electron. Consequently, the resistance of the sensor decreases with the increasing concentration of H_2.

The enhanced gas sensing properties of TiO_2/MoS_2 composites are likely to be a result of several factors. First, TiO_2/MoS_2 nanocomposites have structural characteristics of high surface area and good permeability. The surface area of TiO_2/MoS_2 composites is larger than that of pure TiO_2 nanorods. This will provide more adsorptive sites so that more oxygen molecules can be absorbed and ionized. Meanwhile, the nanostructures lead to effective and rapid gas diffusion toward surface regions, resulting in the strong and quick response.

Second, p-n junction between TiO_2 nanorod and MoS_2 nanosheet is the primary factor to improve sensing response of nanocomposites. Fig. 6 shows energy bending of the nanocomposites without gas molecules adsorption. TiO_2 is a n-type semiconductor with band gap of 2.9eV, and MoS_2 is a p-type semiconductor with band gap 1.9eV., The P-N junction is formed at the interface due to different work functions of two materials, which the electron may transfer from TiO_2 to MoS_2, and hole moves on the contrary direction[34]. Dynamic equilibrium of vacancy and electron appearing and merging around the p-n junction forms a depleted layer. Since the Fermi-level energy is at the same energy level for the two kinds of metal oxide semiconductor (MOS) materials, energy bending occurs at the depleted layer, and leads to greater change of resistance. When a depletion layer is formed, the barrier will increase and the initial resistance will increase, so the resistance will decrease significantly when it reacts with reducing gases such as H_2. Eventually an enhanced response will be observed.

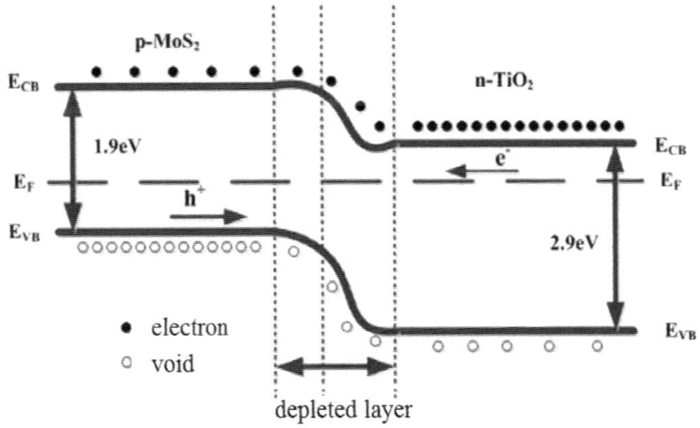

Fig. 6 Energy bending of the TiO_2/MoS_2 nanocomposites without adsorption of gases.

4. Conclusion

TiO_2/MoS_2 composites consisted of TiO_2 nanorods and flowerlike MoS_2 nanosheets were successfully synthesized by hydrothermal method. Based on the nanocomposites, the gas sensor was developed successfully, and it had better sensing properties to hydrogen compared with the sensor based on pure TiO_2 nanorods and MoS_2 nanosheets. Improvement of the gas sensor properties could be primarily attributed to depleted layer, large specific area and heterojunction of nanocomposites.

Acknowledgments

The work was supported by National Nature Science Foundation of China (Grant No. E51675420), Innovation Fund for Small and Medium Technology Based Firms of Shaanxi Province (Grant No. 2017CXZJ09) and 111 Program (Grant No.B12016). We also appreciated the support from the International Joint Laboratory for Micro/Nano Manufacturing and Measurement Technologies.

References

[1] Pritchard DK, Royle M, Willoughby D. (2009) Installation permitting guidance for hydrogen and fuel cell stationary applications. UK Version Health & Safety Laboratory, RR715:63-68.
[2] Kumar MK, Ramaprabhu S. (2007) Palladium dispersed multiwalled carbon nanotube based hydrogen sensor for fuel cell applications. Int J Hydrogen Energy, 32(13):2518–2526.
[3] Shukla S, Agrawal R, Cho HJ, et al. (2005) Effect of ultraviolet radiation exposure on room-temperature hydrogen sensitivity of nanocrystalline doped tin oxide sensor incorporated into micro electromechanical systems device. Journal of Applied Physics, 97 (5):1-13.
[4] Lupan O, Chai G, Chow L. (2008) Novel hydrogen gas sensor based on single ZnO nanorod. Microelectronic Engineering, 85(11):2220-2225.
[5] Miyazaki H, Hyodo T, Shimizu Y, Egashira M. (2005) Hydrogen-sensing properties of anodically oxidized TiO_2 film sensors: effects of preparation and pretreatment conditions. Sens Actuators B, 108:467–72.
[6] G.K. Mor, O.K. Varghese, M. Paulose, K.G. Ong, C.A. Grimes, (2006) Fabrication of hydrogen sensors with transparent titanium oxide nanotube-array thin films as sensing elements, Thin Solid Films, 496:42–48.
[7] E. S, ennik, Z. C, olak, N. Kılınc, Z.Z. (2010) Ozturk, Synthesis of highly-ordered TiO_2 nanotubes for a hydrogen sensor, Int. J. Hydrogen Energy 35:4420–4427.
[8] Varghese, (2003) Hydrogen sensing using titania nanotubes, Sens. Actuators B: Chem. 93:338–344.

[9] Wang LQ, Yang XN, Zhao XL, et al. (2011) Preparation of TiO_2 Nanoparticles in the Solvothermal Method. Key Engineering Materials: 1672-1677.
[10] Lyson-Sypien B, Czapla A, Lubecka M, et al. (2012) Nanopowders of chromium doped TiO_2 for gas sensors. Sensors and Actuators B: Chemical, 175: 163-172.
[11] Hayakawa I, Iwamoto Y, Kikuta K, et al. (2000) Gas sensing properties of metal-organics derived Pt dispersed-TiO_2 thin film fired in NH3. Sensors and Actuators B: Chemical, 67(3): 270–27.
[12] F. Qiu, W. Shin, M. Matsumiya, N. Izu, I. Matsubara, N. Murayama, (2004) Miniaturization of thermoelectric hydrogen sensor prepared on glass substrate with low-temperature crystallized SiGe film, Sens. Actuators B: Chem. 103 (1–2), 252–259.
[13] K. Scharnagl, et al., (2001) Low temperature hydrogen detection at high concentrations: comparison of platinum and iridium, Sens. Actuators B: Chem. 80(3): 163-168.
[14] Gu H, Wang Z, Hu Y. (2012) Hydrogen gas sensors based on semiconductor oxide nanostructures. Sensors; 12: 5517–50.
[15] M. Yang, H. Liu, D. Zhang, X. Tong, (2010) Hydrogen sensing performance comparison of Pd layer and Pd/WO_3 composite thin film coated on side-polished singleand multimode fibers, Sens. Actuators B: Chem. 149:161–164.
[16] M. Ando, T. Kobayashi, S. Iijimac, M. Harutaa, (1997) Optical recognition of CO and H_2 by use of gas-sensitive Au–Co_3O_4 composite films, J. Mater. Chem. 7, 1779–1783.
[17] A.Z. Sadek, W. Wlodarski, K. Shin, R.B. Kaner, K. Kalantar-zadeh, (2008) A polyaniline/WO_3 nanofiber composite-based ZnO/64◦ YX $LiNbO_3$ SAW hydrogen gas sensor, Synth. Metals 15:29–32.
[18] Schedin F, Geim AK, Morozov SV, Hill EW, Blake P,Katsnelson MI, et al. (2007)Detection of individual gas molecules adsorbed on graphene. Nat Mater; 6(9): 652–655.
[19] RobinsonJT, Perkins FK, SnowES,WeiZQ, Sheehan PE. (2008) Reduced graphene oxide molecular sensors. Nano Lett; 8: 3137–3140.
[20] Akhavan O, Ghaderi E, Rahighi R. (2012) Toward single-DNA electrochemical biosensing by graphene nanowalls. ACSNano; 6: 2904–2916.
[21] Radisavljevic, B.; Radenovic, A.; Brivio1, J.; Giacometti, V.;Kis, A. (2011)Single-Layer MoS_2 Transistors. Nat. Nanotechnol, 6, 147–150.
[22] Wang H, Yu L, Lee Y H, et al. (2012) Integrated Circuits Based on Bilayer MoS_2 Transistors[J]. Nano Letters, 12(9):4674-4680.
[23] Jariwala D, Sangwan V K, Late D J, et al. (2013) Band-like transport in high mobility unencapsulated single-layer MoS_2 transistors[J]. Applied Physics Letters, 102(17):699.
[24] Kim, S.; Konar, A.; Hwang, W.; Lee, J. H.; Lee, J.; Yang, J. et al. (2012) High-Mobility and Low-Power Thin-Film Transistors Based on Multilayer MoS_2 Crystals. Nat. Commun, 3(8):1011.
[25] Matte, H. S. S. R.; Gomathi, A.; Manna, A. K.; Late, D. J.; Datta,R.; Pati, S. K. (2010)MoS_2 and WS_2 Analogues of Graphene. Angew. Chem., Int. Ed., 49:4059–4062.
[26] Luo, Y.F. and C. Zhang, (2018) Pt-activated TiO_2-MoS_2 nanocomposites for H-2 detection at low temperature[J]; Journal of Alloys and Compounds,747: 550-557.
[27] Late, D. J.; Liu, B.; Matte, H. S. S. R.; Rao, C. N. R.; Dravid, V. P. (2012) Rapid Characterization of Ultrathin Layers of Chalocogenides on SiO_2/Si Substrates. Adv. Funct. Mater, 22, 1894–1906.
[28] Late, D. J.; Liu, B.; Luo, J.; Yan, A.; Matte, H. S. S. R.; Grayson. M. (2012) GaS and GaSe Ultrathin Layer Transistors. Adv. Mater. 24, 3549–3554.
[29] Fang, H.; Chuang, S.; Chang, T. C.; Takei, K.; Takahashi, T.;Javey, A. (2012)High-Performance Single Layered WSe_2 p-FETswith Chemically Doped Contacts. Nano Lett. 12, 3788–3792.

[30] Liu, W.; Kang, J.; Sarkar, D.; Khatami, Y.; Jena, D.; Banerjee, K. (2013) Role of Metal Contacts in Designing High-Performance Monolayer n-Type WSe_2 Field Effect Transistors. Nano Lett. 13, 1983–1990.

[31] S.R. Shakil1, N. Tarannum1, M.K. Rhaman, (2017) A Comparative Study of Increasing Sensitivity Based on MoS_2 Gas Sensor, Journal Of Nano- And Electronic Physics, 9(6)06003:1-4.

[32] Koh E W K, Chiu C H, Lim Y K, et al. (2012) Hydrogen adsorption on and diffusion through MoS_2 monolayer: First-principles study. International journal of hydrogen energy, 37(19): 14323-14328.

[33] Gautam M, Jayatissa AH. (2012) Adsorption kinetics of ammonia sensing by graphene films decorated with platinum nanoparticles. J Appl Phys; 111: 094317–29.

[34] Miller, D.R., S.A. Akbar, and P.A. Morris, (2014) Nanoscale metal oxide-based heterojunctions for gas sensing: A review; Sensors & Actuators B Chemical. 204: p. 250-272.

IOP Conf. Series: Materials Science and Engineering **799** (2020) 012022 doi:10.1088/1757-899X/799/1/012022

Distribution Characteristics and Edge Detection of Electromagnetic Field in Two Beds Formation Based on Analytical Solution

Jie Wu* and Ye Li

Departments of Electronic Engineering, University of Xi'an Shiyou, Xi'an, Shaanxi, 710000, China

wujie@xsyu.edu.cn

Abstract. After decades of development of the Chinese petroleum industry, complex oil and gas reservoirs have become the main battlefield for oil and gas field detection and development, which pose severe challenges to geosteering drilling technology that uses integrated logging and conventional logging-while-drilling tools to locate and predict formation interfaces. Based on the developed software for the distribution of forward and lateral detection electromagnetic fields, this paper calculates and analyzes the distribution characteristics of electromagnetic fields at different distances in front, and studies the lateral detection characteristics based on the half-coil of the electric field. The results show that the influence of boundary on the field is revealed according to the distribution characteristics of electromagnetic fields at different distances in front. The 1-3m receiving-transmitting short-pitch half-coil have approximately the same edge detection distance, and when the frequency is less than 20 kHz, the edge detection distance is more than 30m in two beds formation of 0.01S/m and 1.0S/m.

1. Introduction

In the field of oil drilling, logging-while-drilling (LWD) technology is the key to geosteering and real-time formation comparison evaluation in intelligent drilling, and it is an important development direction of logging technology. The geosteering should address two key technologies of drill bit front and side detection [1]. Forward detection can provide effective help for determining geological reservoirs and adjusting drilling trajectories by understanding the formation structure amplitude, reservoir distribution range, and formation properties, etc. [2,3]. Lateral detection, on the one hand, confirms the upper and lower boundaries of the reservoir and guides drilling along the best trajectory of the reservoir [4,5]; on the other hand, determines the geological reserves and oil-bearing areas of the oil and gas reservoirs, and recognizes and evaluates the oil and gas reservoirs [6]. Currently the electromagnetic remote detection technology is mature. Schlumberger respectively introduced electromagnetic wave resistivity instruments such as VDR in 2003, PeriScope HD in 2008, and DDR in 2010, which used tilt or quadrature coil technology to achieve the formation interface and orientation within a range of 1 to 6 m near the wellbore probe. VDR does not have the capability of azimuth measurement. PeriScope HD and DDR have real-time mapping function of multiple interfaces above and below the borehole, whose formation boundary detection distance can reach 30 m, and have a certain forward-looking prospecting capability. In 2014, Schlumberger launched a new generation of Geosphere reservoir mapping and other high-end logging-while-drilling technologies and services, which achieved well logging Long-range detection of formation and geological structures 30m near the eye by adopting long-distance drilling-while- drilling electromagnetic resistivity logging tools DDR, using multi-frequency multi-component array antennas and segmented modular design as well as real-time inversion algorithms[7,8]. The half-coil has the capability for a long-distance detection, and its open-loop half-coil design is used to realize the electric field measurement under the condition of drilling. Based on the software developed for forward and lateral detection of electromagnetic field distribution characteristics, this paper calculates and analyzes the electromagnetic field distribution characteristics at different distances in front studies the lateral detection characteristics based on the electric field half-coil, and thus provides theoretical basis for the design and application of electromagnetic remote detection instrument.

2. Methods

2.1. Forward detection electromagnetic field distribution theory

In order to study the distribution characteristics of the forward detection electromagnetic field, taking two beds as an example, considering the ideal case, the transmitting coil faces the boundary of the target intervals in front of the detection. The origin of the cylindrical coordinate system is taken on the interface, upwards are positive, the upper dielectric conductivity is σ_2 and lower is σ_1 , the permeabilities are μ_2 and μ_1 respectively, the dielectric constants are ε_2 and ε_1 respectively, the transmission and reception distance is L, and the distance from the transmitting coil to the interface is h. Suppose that the transmitting coil is in σ_1 medium, and the simplified schematic diagram of the two beds formation vertical coil system is shown on the left of Figure 1.

The coil source is a magnetic dipole. The magnetic dipole moment $\mathrm{M} = N_T I_T A_T$, N_T is the number of turns of the transmitting coil, I_T is the current of the transmitting coil, and A_T is the cross-sectional area of the transmitting coil. The coil source is rotationally symmetric, so only E_φ and H_z in the cylindrical coordinate system are analyzed. The expression of electric field and magnetic field in each bed of medium is given in Reference 9. Based on the expression, "the analysis software for detecting the distribution of electromagnetic fields in front of two beds of formation V1.0" is developed, which can analyze the distribution of electromagnetic fields at different positions from the front boundary.

$$E_{1\varphi} = -\frac{j\omega\mu M}{4\pi}\int_0^\infty \frac{\xi^2}{\lambda_1}\left[e^{-\lambda_1|z+h|} - F_{12}e^{-\lambda_1(h-Z)}\right]\mathrm{J}_1(\xi\rho)\,d\xi \quad (1)$$

$$E_{2\varphi} = -\frac{j\omega\mu M}{4\pi}\int_0^\infty \frac{\xi^2}{\lambda_1}(1 - F_{12})e^{-\lambda_1 h-\lambda_2 Z}\cdot \mathrm{J}_1(\xi\rho)\,d\xi \quad (2)$$

$$H_{1z} = \frac{M}{4\pi}\int_0^\infty \frac{\xi^2}{\lambda_1}\left[e^{-\lambda_1|z+h|} - F_{12}e^{-\lambda_1(h-Z)}\right]\xi\mathrm{J}_0(\xi\rho)\,d\xi \quad (3)$$

$$H_{2z} = \frac{M}{4\pi}\int_0^\infty \frac{\xi^2}{\lambda_1}(1 - F_{12})e^{-\lambda_1 h-\lambda_2 Z}\xi\mathrm{J}_0(\xi\rho)\,d\xi \quad (4)$$

Where $\lambda = \sqrt{\xi^2 - k^2}$, $F_{12} = \frac{\lambda_2-\lambda_1}{\lambda_2+\lambda_1}$, F_{12} is the reflection coefficient of the electromagnetic wave when it is transmitted from the medium σ_1 to σ_2, $(1 - F_{12})$ is the transmission coefficient when the electromagnetic wave is transmitted from the medium σ_1 to σ_2.

2.2. Theory of electromagnetic field distribution in lateral detection

Compared with the calculation of the vertical magnetic dipole response, the calculation of the horizontal magnetic dipole requires the use of a cylindrical coordinate system and a rectangular coordinate system. The horizontal magnetic dipole points to the x-axis direction at (0,0,0). The two dielectric conductivities are σ_1 and σ_2 respectively, the permeabilities are μ_1 and μ_2, the dielectric constants are ε_1 and ε_2, and the distance from the transmitting coil to the interface is h. Assuming that the transmitting coil is in the σ_1 medium, a simplified schematic diagram of a two beds formation horizontal coil system is shown on the right of Figure 1.

IOP Conf. Series: Materials Science and Engineering **799** (2020) 012022 doi:10.1088/1757-899X/799/1/012022

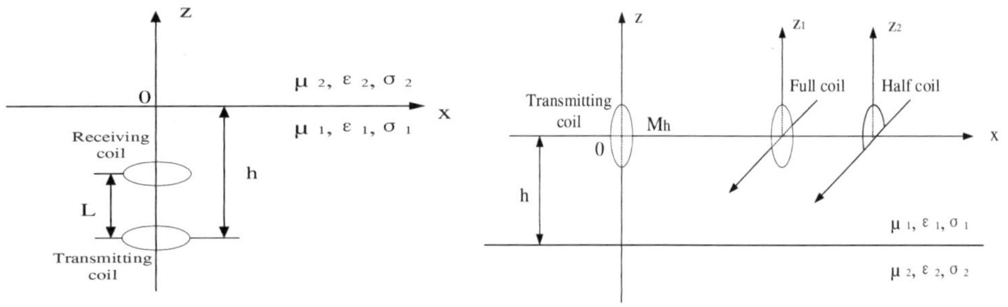

Figure 1. Diagram of two beds formation vertical coil system(left)and horizontal coil system(right).

In horizontal wells, formations do not have rotational symmetry. Therefore, the expressions of E_ρ, E_φ and H_z in the two beds medium are given in reference 10. Based on the expressions, the "three-beds formation horizontal well electromagnetic field distribution characteristics analysis software V1.0" is developed, which can analyze distribution characteristics of electromagnetic fields at different positions. The formulas for calculating the electromagnetic field distribution characteristics of the two beds formation horizontal coil system is shown in Table 1. The parameters of the formulas for calculating the electromagnetic field distribution characteristics of the two beds formation horizontal coil system are shown in Table 2.

Table 1. Calculation formulas of the electromagnetic field distribution characteristics of the horizontal coil system in two beds.

	Range	Formula
$E_{\rho 1}$	z>h	$-i\omega\mu_1 sin\emptyset M_h(4\pi)^{-1}\left[\int_0^\infty[\xi_1 P_1 e^{-\xi_1(z-h)}]\lambda J_0(\lambda\rho)d\lambda+\rho^{-1}\int_0^\infty S_1 e^{-\xi_1(z-h)}\lambda J_1(\lambda\rho)d\lambda\right]$
	|z|<h	$-i\omega\mu_1 sin\emptyset M_h(4\pi)^{-1}\left[\int_0^\infty[z(\|z\|)^{-1}e^{-\xi_1\|z-z_0\|}+\xi_1 P_2 e^{-\xi_1 z}]\lambda J_0(\lambda\rho)d\lambda+\rho^{-1}\int_0^\infty[S_2 e^{-\xi_1 z}+T_2 e^{+\xi_1 z}]\lambda J_1(\lambda\rho)d\lambda\right]$
$E_{\rho 2}$	z<-h	$-i\omega\mu_2 sin\emptyset M_h(4\pi)^{-1}\left[\int_0^\infty[-\xi_2 Q_3 e^{+\xi_2(z+h)}]\lambda J_0(\lambda\rho)d\lambda+\rho^{-1}\int_0^\infty T_3 e^{+\xi_2(z+h)}\lambda J_1(\lambda\rho)d\lambda\right]$
$E_{\varphi 1}$	z>h	$-i\omega\mu_1 cos\emptyset M_h(4\pi)^{-1}\left[\int_0^\infty[(\xi_1 P_1+\lambda S_1)e^{-\xi_1(z-h)}]\lambda J_0(\lambda\rho)d\lambda-\rho^{-1}\int_0^\infty S_1 e^{-\xi_1(z-h)}\lambda J_1(\lambda\rho)d\lambda\right]$
	|z|<h	$-i\omega\mu_1 cos\emptyset M_h(4\pi)^{-1}\left[\int_0^\infty[z(\|z\|)^{-1}e^{-\xi_1\|z-z_0\|}+(\xi_1 P_2+\lambda S_2)e^{-\xi_1 z}+\lambda T_2 e^{+\xi_1 z}]\lambda J_0(\lambda\rho)d\lambda-\rho^{-1}\int_0^\infty[S_2 e^{-\xi_1 z}+T_2 e^{+\xi_1 z}]\lambda J_1(\lambda\rho)d\lambda\right]$
$E_{\varphi 2}$	z<-h	$-i\omega\mu_2 cos\emptyset M_h(4\pi)^{-1}\left[\int_0^\infty[(\lambda T_3-\xi_2 Q_3)e^{+\xi_2(z+h)}]\lambda J_0(\lambda\rho)d\lambda-\rho^{-1}\int_0^\infty T_3 e^{+\xi_2(z+h)}\lambda J_1(\lambda\rho)d\lambda\right]$
H_{z1}	z>h	$cos\emptyset M_h(4\pi)^{-1}\int_0^\infty[(\xi_1\lambda P_1+{\xi_1}^2 S_1+{k_1}^2 S_1)e^{-\xi_1(z-h)}]\lambda J_1(\lambda\rho)d\lambda$
	|z|<h	$cos\emptyset M_h(4\pi)^{-1}\int_0^\infty\left[\frac{\lambda z}{\|z\|}e^{-\xi_1\|z-z_0\|}+(\xi_1\lambda P_2+{\xi_1}^2 S_2+{k_1}^2 S_2)e^{-\xi_1 z}+({\xi_1}^2 T_2+{k_1}^2 T_2)e^{+\xi_1 z}\right]\lambda J_1(\lambda\rho)d\lambda$
H_{z2}	z<-h	$cos\emptyset M_h(4\pi)^{-1}\int_0^\infty[-(\xi_2\lambda Q_3-{\xi_2}^2 T_3-{k_2}^2 T_3)e^{+\xi_2(z+h)}]\lambda J_1(\lambda\rho)d\lambda$

Table 2. Parameters for calculating the electromagnetic field distribution characteristics of the horizontal coil system in two beds.

Parameter	Formula	Parameter	Formula
Δ_v	$2({\xi_1\xi_2}^{-1}+{\mu_1\mu_2}^{-1})e^{2\xi_1 h}$	Δ_H	$2[{k_1}^2{k_2}^{-2}+\mu_1\xi_1(\mu_2\xi_2)^{-1}]e^{2\xi_1 h}$
$\boldsymbol{P_{21}}$	${\xi_1}^{-1}e^{-\xi_1(h-z_0)}+P_2e^{-\xi_1 h}-P_1$	$\boldsymbol{P_{32}}$	$Q_3-{\xi_1}^{-1}e^{-\xi_1(h+z_0)}-P_2e^{\xi_1 h}$
$\boldsymbol{P_1}$	$2(\Delta_H\xi_1)^{-1}\{[{k_1}^2{k_2}^{-2}+\mu_1\xi_1(\mu_2\xi_2)^{-1}]e^{\xi_1(h+z_0)}+[\mu_1\xi_1(\mu_2\xi_2)^{-1}-{k_1}^2{k_2}^{-2}]e^{-\xi_1(h+z_0)}\}$		
$\boldsymbol{P_2}$	$2[\mu_1\xi_1(\mu_2\xi_2)^{-1}-{k_1}^2{k_2}^{-2}]e^{-\xi_1 z_0}(\Delta_H\xi_1)^{-}$	$\boldsymbol{Q_3}$	$4\mu_1\xi_1{k_1}^2(\mu_2\xi_2{k_2}^2)^{-1}e^{\xi_1(h-z_0)}(\Delta_H\xi_1)^{-1}$
$\boldsymbol{S_1}$	$\lambda P_{21}(\xi_1\Delta_v)^{-1}[({\xi_1\xi_2}^{-1}+{\mu_1\mu_2}^{-1})e^{+2\xi_1 h}-({\xi_1\xi_2}^{-1}-{\mu_1\mu_2}^{-1})e^{-2\xi_1 h}]+2\lambda P_{32}(\xi_2\Delta_v)^{-1}$		
$\boldsymbol{S_2}$	$2\lambda P_{32}(\xi_2\Delta_v)^{-1}e^{+\xi_1 h}-\lambda P_{21}(\xi_1\Delta_v)^{-1}({\mu_1\mu_2}^{-1}-{\xi_1\xi_2}^{-1})e^{-\xi_1 h}$		
$\boldsymbol{T_2}$	$\lambda P_{21}(\xi_1\Delta_v)^{-1}({\xi_1\xi_2}^{-1}+{\mu_1\mu_2}^{-1})e^{+\xi_1 h}$	$\boldsymbol{T_3}$	$2\lambda\mu_1(\xi_2\mu_2\Delta_v)^{-1}(P_{32}e^{2\xi_1 h}+P_{21})$

3. Results

3.1. Distribution characteristics of forward detection electromagnetic field

Heterogeneous formation parameters are showed as follows: the conductivity of the target intervals is 0.01S/m, the conductivity of the surrounding rock is 1 S/m, the frequency is 200kHz, the emission current is 1A, the number of coil turns is 1 turn, the coil radius is 0.1m, The distance from the coil to the interface is 2m.Draw the electromagnetic field distributions of homogeneous formations with electrical conductivity of 0.01 S/m, other parameters are the same as above, and compare them with heterogeneous formations. In the figure below, the real part is on the left and the imaginary part is on the right. Two pictures on the left shows the distribution of electromagnetic fields in heterogeneous formations, two pictures on the right shows the distribution of electromagnetic fields in homogeneous formations with a conductivity of 0.01S/m.

3.1.1. E_φ distribution characteristics of forward detection.

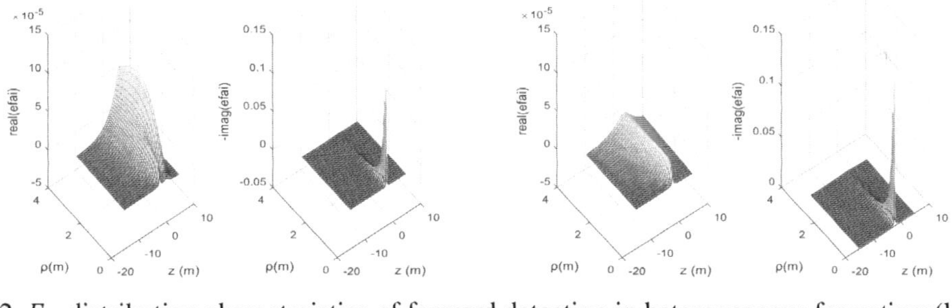

Figure 2. E_φ distribution characteristics of forward detection in heterogeneous formations(left) and homogeneous formation with conductivity of 0.01S/m(right).

By comparing, it can be seen that the real part of the heterogeneous formation E_φ shows significant numerical changes at the interface boundary (mainly due to the change in boundary conductivity) besides changing obviously at the source emission position. But There is almost no difference in the trend and amplitude of imaginary part change (the main reason is that the direct coupling signal is strong and the small changes are submerged).

3.1.2. H_z distribution characteristics of forward detection.

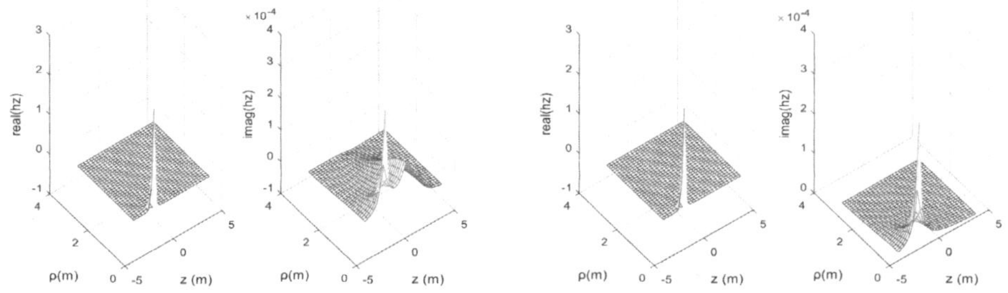

Figure 3. H_z distribution characteristics of forward detection in heterogeneous formations(left) and homogeneous formation with conductivity of 0.01S/m(right).

By comparison, we can see that the imaginary part of H_z in heterogeneous formations changes significantly at the source emission position as well as at the interface boundary, while the real part has almost no difference in trend and amplitude.

3.2. Distribution characteristics of lateral detection electromagnetic field

3.2.1. E_φ distribution characteristics of lateral detection.

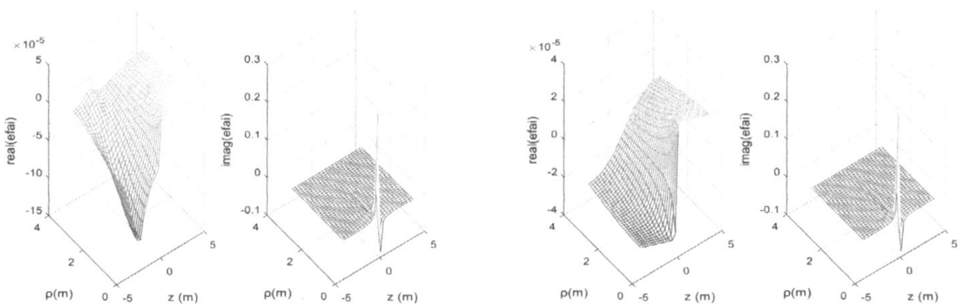

Figure 4. E_φ distribution characteristics of lateral detection in heterogeneous formations(left) and homogeneous formation with a conductivity of 0.01S/m(right).

By comparing, it can be seen that the real part of the heterogeneous formation E_φ shows a significant value change at the source emission position as well as at the interface boundary. But there is almost no difference in the trend and amplitude of imaginary part change.

3.2.2. E_ρ distribution characteristics of lateral detection.

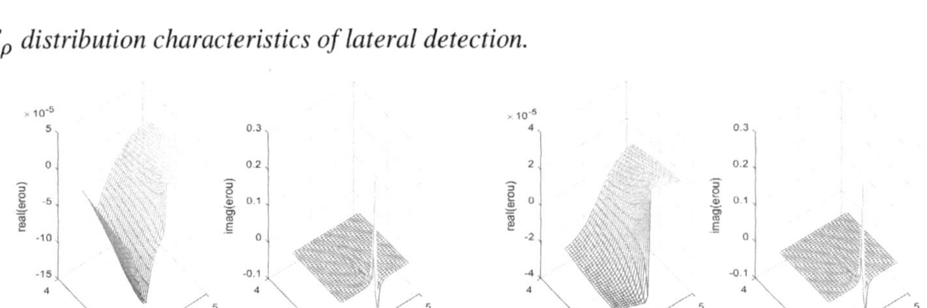

Figure 5. E_ρ distribution characteristics of lateral detection in heterogeneous formations(left) and homogeneous formation with a conductivity of 0.01S/m(right).

By comparison, we can see that the real part of the heterogeneous formation E_ρ has significant changes in value besides at the source emission position, and also shows significant numerical changes at the boundary of the interface. Compared with the lateral detection E_φ, the change trend and amplitude of the two are almost as same as each other.

3.2.3. H_z distribution characteristics of lateral detection.

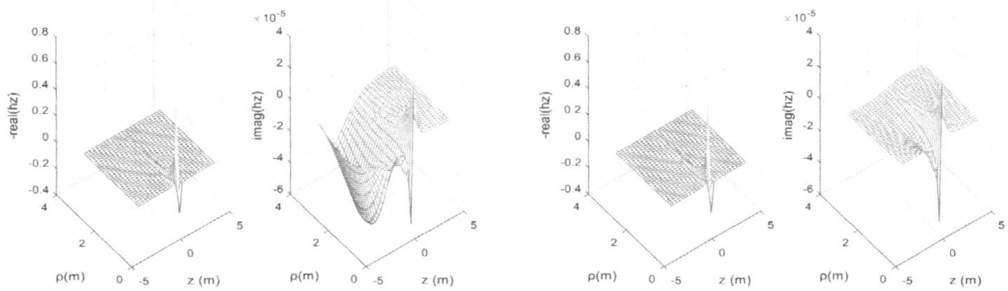

Figure 6. H_z distribution characteristics of lateral detection in heterogeneous formations(left) and homogeneous formation with a conductivity of 0.01S/m(right).

By comparing, it can be seen that the imaginary part of H_z in heterogeneous formations changes significantly at the source emission position as well as at the interface boundary, while the change trends of the real part of H_z are almost as same as each other.

4. Discussion

The research is based on the electric field half-coil edge detection characteristics. In this paper, it is to study the effect of changes in edge detection distance and transmit and receive distance on apparent conductivity. The x-axis is the edge detection distance, the y-axis is the receiving and transmitting distance, and the z-axis is the apparent conductivity. The left part is the real part of apparent conductivity, and the right part is the imaginary part of apparent conductivity. The parameters are shown in Table 3.

Table 3. Research parameters based on electric field half-coil edge detection.

Parameter	First group	Second group	Third group
Frequency (kHz)	200	200	20
Target layer conductivity (S/m)	0.01	0.1	0.01
Conductivity of surrounding rock (S/m)	1	1	1
Probe distance (m)	1,1.5,2,2.5,3,3.5,4,4.5,5,6,7,8,9,10,15,20,25,30,35,40		
Receiving and transmitting distance (m)	1,1.2,1.4,1.6,1.8,2,2.2,2.4,2.5		

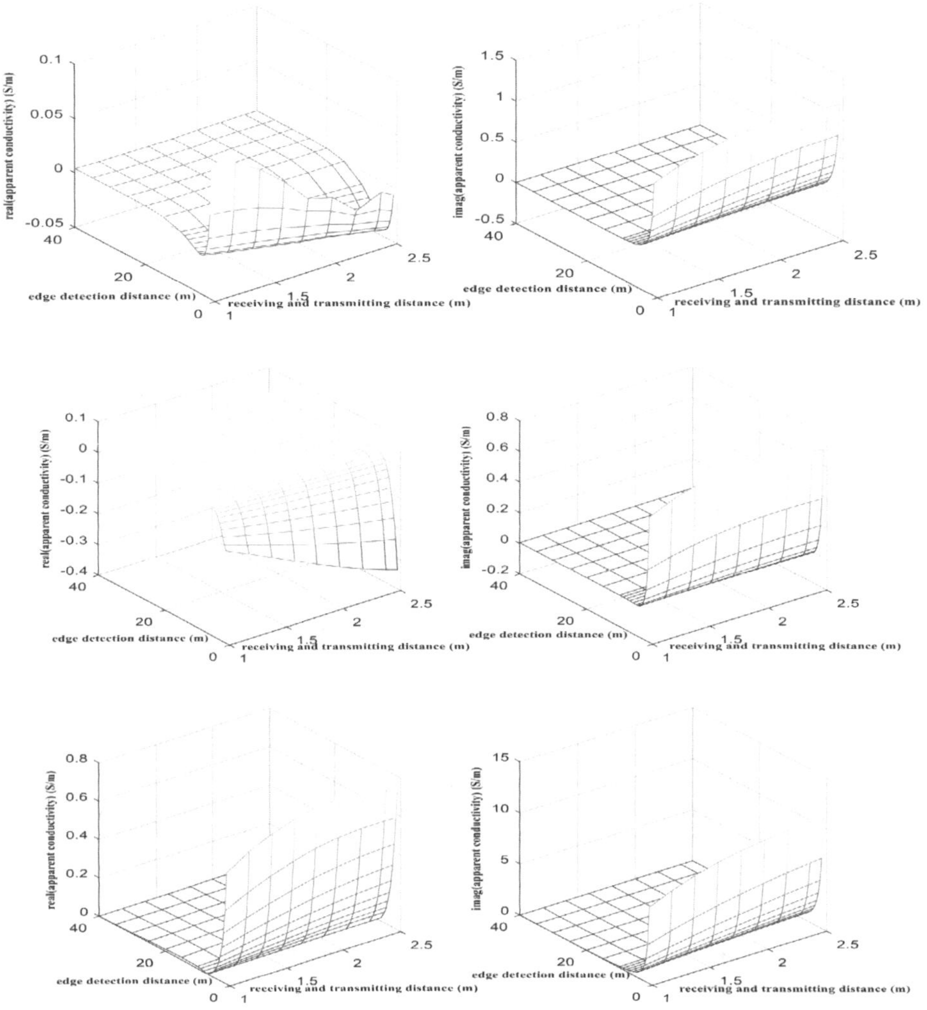

Figure 7. Three groups of data graphs.

It can be seen from the first two sets of data graphs that when the frequency is 200 kHz, the real part of the apparent conductivity shows a negative value, and the edge detection distance is very short. Therefore, it is best to work at low frequency; According to the third set of data graphs, when the frequency is 20kHz, the change of the transmitting-receiving distance has little effect on the apparent conductivity amplitude, while the apparent conductivity varies monotonically with the edge detection distance. In general, when the formation conductivity is less than 10^{-3}S/m (the resistivity is greater than 1000Ω·m), the measurement signal is very small and cannot be accurately measured. If the apparent conductivity is equal to 10^{-3}S/m, the minimum measurement value is 10^{-3}S/m, it can be seen from the calculation data that under the condition of a frequency of 20kHz, the maximum detectable distance is 35m, and the receiving and transmitting distance can be realized at 1-3 m, which shows that a short distance half-coil of 1m can realize a long-distance detection of 30m.

5. Conclusions

(1) The distribution characteristics of E_φ real part and H_z imaginary part at different distances from the two beds formation reveal the field change regularity in the boundary area of the front formation, which can be used to study the influence regularity of electromagnetic forward detection instruments,

IOP Conf. Series: Materials Science and Engineering **799** (2020) 012022 doi:10.1088/1757-899X/799/1/012022

judge the state of the front formation, and provide theoretical basis for determining geological reservoirs and adjusting drilling trajectories;

(2) The E_φ real part, E_ρ real part and H_z imaginary part of the horizontal coil system in the two beds formation reveal the change characteristics of the lateral boundary region, which can be used to study the lateral formation influence regularity of the edge detection instrument and guide drilling along the optimal trajectory of the reservoir;

(3) In the half-coil edge detection instrument, the edge detection distance is little affected by the receiving and transmitting distance, and the frequency has a great influence on the edge detection distance. For the formation with 1S/m surrounding rock and 0.1S/m target intervals, if the minimum apparent conductivity is 0.001S/m when the frequency is 20kHz, the receiving-transmitting distance is 1-3m and the edge detection distance can reach up to 35 m.

References

[1] Du S.L. (2016) Domestic LWD geosteering survey technology and application.Enterprise Technology Development, 35: 2-3.

[2] Wei B.J., Wu K.K., Lin N., Dang F., Ren C.(2019) Signal decoupling of electromagnetic measurement instrument while drilling and its application in forward detection of drill bit.Journal of China University of Petroleum (Edition of Natural Science), 43: 33-41.

[3] Wu J., Hu X.Y., Li H.D., Tang Z. (2014) Application of LWD technology in geosteering of horizontal wells. Drilling & Production Technology, 37: 26-30+7.

[4] Wang H.B., Liu R.F. (2019) Design of small diameter dual mode array induction logging tool. Well Logging Technology, 43: 268-272.

[5] Li Q.M., Omeragic D., Chou L., et al. (2005) New directional electromagnetic tool for proactive geosteering and accurate formation evaluation while drilling. In: SPWLA 46th Annual Logging Symposium 2005. New Orleans, LA, United states.

[6] Li G.Q., Liu Z.L., Zhang M.Y. (2004) Application of edge detection test in detection and development of Liuzhuang oil and gas field. Fault Block Oil and Gas Field, 11: 86-88+94.

[7] Dupuis C, Mendoza V. (2014) Avoid Pilot Holes, Land Wells, and Optimize Well Placement and Production with Deep Directional Resistivity Logging-while-drilling. Society of Petroleum Engineers,29: 473-480.

[8] Constable M.V., et al. (2016) Looking Ahead of the Bit While Drilling: From Vision to Reality. In: SPWLA 57th Annual Logging Symposium 2016. Reykjavik, Iceland.

[9] Hu Q., Wu J.(1990) Induction logging theory. Shanxi People's Education Press,xi'an.

[10] Hardman R.H., Shen L.C.(1986) Theory of induction sonde in dipping beds. Geophysics, 51: 800-809.

IOP Conf. Series: Materials Science and Engineering **799** (2020) 012023 doi:10.1088/1757-899X/799/1/012023

Borehole transient electromagnetic array based relative attitude determination for relief well drilling

Yang Ling[1], Dang Bo[1*], Liu Changzan[2], Xu Linkang[1], Ren Zhiping[1] and Dang Ruirong[1*]

[1]Key Laboratory of Education Ministry for Photoelectric Logging and Detecting of Oil and Gas, Xi'an Shiyou University, Xi'an, 710065, Shaanxi, China

[2]School of Marine Science and Technology, Northwestern Polytechnical University, Xi'an, 710072, Shaanxi, China

[*]Corresponding author's e-mail: dangrr@xsyu.edu.cn & bodang521@126.com

Abstract. Drilling relief well is one of the most effective way to control blowout well at present. In order to improve the connection rate, the relief well needs to trace a short distance and then gradually approaches the accident well, where the relative space attitude determination of relief well and blowout well is the key to improve the drilling efficiency. In this paper, a symmetry multi-coil array-based borehole transient electromagnetic (TEM) system for relative attitude determination of relief well drilling is presented. On the basis of the borehole TEM theory, the multi-coil array weighting is applied to improve the detection performance in the case that the blowout well is parallel to the relief well. Moreover, in order to determine the relative attitude, a symmetrical borehole TEM array structure is proposed for the short-range detection. Experiments were conducted and the results demonstrate the effectiveness of the proposed method for improving the short-range detection performance, which would offer significant information for the tracing phase of relief well drilling.

1. Introduction

In the process of offshore and onshore drilling, blowout out and leakage in oil and gas well will not only cause economic losses, but also make serious environmental pollution [1]. Drilling relief well is one of the most effective methods to solve blowout accidents. However, due to the influence of error ellipsoid, it is difficult to hit the target directly only by relying on the trajectory information of relief well and accident well. The relative distance and orientation should be measured continuously in the process of drilling to gradually guide the connection of two wells [2]. Therefore, the location of the blowout well from the relief well is the key to determine the connection [3].

At present, there have been three typical ways for the location of blowout wells, including the geomagnetic disturbance (GMD) method, the current injection method and the transient electromagnetic (TEM) method. Using the measurement-while-drilling's own fluxgate, the GMD techniques measure the geomagnetic field disturbed by the casing to determine the location of the accident well, where the detection range is limited and the accuracy will be greatly affected by the residual magnetic field of the casing [4,5]. Compared with the GMD detection, the current injection method [6] injects large current into the formation by employing the downhole electrode to form the converging current on the casing of the accident well, which can obtain a larger detection range. The relative distance between the relief well and the blowout well can be obtained by measuring the magnetic field generated by the converging current [7-9]. However, this method is too difficult to be implemented due to the high power of the injected current, especially for an offshore drilling. Employing the feature that the resistivity of the metal casing is 6-7 orders lower than that of the formation, the TEM system [10-13] determines the relative distance between the relief well and the accident well by detecting the response of the anomaly due to the metal casing in uniform formation [14]. The above methods have achieved some positive results for measuring the relative position of two wells. However, in the tracing phase with respect to short-range detection, the responses of all three methods contain more information of the metal casing in longitudinal direction, which means the blowout well cannot be regarded as a point target any longer, and will lead to a serious distortion when detecting the distance and orientation. As a result, the relative attitude of the blowout well to relief well cannot be determined accurately, and it is not conducive to the high-precision detection in the tracing phase.

In this paper, we present borehole TEM array based relative attitude determination for relief well drilling. Using a symmetry multi-coil array structure, the response of the casing within the detection range in the accident well is analyzed on the basis of the symmetric characteristics of TEM array and the relative attitude between two wells can be determined. The effectiveness of the proposed method was verified by applying it to experiments and the results show that the short-range detection performance can be effectively improved, which may offer significant information for the tracing phase in relief well drilling.

2. TEM Signal Model for Relief Well

Relief well drilling consists of three stages: approaching, tracing and connecting [2]. The design of relief well trajectory should give priority to satisfying the utilization of measurement tool to improve the connection rate. There have been three common trajectories of relief well: gradually approaching, directly connecting and bypass method [3,14], as shown in Figure 1.

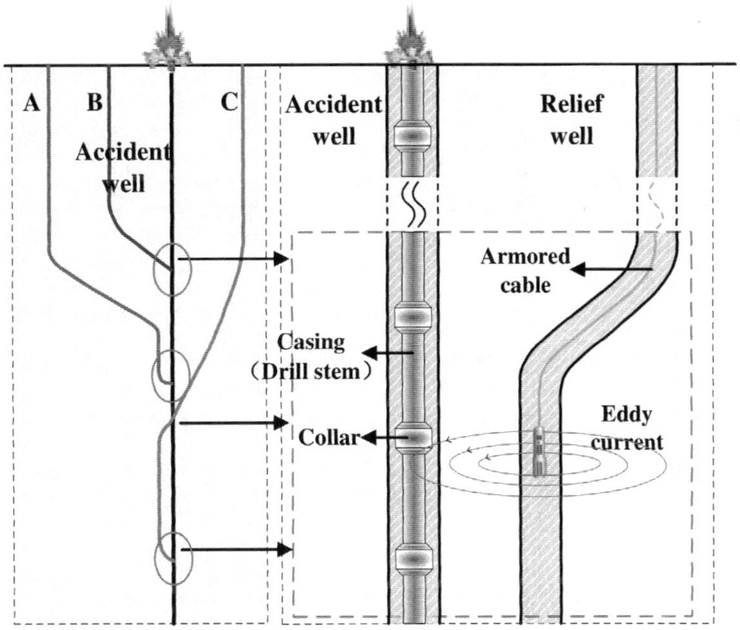

Figure 1. Transient electromagnetic (TEM) detection system of relief well

As shown in Fig. 1, the bypass method of relief well C eliminates the error ellipse by approaching the accident well two times. However, if the vertical depth of the connection point for the trajectory cannot meet the requirement of bypass method, the gradually approaching of relief well A should be given priority. At last, if the above two methods cannot be implemented due to the large dogleg, the direct connection method of relief well B can be utilized. The tracing phase of the direct connection method is suitable for the case that the connection point is very shallow, but the connection rate is lowest. Notably, no matter which kind of the above methods is adopted, the localization of the blowout well in relief well should be made several times in different depths to ensure high connection rate, especially in short-range detection. The detection scheme is shown in red dotted box of Fig. 1. However, long-time tracing will not only be inconvenient for engineering implementation, but also reduce the drilling efficiency of relief well, which is not conducive to minimize the loss of the accident well.

To solve this problem, the borehole TEM array-based short-range detection for relief well including the electrical and geometrical parameters of the jth layer (μ_j, ε_j, σ_j, and r_j with $j = 1,2,\ldots,J$) are illustrated in Figure 2. We consider the magnetic core to be the innermost layer. The transmitter and the M receivers are located in the second layer, with their number of turns given by N_T and N_R, respectively, where each receiver has the same number of turns. The inter-element spacing is Δz and the transmitter-receiver distance (TRD) is z_M, $m=0,1,\ldots,M$, where the receiver with a TRD of zero is collocated with the transmitter.

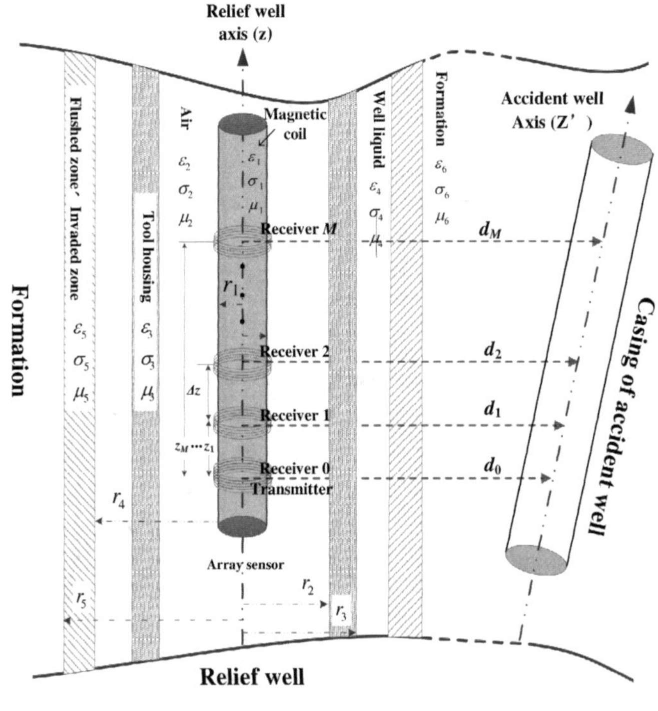

Figure 2. TEM detection model of a borehole

With the introduction of variables x_j and λ that satisfy $x_j^2 = \lambda^2 - k_j^2$ and $k_j^2=\mu_j\varepsilon_j\omega^2-i\mu_j\sigma_j\omega$, the vertical component of the magnetic field in the mth receiver of multi-coil array with a TRD of z_m can be obtained by solving the homogeneous and inhomogeneous Helmholtz equations [12]

$$H_{z1}\left(\omega, z_m, d_m\right) = \frac{N_T r_1 I_T}{\pi}\int_0^{\infty} x_1 C_1\left(\omega, d_m\right) I_0\left(x_1 r\right)\cos(\lambda z_m) d\lambda \quad (1)$$

where ω denotes the angular frequency, I_T denotes the transmitter current and $I_0(\cdot)$ is a modified Bessel function of the first kind of order zero. C_1 denotes the reflection coefficient, which is related to the geometrical and electrical parameters of all the layers. The distances d_m between the casing of the accident well and the receivers are also included in C_1. Then the induced electromagnetic force (EMF) in the frequency domain of the mth receiver can then be expressed as

$$U_m(\omega, z_m, d_m) = -i\omega\xi \int_0^{r_1} \int_0^{\infty} 2\pi r f(\lambda, r, \omega, d_m) \cos(\lambda z_m) d\lambda dr \tag{2}$$

where $\xi = \mu_l N_R N_T I_T / \pi$ and $f(\lambda, r, \omega, d_m) = x_1 C_1(\omega, d_m) I_0(x_1 r)$. Assuming that the detected area is longitudinally uniform, the distance, shape and other parameters of the anomaly can be obtained by analysing the decay speed of the TEM response. It should be noted that the borehole TEM system is suitable for the multi-cylindrically layered structures and the accident well is actually located on one side of the relief well. Therefore, the d_m in the TEM signal model in relief well is different from the real distance, where the parameters need to be corrected with gradation. Besides, since the cylindrical symmetry model cannot reflect the azimuth information of the accident well relative to the relief well, it is necessary to obtain it through multiple detection in different vertical depths based on the known well trajectory [14]. In this paper, the relative attitude of relief well and blowout well is determined according to the TEM response in the above model.

3. TEM array-based signal processing method for relief well

The traditional relief well detection system uses a single receiving coil for blowout well location. However, when the relative distance between the relief well and the accident well casing is too close, the response of the receiver will contain more longitudinal information of the casing. No matter the relative attitude of the two wells is parallel or oblique, the accident well casing is regarded as a point target, thereby introducing serious measurement error. In this paper, in order to improve the detection performance of the relief well, the multi-coil array-based TEM system with one transmitter and multiple receivers is adopted to detect the metal casing of the accident well several times in different depths.

Given a ramp signal with a turn-off time of t_{of}, the two integration in Equation (2) can both be expanded as multi-stage Legendre polynomials with respect to the number of stages P and Q respectively. Using the Gaver–Stehfest inverse Laplace transform with S stages, the induced EMF $U_{m,l}(t_l, z_m, d_m)$ can be obtained by converting Equation (2) into the time domain, and can be expressed as

$$U_{m,l}(t_l, z_m, d_m) = -\frac{\xi\chi}{t_{\text{of}}} \sum_{s=1}^{S} \sum_{q=1}^{Q} \sum_{p=1}^{P} g_{s,q,p}(t_l, d_m) \cdot \cos(\lambda_p z_m) \tag{3}$$

where

$$g_{s,q,p}(t_l, d_m) = D_s \frac{e^{(-s\ln 2t_{\text{of}})/t_l} - 1}{s} \cdot A_q \frac{B_q + 1}{2} A_p f(\lambda_0 \frac{B_p + 1}{2}, r_1 \frac{B_q + 1}{2}, s\ln 2 / it_l, d_m) \tag{4}$$

$\chi = \pi r_1^2 \lambda_0 / 2$, A and B denote the quadrature coefficient and zero-point, respectively. λ_0 is the upper limit of the infinite integral. In matrix form, Equation (3) can be expressed as

$$U_{m,l}(t_l, z_m, d_m) = -\frac{\xi\chi}{t_{\text{of}}} \cdot \mathbf{x}(z_m) \cdot \mathbf{g}^{\mathrm{T}}(t_l, d_m) \tag{5}$$

where

$$\mathbf{x}(z_m) = \Big[\underbrace{\overbrace{\cos(\lambda_1 z_m), \ldots, \cos(\lambda_P z_m)}^{P} \ \cdots \ \overbrace{\cos(\lambda_1 z_m), \ldots, \cos(\lambda_P z_m)}^{P}}_{SQP} \Big] \tag{6}$$

$$\mathbf{g}(t_l, d_m) = \left[g_{1,1,1}(t_l, d_m), \ldots, g_{S,Q,P}(t_l, d_m) \right]_{1\times SQP} \tag{7}$$

It can be seen from Equation (5) that the response of TEM system is mainly related to the TRD, the sampling time t and the relative distance d between the relief well and the accident well, where the term including TRD z is a cosine function and can be separated. On the one hand, when the accident well is parallel to the relief well, the geometrical–electrical parameters $\mathbf{g}(t_l, d_m)$ of the multi-cylindrical structure along the borehole axis is the same, and the signals can be weighted according to the array weighting method in Reference [12], so as to improve the signal-to-noise (SNR) ratio and short-range detection resolution.

IOP Conf. Series: Materials Science and Engineering **799** (2020) 012023 doi:10.1088/1757-899X/799/1/012023

Assuming that the induced EMFs of each coil are discretely sampled with a sampling length L and a sampling interval Δt, then, the induced EMF of the multi-coil array with sampling time t_l and TRD z_m can be written as

$$\mathbf{U}_{0-M,1-L} = -\frac{\xi\chi}{t_{\mathrm{of}}}\cdot\begin{bmatrix} \mathbf{x}(z_0)\cdot\mathbf{g}^{\mathrm{T}}(t_1,d_0) & \mathbf{x}(z_0)\cdot\mathbf{g}^{\mathrm{T}}(t_2,d_0) & \cdots & \mathbf{x}(z_0)\cdot\mathbf{g}^{\mathrm{T}}(t_L,d_0) \\ \mathbf{x}(z_1)\cdot\mathbf{g}^{\mathrm{T}}(t_1,d_1) & \mathbf{x}(z_1)\cdot\mathbf{g}^{\mathrm{T}}(t_2,d_1) & \cdots & \mathbf{x}(z_1)\cdot\mathbf{g}^{\mathrm{T}}(t_L,d_1) \\ \vdots & \vdots & \ddots & \vdots \\ \mathbf{x}(z_M)\cdot\mathbf{g}^{\mathrm{T}}(t_1,d_M) & \mathbf{x}(z_M)\cdot\mathbf{g}^{\mathrm{T}}(t_2,d_M) & \cdots & \mathbf{x}(z_M)\cdot\mathbf{g}^{\mathrm{T}}(t_L,d_M) \end{bmatrix}_{(M+1)\times L} \tag{8}$$

In (8), if the accident well is parallel to the relief well with $d_1 = d_2 = \ldots = d_M = d_0$, a receiving weight vector $\mathbf{W}$ can be applied to optimize the array data of the lth sampling time [12].

$$y_l = -\frac{\xi\chi}{t_{\mathrm{of}}}\mathbf{W}_{0\text{-}M}^{\mathrm{T}}\begin{bmatrix} \mathbf{x}(z_0) \\ \mathbf{x}(z_1) \\ \vdots \\ \mathbf{x}(z_M) \end{bmatrix}\mathbf{g}^{\mathrm{T}}(t_l,d_0) \tag{9}$$

Using this property, the array weighting can be extended by applying the weight to the received signals of the multi-coil array for all L sampling times such that

$$\mathbf{Y}_{0\text{-}M,1\text{-}L} = \begin{bmatrix} y_1 & y_2 & \cdots & y_L \end{bmatrix} \tag{10}$$

On the other hand, considering the symmetry of cosine function, the responses of two symmetric receivers with $\pm z_m$ should be identical if the distance between the accident well and the relief well is the same. On the contrary, if the signals of the symmetric receivers are not identical, it reveals that the accident well is not parallel to the relief well, and it is necessary to make a further determination on the relative attitude of two wells. Thereby, we can use the response of the symmetric TEM array to determine the relative attitude of two wells by using the symmetric characteristic of cosine function.

4. Symmetric TEM array-based relative attitude determination for relief well

When the blowout well is parallel to the relief well, the relative distance between two wells is the same along the borehole axis; otherwise, it is necessary to determine the relative attitude of two wells according to the responses of the receivers in different observation places. In order to solve this problem, a symmetric TEM array detection method is proposed in this paper. The array structure and the relative attitude of relief well and accident well are shown in Figure 3.

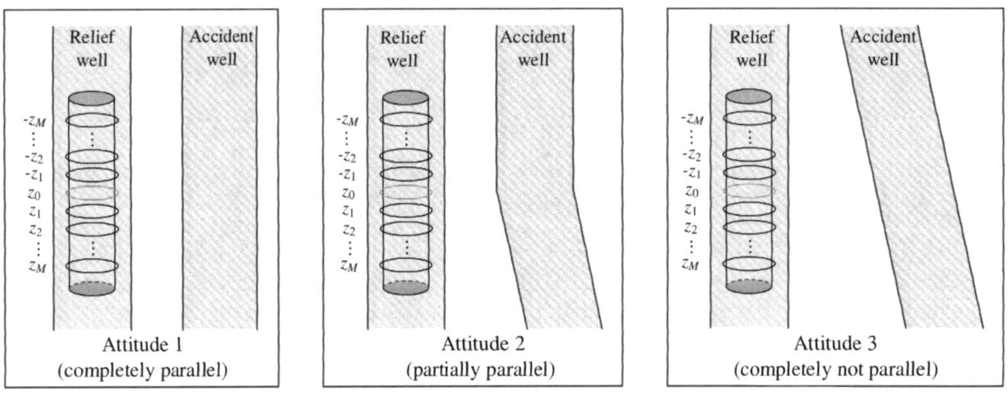

Figure 3. Three attitudes of the relief well and the accident well

The array structure consists of one transmitter and $2M+1$ receivers, where the receiver z_0 is collocated with the transmitter and the rest $2M$ receivers are symmetric with respect to the transmitter (the centre receiver, z_0). Since the TRD z is included in the cosine function, if the distance between the blowout well and the relief well along the borehole axis is the same, the response of the symmetric element of array should be identical. Then, the relative attitude of two wells can be determined by analysing the response of each element of symmetric array.

Firstly, assuming the threshold δ and taking m from $-M$ to M, if there exists

$$\left\|\mathbf{Y}_{(-m)-m,1-L}-\mathbf{U}_{0,1-L}\right\|_{\mathrm{F}}^{2}\leq\delta \tag{11}$$

the weighted array output from $-z_m$ to z_m is nearly the same as the response of a receiver with a TRD of zero, which means the accident well is parallel to the relief well. Conversely, if there is no symmetry of the response of any pair of symmetric receivers, the accident well should be not parallel to the relief well within the detection area. Then, it is necessary to determine whether the accident well is partially parallel to the relief well or not. By dividing the multi-coil array into two parts (termed as the upper part and the lower part) without the centre receiver, and taking m from 1 to M and from -1 to $-M$, respectively, if there exists parallel part, we have

$$\left\|\mathbf{Y}_{1-m,1-L}-\mathbf{U}_{0,1-L}\right\|_{\mathrm{F}}^{2}\leq\delta \text{ or } \left\|\mathbf{Y}_{-1-(-m),1-L}-\mathbf{U}_{0,1-L}\right\|_{\mathrm{F}}^{2}\leq\delta \tag{12}$$

The parallelism of the two wells correspond to the two parts of the multi-coil array can be analysed from Equation (12), as shown in attitude 2 in Fig. 3(b). On the contrary, if both the upper part and the lower part of the multi-coil array do not satisfy Equation (12), the accident well is completely not parallel to the relief well within the detected area. Assuming that there is no anomaly except the metal casing of the accident well in the formation, the shorter the distance of two wells is, the better the performance of the TEM response can be achieved when the conductivity of the metal casing is a constant. Therefore, the trajectory of the casing in the accident well relative to the relief well can be determined by analysing the response of the M symmetrical receivers one by one. In addition, considering the limitation of the dogleg during drilling, it is impossible to have two inflexion points in accident well in a short distance. In this paper, the responses of the multi-coil array are differential accumulated as E to evaluate improve the SNR and the accuracy of the TEM system for relative attitude determination such that

$$E=\sum_{l=1}^{L}\sum_{m=1}^{M}\left(U\left(t_l,z_m,d_m\right)-U\left(t_l,-z_m,d_m\right)\right) \tag{13}$$

The relative attitude determination process is shown in Figure 4.

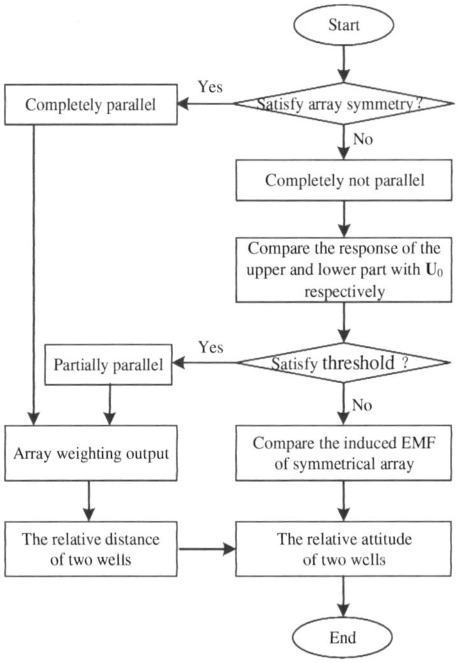

Figure 4. Relative attitude determination process of two wells

5. Experiment and Analysis

In this paper, the standardized $5^{1/2}$-inch metal casings are utilized as the accident well to verify the validity of the TEM array-based relative attitude determination method for relief well drilling. The distance in normal direction of transmitter is used to represent the distance between the accident well and the relief well, and the relative distance of two wells is 3 meters. The relative attitudes of the two wells are shown in Fig. 3, in which the inclination angle of the casing and the relief well in non-parallel area is 15 degrees. The parameters of the multi-coil array sensor and the experiment are shown in Table 1.

Table 1. Parameters of the multi-coil array sensor and the field experiment

Parameter	Value
Radius of the multi-coil array sensor	65 mm
Length of the sensor	1300 mm
Number of transmitting coils	1
Number of receiving coils	9
Number of transmitting coil turns	80
Number of receiving coil turns	300
Wire diameter of transmitting coil	0.46 mm
Wire diameter of receiving coil	0.18 mm

For the convenience of description, the TEM array consists of 9 receivers from $-z_4$ to z_4. The induced EMFs at 30 ms and 50 ms corresponding to three attitudes of Fig. 3 are shown in Figure 5 and 6.

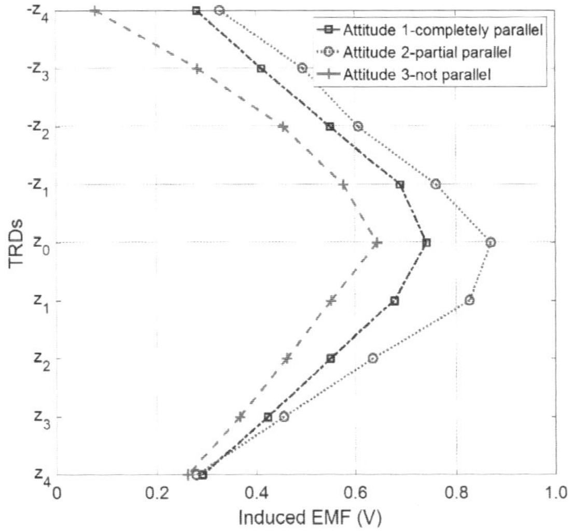

Figure 5. Induced EMF in the symmetric array at 30 ms

It can be seen from Fig. 5 and 6 that for both early and late sampling times of 30 ms and 50 ms, respectively, the signal of multi-coil array for attitude 1 shows typical symmetrical characteristic, which indicates that the accident well is completely parallel to the relief well. The signals of the symmetrical receivers are not identical for attitude 2 and 3 in Fig.4. By using the relative attitude determination method in Section 4 and analysing the characteristic of the signal in upper and lower part of multi-coil array, we can conclude that the upper part in accident well is parallel to the relief well whereas the lower part is asymmetric for attitude 2. The accident well is not parallel to the relief well for attitude 3, but the distance of the upper part to relief well is smaller than that of the lower part.

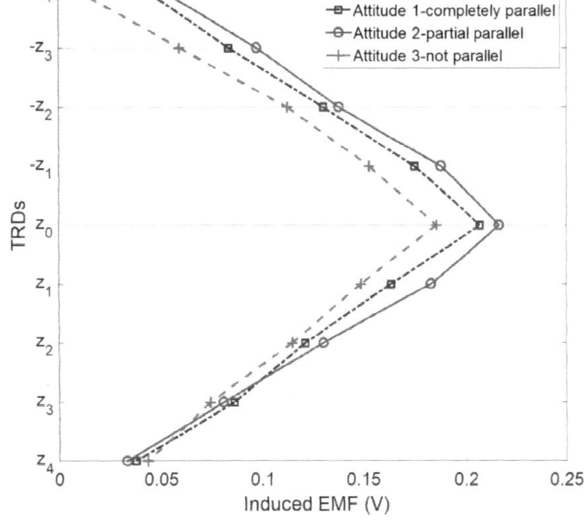

Figure 6. Induced EMF in the symmetric array at 50 ms

For different attitudes, on the one hand, by analysing the symmetric or asymmetric relationship of the signal of the upper and lower receivers, the determination whether the accident well and the relief well are completely parallel or not can be made; on the other hand, if only the upper part or the lower part of the multi-coil array is considered, the induced EMF will change with different observation places because the distance between each receiver and the accident well casing is not the same.

Therefore, we can determine the partially parallel area through any half signal of the multi-coil symmetrical array.

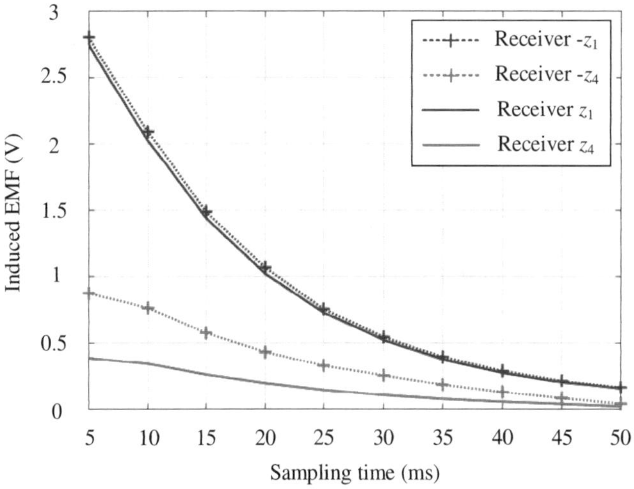

Figure 7. Comparison of the symmetrical array signal for attitude 3

In addition, the incline direction can be obtained by analysing the signal of multi-coil array for attitude 3 in Fig.4. As shown in Fig. 7, the signal of the upper part in the symmetrical array is significantly larger than that of the lower part, which indicates that the casing of the upper part of the accident well is closer to the relief well. Notably, Since the response of each receiver contains more information in longitudinal direction of the accident well, the smaller the element spacing of symmetrical receivers is, the lower resolution of response can be obtained by considering the influence of noise and detection error, and the difference of the response of $\pm z_1$ is significantly less than $\pm z_4$ as shown in Fig. 7. Notably, the array signal processing method proposed in this paper can only calibrate the general attitude, and the accurate calculation method of the relative angle of the accident well and the relief well needs to be investigated further in future studies.

6. Conclusion

In this paper, a symmetry TEM array based relative attitude determination for relief well drilling is proposed to improve the detection performance in the short-range tracing phase. Based on the TEM signal model, the array weighting is applied and the symmetrical TEM array structure is employed, where the relative attitude of two wells can be determined by analysing the symmetry of the response of the symmetrical array. The experiment results of three different attitudes demonstrate that the proposed method can effectively determine the relative attitude between the accident well and the relief well, which is of great significance to improve the short-range detection performance and drilling efficiency of the relief well.

Acknowledgements

This work was supported by the National Science and Technology Major Project (Grand No. 2016ZX05028-001), the National Natural Science Foundation of China (Grant No. 51974250, 41874158), the Natural Science Basic Research Plan in Shaanxi Province of China (Grant No. 2018JQ5133), the Natural Science Foundation of Shaanxi Provincial Department of Education (Grant No. 17JS106), and the Postgraduate Innovation and practical ability training program of Xi'an Shiyou University (Grant No. YCS19211002).

IOP Conf. Series: Materials Science and Engineering **799** (2020) 012023 doi:10.1088/1757-899X/799/1/012023

References

[1] Gilian Schout, Niels Hartog, S. Majid Hassanizadeh, et al. (2017) Impact of an historic underground gas well blowout on the current methane chemistry in a shallow groundwater system. Proceedings of the National Academy of Sciences, 115(2):1-6.
[2] Goobie R B, Allen W T, Lasley B M, et al. (2015) A Guide to Relief Well Trajectory Design using Multidisciplinary Collaborative Well Planning Technology. In: Drilling Conference and Exhibition. London. pp. 1-5.
[3] Li F F, Jiang S Q, Zhou J L, et al. (2017) Discussion on the Design of Well Trajectories in Deepwater Relief Wells. Petroleum Drilling Techniques, 45(1): 21-26.
[4] Dittrick P. BP spill. (2010) Drilling ahead of schedule for first relief well. Oil & Gas Journal, 108(23): 20-22.
[5] Li F F, Ye J H, Yang W X. (2015) Application of electromagnetic detection and location system in the design of relief well, Oil Drilling & Production technology, 37(1): 154-159.
[6] Li C, Gao D L. (2013) Preliminary Research on Detection Method for Connection Relief, Petroleum Drilling Techniques, 41(3): 56-61.
[7] Wu Y, Mao J L, Li F F, et al. (2014) Research on precise detection technology of relief wells for deepwater petroleum exploration, Oil Drilling & Production technology, 36(4): 26- 29.
[8] Li C, Gao D D, Diao B B, et al. (2013) A Detection System based on Three-electrode Array for Connecting a Relief Well to Failure Well, Acta Petrolei Sinica, 34(6): 1181-1188.
[9] Jia H Q, Liu R, Liu M, et al. (2016) Analysis and simulation of relief well detection model based on point source, Oil Drilling & Production technology, 38(4): 422-426.
[10] Wang H Y, Xue G Q, Guo H, et al. (2018) Numerical analysis for direct response of the transient electromagnetic field in the time domain in a homogeneous half space. Chinese J. Geophys, 61 (2): 750-755.
[11] Xue G Q, Li X, Di Q Y. (2007) The progress of TEM in theory and application, Progress in Geophysics, 22(4): 1195-1200.
[12] Dang B, Yang L, Liu C Z, et al. (2018) A Uniform Linear Multi-Coil Array-Based Borehole Transient Electromagnetic System for Non-Destructive Evaluations of Downhole Casings[J]. Sensors, 18(8): 1-9.
[13] Yang L, Dang B, Liu R, et al. (2018) Modeling and Simulation of Transient Electromagnetic Detection Sensor in Relief Well[J], Transducer and Microsystem Technologies, 37(11): 37-39+43.
[14] Liu R. (2016) Study on the theory and method of relief well electromagnetic detection, Xi'an Shiyou University.

IOP Conf. Series: Materials Science and Engineering **799** (2020) 012024 doi:10.1088/1757-899X/799/1/012024

Radiation Characteristics Test of FOG Circuit Magnetic Field

Jun Lu, Zhirui Guo* and Lei Liu

School of Electrical and Electronic Engineering, Shijiazhuang Tiedao University, Shijiazhuang, Hebei Province,050043, China

Corresponding author's e-mail: lj19752019@163.com

Abstract: Aiming at the magnetic field issue in the engineering application of fiber optic gyroscope (FOG), this paper explores the distribution law of magnetic field generated by the internal circuit of gyroscope and builds an electromagnetic field distribution scanning system to complete the test. The experimental results show that the overall radiation of FOG circuit magnetic field is low, and the front circuit radiation intensity is higher than that of bottom circuit. The front radiation is strongest at 8.76MHz, approximately 0.01mGs. It is of referential significance for the study of FOG magnetic sensitivity.

1. Introduction

Since the fiber optic gyroscope (FOG) came out in 1976, it has become a research hotspot in the field of domestic inertial instrument in recent years. FOG features all solid state structure without rotatable parts, large dynamic range, fast response speed, low power consumption, resistance to shock vibration, short start-up process, long service life and other advantages. It can be widely applied in military and civilian fields [1]. The operating principle of FOG is based on the Sagnac effect of optical path. The sensing component is a fiber optic ring. The fiber optic ring is not only sensitive to the non-reciprocal phase shift generated by Sagnac effect, but also sensitive to a variety of physical quantities. Therefore, the external environment is easy to produce a variety of nonreciprocal errors in FOG, and the error is superimposed on the Sagnac effect of FOG, which will lead to the change of the zero deviation of FOG output [2-4].

In the practical application of FOG, magnetic field is an inevitable interference source. So, it is of great significance to study the FOG magnetic field to avoid interference and ensure the precision of FOG. The magnetic fields faced by FOG include external magnetic field and internal magnetic field. External magnetic field includes earth magnetic field, power frequency interference and other environmental magnetic fields. Internal magnetic field is mainly generated by circuit boards, wires and photoelectric components [5-6]. The geomagnetic field belongs to the static and uniform magnetic field. This paper mainly analyzes the distribution law of the magnetic field generated by the internal circuit of gyroscope.

2. Test of the Radiation Magnetic Field of FOG Internal Circuit

2.1. Test Equipment

The test system consists of electromagnetic field scanner, spectrum analyzer, computer control and display software. The circuit board of FOG is scanned for electromagnetic field radiation characteristics, as shown in Figure 1. Electromagnetic field scanner adopts Sweden DETECTUS' RX644EH electromagnetic field distribution scanner for thermal distribution, and spectrum analyzer

IOP Conf. Series: Materials Science and Engineering **799** (2020) 012024 doi:10.1088/1757-899X/799/1/012024

adopts Agilent's E4440A spectrum analyzer to scan the circuit board of FOG for electromagnetic radiation characteristics. The scanning system can accurately measure the electromagnetic and thermal radiation of circuit board, cable and whole device to determine the intensity and location of component and radiation source. The measurement results can be presented in the form of two-dimensional or three-dimensional color graphics.

Figure 1. Electromagnetic Field Distribution Scanning System

2.2. Circuit Scanning

The main circuit of FOG is a double-sided circuit board, as shown in Figure 2 and 3. Electronic components and circuit board wiring are different on both sides. Hence, the radiation characteristics on both sides will be different. Therefore, the magnetic field radiation characteristics of circuit board are conducted scanning and test on both sides separately.

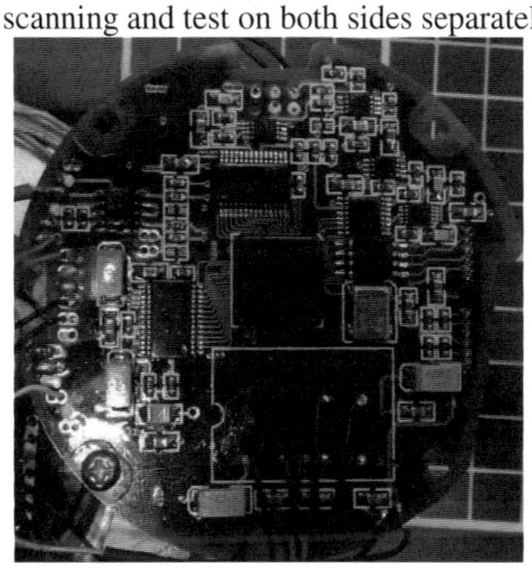

Figure 2. Front Circuit

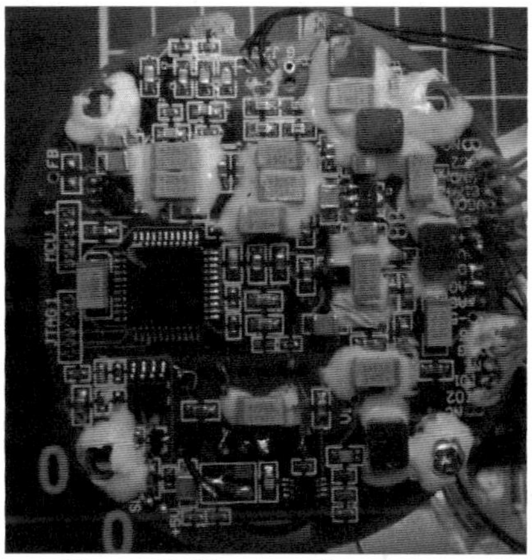

Figure 3. Bottom Circuit

According to the size of circuit, the scanning range of scanner is 0~70mm on the horizontal axis and 0~70mm on the vertical axis. The probe height is 10mm higher than that of the circuit board. The probe movement step size is 5mm. The antenna ring of magnetic field probe is placed horizontally, and the sweep frequency range of spectrometer is set as 0~50MHz through software, the resolution bandwidth is 9kHz, video bandwidth is 30kHz and the scanning frequency is 5 times. The scan is conducted twice: pre-scan and precision scan. First, the whole circuit is pre-scanned to obtain the frequency points with high radiation intensity of circuit board. According to the results of pre-scanning, the frequency points with high radiation intensity are targeted and then the positioning and accurate scanning are conducted to obtain the distribution map of magnetic field radiation intensity of each part of circuit board with different frequencies.

According to the pre-scan results, the frequency points with large radiation are 8.761mhz,

35.047mhz, 17.524mhz, 43.808mhz and 26.285mhz. Of which, the fundamental frequencies with the strongest radiation are 8.761mhz, 35.047mhz, 17.524mhz, 43.808mhz and 26.285mhz, and they are the higher harmonics of the maximum radiation frequency point 8.7612mhz.

The distribution of magnetic field radiation intensity of partial accurate scanning is shown in Figure 4. By comparing the radiation intensity distribution diagram with Figure 2, it is found that the magnetic field radiation of front circuit is higher mainly in the vicinity of digital-to-analog converter DAC, FPGA chip, active crystal oscillator and PIN-FET detector (photoelectric converter).

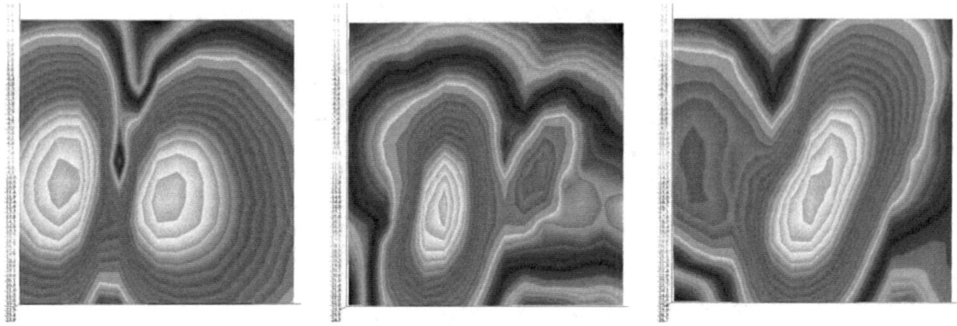

(a) 8.761MHz (b) 35.047MHZ (c)17.524MHz

Figure 4. The Magnetic Field Radiation Intensity Distribution of Front Circuit at Different Frequencies

According to the test method of abovementioned front circuit, the radiation characteristics of FOG bottom circuit are tested. During the test, the setup of software and spectrometer and the placement of probe are exactly the same as that of front circuit. Firstly, the whole circuit is prescanned. According to the results of pre-scan, the frequency points with greater radiation are 8.761mhz, 17.524mhz, 26.285mhz and 35.047mhz. Among them, 17.524mhz, 26.285mhz and 35.047mhz are the higher harmonics of the maximum radiation frequency point 8.7612mhz. This indicates that the fundamental frequency with the strongest radiation is 8.761mhz, which is basically consistent with the pre-scan results of the front circuit. According to the results of the pre-scan, the accurate positioning scan is conducted. The radiation intensity distribution of partial magnetic field is shown in Figure 5. By comparing the radiation intensity distribution diagram with Figure 3, it is found that the area with strong magnetic field radiation in the bottom circuit is mainly near the PIN-FET detector (photoelectric converter)'s pin and individual capacitance.

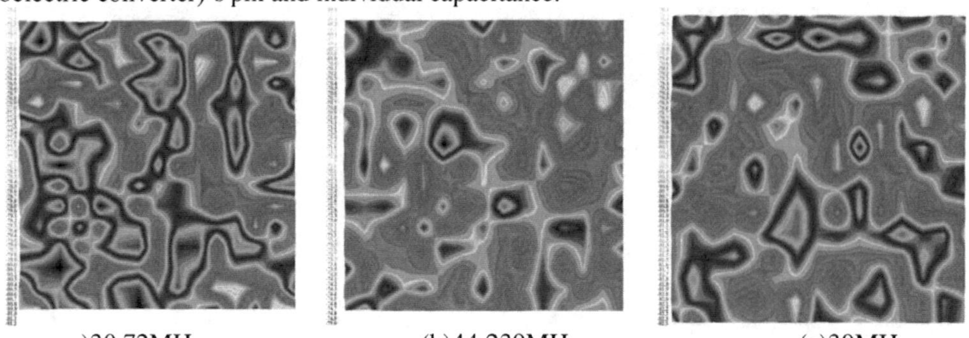

a)30.72MHz (b)44.239MHz (c)30MHz

Figure 5. Distribution of Magnetic Field Radiation Intensity of Bottom Circuit at Different Frequencies

3. Experimental Results

As the near-field probe of field distribution scanning system is not calibrated, only relative value of circuit board radiation intensity can be given. So, it is necessary to calibrate the field probe of the scanning equipment with other field intensity meters and standard radiation sources. Based on the calibrated probe antenna coefficient and the values read by the scanner at time of circuit board measurement, the magnetic field intensity and magnetic induction intensity (air medium) radiated by the gyroscope circuit can be calculated, as shown in Table 1 and Table 2.

Table 1. Magnetic Field Radiation Intensity of Front Circuit of Gyroscope

Frequency (MHz)	Spectrometer Readings (dBμV)	Magnetic Field Intensity (A/m)	Magnetic Induction Intensity (mGs)
2.098	0.7	0.000146	0.001838
4.195	-4.1	0.000427	0.00536
8.76	8.8	0.000796	0.010004
17.52	2.6	0.000279	0.003503
26.28	0.9	0.000177	0.002225
35.05	3.2	0.000363	0.004563
43.8	0.9	0.000165	0.002078

Table 2. Magnetic Field Radiation Intensity of Bottom Circuit of Gyroscope

Frequency (MHz)	Spectrometer Readings (dBμV)	Magnetic Field Intensity (A/m)	Magnetic Induction Intensity (mGs)
2.098	0.8	0.000148	0.001859
4.195	-5.6	0.000359	0.00451
8.76	0.5	0.000306	0.003847
17.52	-3.6	0.000137	0.001716
26.28	-10.7	0.000047	0.000585
35.05	-7.7	0.000104	0.001301

From the tables, it can be found that the overall radiation of gyroscope circuit is low, while the radiation intensity of front circuit is greater than that of the bottom circuit, which indicates that the overall electric field radiation of bottom circuit is weak. The front radiation is strongest at 8.76MHz, approximately 0.01mgs. According to the above test results and for convenience of comparison, they can be summarized as the following table:

Table 3. Scan and Calibration Results

Magnetic Field	Front Circuit	Bottom Circuit
Frequency Points with Larger Radiation	Strong → Weak: 8.761MHz, 4.195MHz, 35.047MHZ, 26.28MHz, 17.524MHz, 43.80MHz, 2.098MHz	Strong → Weak: 4.195MHz, 8.761MHz, 2.098MHz, 17.52MHz, 35.05MHZ, 26.28MHz
Generated position	DAC、FPGA chip, active crystal oscillator and PIN-FET detector	PIN-FET detector pins, individual capacitors
Strongest Frequency Point	8.762MHz (Calibrated magnetic field intensity: 0.01mGs)	

4. Conclusions

Magnetic sensitivity is one of the research contents of FOG environmental adaptability. In this paper, the radiation characteristics of magnetic field generated by gyroscope signal detection circuit are studied. The experimental results show that the overall radiation of gyroscope circuit is low, the radiation intensity of front circuit is higher than that of bottom circuit, and the front radiation is the strongest at 8.76MHz, approximately 0.01mGs. In addition, the driving circuit of light source is also one of the components of gyroscope circuit, and the radiation characteristic of generated magnetic field is the next research content.

Author

Author's Information: Lu Jun (1975-), male, lecturer, PhD, mainly engage in the research on optical fiber sensing technology. E-mail: lj19752019@163.com

Acknowledgments

This work was financially supported by Natural Science Foundation of Hebei Province (F2016506014).

References

[1] Zhang Weixu. Fiber Optic Gyroscope and Its Application [M]. Beijing: National Defense Industry Press, 2008:1-3.
[2] Wang Xiaxiao, Song Ningfang, Zhang Chunxi, et al. Experimental Study on the Magnetic Sensitivity of Fiber Optic Gyroscope [J]. Journal of Beijing University of Aeronautics and Astronautics, 2005,31 (10) : 1116-1120.
[3] Xiao Cheng, Liu Jun, Yin Jianling, Lu Jun. Research on Magnetic Sensitivity of Fiber Optic Gyroscope in Spherical Inhomogeneous Magnetic Fields[J]. Optical Instruments, 2014, 36 (5): 426-431.
[4] Yin Jianling, Lu Jun, Li Li, Chen Yudan. Effect on FOG Bias Error of Cylindrical Nonuniform Distribute Weak Magnetic Field, 2017, 38 (6) : 872-876.
[5] Hotate K, Tabe K. Drift of an fiber gyroscope caused by the Faraday effect: influence of the earth's magnetic field [J]. Appl. Opt., 1986, 25 (7): 1086-1092.
[6] Boehm K, Petermann K, Weidel E. Sensitivity of a fiber optic gyroscope to environmental magnetic fields [J]. Opt. Lett. , 1982, 7 (4): 180-183.

IOP Conf. Series: Materials Science and Engineering **799** (2020) 012025 doi:10.1088/1757-899X/799/1/012025

Analysis of Key Security Technologies for Power Dispatch Control System of Power Dispatch Support System

Zhou Siming[1], Liu Guanqiang[2], Qin Hai[1], Xia Tian[1]

[1]Power dispatching control center of Guizhou Power Grid Co., Ltd., Guiyang 550007, Guizhou

[2]NARI-TECH Nanjing Control Systems Ltd., Nanjing, Jiangsu 211106

Abstract. The scale of China's power grid is continuously expanding, the amount of collected data is increasing rapidly, and data processing and analysis services are developing towards clustering. Therefore, at present, task management in a stand-alone mode faces severe challenges in ensuring the high efficiency and reliability of distributed tasks. Information security risks in cyberspace can cause fatal threats to smart grid entities through the destruction of grid dispatching control systems and communication networks. Security check is one of the application functions of intelligent power dispatching control system. It is an important security defense line to ensure the stable operation of power grid and provides security check service for power grid dispatching plan and power grid operation. The existing operation and maintenance management mode of the power grid dispatching control system is distributed, and the data of each system and unit is not effectively integrated and connected, and lacks overall consideration. Based on the development history of power grid dispatching automation, this paper comprehensively analyzes the current status of the overall structure of the power grid dispatching system, and summarizes the key technology innovations and application effects.

1. Introduction

In the new era of rapid development and excessive development of China's UHV power grid, in order to effectively promote the safe and stable operation of China's power grid system, effective measures should be put forward to make certain adjustments to the structure and current flow of the grid [1]. Information security risks in cyberspace can cause fatal threats to smart grid entities through the destruction of grid dispatch control systems and communication networks. With the continuous expansion of the scale of the interconnected power grid, the electrical connections of the entire network are getting closer, the cross-section coupling relationship is more complicated, the shape and characteristics of the power grid are facing profound changes, and the level of security and stability are mutually restricted [2]. The concept of cyber security has evolved from traditional information system security protection to cyberspace confrontation. In fact, the network has become the fifth combat space after land, sea, sky and space. On the basis of integrated development, the regional line network constructed in China has clearly increased demand and exhibited differentiated usage characteristics. However, in actual construction, information and scale still do not fit the management model [3]. For distributed tasks, the current single-machine task management cannot manage such tasks across nodes, that is, tasks cannot be automatically deployed to multiple nodes, and computing performance cannot be improved by making full use of the resources of all nodes [4]. Security check is one of the application functions of smart grid dispatching and control system. It is an important security defense line to ensure the stable operation of power grid and provides security check service for power grid dispatching plan and power grid operation [5].

Each specialized system of the dispatching control center has the application requirement of safety check, and the separate construction of safety check function in each specialized system will lead to waste of resources and difficulty in coordination [6]. Dispatching agencies at or above the provincial level shall set up automatic departments to undertake the operation, construction, maintenance and related technical management of the system [7]. The State Grid Corporation of China has carried out the construction and application of the smart grid dispatching and control system pilot project in dispatching and control centers above the provincial level. It has completed and put into operation the largest and most powerful grid dispatching and control system in the world, ensuring the safe operation of large power grids [8]. The operation and maintenance management mode of the existing power grid dispatching and control system is distributed and not centralized enough, and the data of each system and unit are not effectively integrated and linked, thus lacking overall consideration [9]. This paper analyzes the service bus and parallel computing service based on smart grid dispatching and control system to realize service-oriented security check service of power grid, which provides customizable and multi-task parallel security service for power grid service. So as to ensure the safety and reliability of power grid operation.

2. Analysis of Security Check Service Architecture

2.1 Security Check Service Based on Service Bus

With the improvement of automation level of power monitoring system, the enrichment of functions, the extension of coverage of dispatching data network and the increase of users, the sources of information security threats in power monitoring system are becoming more diversified. The power flow corresponding to the check section is formed for the maintenance plan of the power grid system function and the operation of the power grid, so that the safety check can be carried out for the faults and problems occurring in the operation of the power grid [10]. The information between the security check service process and each port is implemented interactively by means of interface functions, which can effectively meet the requirements of different application functions for the query and location services of the security check service. After the safety check of the system is completed, auxiliary decision-making and margin assessment calculations need to be carried out. This can effectively analyze the problems in the safety and stability of the power grid in the dispatch plan and operation, and propose correct judgments. The smart grid dispatch control system is gradually put into practical operation at all levels of dispatch agencies. In order to enable the smart grid dispatch control system to play a better supporting role in the dispatch business system, research and application of the smart grid dispatch control system operation and maintenance plan and key technologies are required. The grid system requires electrical energy to be balanced at all times, so the construction of smart grids is an inevitable choice for the current development of the power industry.

2.2 Parallel Computation of Dynamically Assigned Tasks.

Security check service uses grid dispatching control system to realize parallel computing service, and establishes effective connection between standard interface and cluster computing resources, thus realizing interaction between information resources. In the process of developing and researching the power grid dispatching control system, China has carried out brand-new reforms on the regulation of operating costs of various power grids and the realization of energy conservation and emission reduction. Security check calls the parallel computing service of smart grid dispatching control system, and realizes the interaction with cluster computing resources through standard interfaces. By receiving the optimized real-time data and equipment alarm information of the power grid and using tools such as remote browsing, the reliable operation monitoring of the system can be realized, and the safe and stable operation of the system can be guaranteed [11].

The database of the operation and maintenance service support platform adopts the combination of real-time database and relational database, and uses the real-time database to meet the requirement of providing fast real-time data access, and organically combines the real-time database and relational database [12]. Parallel computing service supports pre allocation and dynamic allocation. The calculation amount of security check changes dynamically according to the application requirements, and dynamic allocation is adopted. After receiving the calculation request received by the service port, the security verification service terminal will select the calculation method according to the specific calculation content, and evaluate the accuracy of the calculation results[13]. As shown in Figure 1, the scanning speed modulation architecture of the power prediction model.

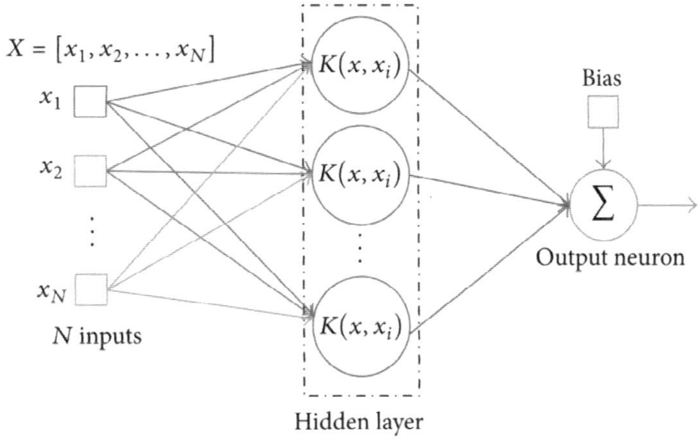

Figure 1 Scanning speed modulation architecture of the power prediction model

2.3 Requirements of Distributed Task Management in Power Grid Dispatching Control System

The connection between the electrical systems of the whole network is continuously strengthened, and the coupling relationship between the sections is becoming more complicated. The cross-security and stability of different levels of power grids are also receiving increasing attention, and their security problems are becoming increasingly prominent. Distributed task management is responsible for managing the entire life cycle of a job, from deployment, startup, run to exit. At present, single-machine task management can only manage the running status of task processes on this node, and cannot manage all tasks subordinate to the job across nodes. Distributed task management of power grid dispatching control system should be able to automatically deploy tasks to each node according to the system's resource usage, and ensure that tasks are evenly distributed according to node load. China's power grid has always adhered to the principles and standards of unified dispatching and hierarchical management during its operation. City and county-level power grids are responsible for the management and supervision of planning and safety check in the region[14].

The security check service system of the power dispatching system puts forward a unified model and a joint check scheme, which effectively ensures the safe and stable operation of the power grid. In addition, the implementation of the multi-level auxiliary dispatching plan plays a key role. In order to make full use of computing resources and improve the computing performance of tasks, distributed task management should monitor the resource usage of nodes in real time. The dynamic part of the unified information model mainly stores a message format template of real-time collected data by sensor nodes, and analyzes and updates the real-time collected data according to the template to form data consistent with the storage format of the data service system. The test preparation data for the delivery accuracy of screened power dispatching data, the delivery speed of power dispatching data, and the safety level detection of power dispatching data are shown in Table 1.

Table 1 System sampling analysis information

Project	First time	Second time	Third time
Detection of power dispatching data security level	36%	38%	40%
Delivery accuracy of power dispatch data	64%	62%	71%
Power dispatch data delivery time	7	8	6

Since the power station does not need to consume fuel, the power company should first dispatch all the power[15]. The goal of dynamic economic dispatch of power system including power station is to minimize the generation cost of traditional generating units. The objective function can be written as follows:

$$dR_t = \chi \cdot a(t) \cdot dt + \delta \cdot a(t) \cdot dw_t \tag{1}$$

Power balance constraint:

$$R_t^{\pi^*} = x + \int_0^t \xi \cdot a(R_s^{\pi^*}) \cdot ds + \int_0^t \zeta \cdot a(R_s^{\pi^*}) \cdot dw_s - U_t^{\pi^*} \tag{2}$$

When the operation cost is not considered, based on the original startup and shutdown state, due to the randomness of wind power, the power generation cost of conventional units is also a random variable, so the objective function is the minimum expected value of power generation cost:

$$dR_t^{\pi} = \xi \cdot a_{\pi}(t) dt + \zeta \cdot a_{\pi}(t) \cdot dw_t - dU_t^{\pi} \tag{3}$$

Like the two non-crossing guard rings of Figure 2, "2" on the link indicates that the link is composed of two fibers in opposite directions. When the working path of the service with bandwidth W passes through the link, the working link A or B belongs to only one ring network, and only one ring network protects the link of the service delivery node. If the service working link is a, the protection capacity w of the ring network 1a and the protection capacity of the ring network 1b are 0.

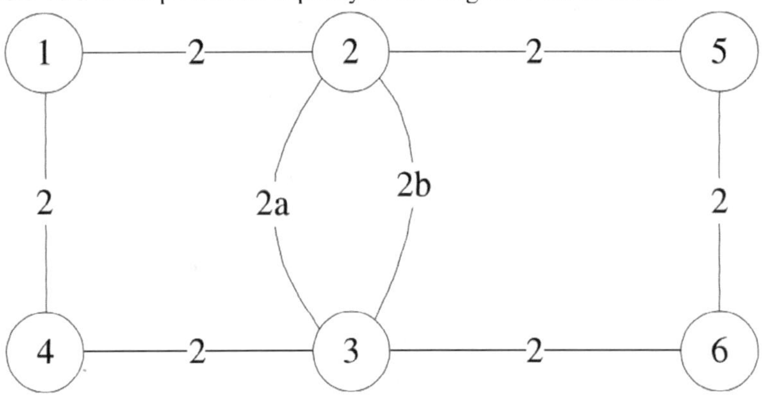

Figure 2 Non-intersecting rings

In the calculation of the power flow of the power dispatching plan, it should be considered that the power demand of the inter-regional tie lines and inter-provincial cross-section power needs is controlled within the planned value, and such considerations can effectively avoid internal plans formulated by different regional and provincial grids Power spreads out of balance. In distributed task management, in order to achieve the recovery of faulty tasks across nodes, one master and one standby redundant fault tolerance technology is used. In this technology, a backup task is configured for each main task on other nodes. When the main task fails, the backup task is immediately switched to the main task [16]. A distributed task management system needs to monitor the running status of each node in the system in real time. When a new node is found to be operational, tasks on a high-load node must be migrated to that node, making full use of idle resources to improve operating efficiency. The reactive power compensation measures shall be timely adjusted according to the node conditions in the specific report of power operation in the whole power grid dispatching control system, so as to ensure that the voltage of each hub node can be within the range specified by the planned value. With the continuous application of modern science and technology, the function of safety check service is continuously developing and improving, and is being promoted and applied in dispatching control centers at all levels. In order to ensure the reliability of the task management node, multiple backup machines are usually configured for the task management node[17]. When the host fails, the backup machine is immediately switched to the host to be responsible for task management.

3. Summary

According to the latest research results at home and abroad, the power grid dispatching control system should continuously improve its own safety check service function in the future development to ensure the safe and effective implementation of the power dispatching plan. With the rapid development of new energy power generation, its randomness and volatility have brought challenges to the dispatching and operation of the power grid, and the accuracy of planned power flow and safety check in some areas with surplus new energy resources have been greatly affected. The operation mode of the power network and the actual distribution of the power network are the keys to guide the dispatching and control technology of the power network and the dispatching and operation of the power network. In the future, the development technology of China's smart grid dispatch control system is mainly reflected in the technical optimization of multi-period and multi-level short-term power markets, self-description of system operation methods, and dynamic analytical development. Uncertainty theoretical analysis should be incorporated into the power grid security checking service to adapt to the security check service level of the power grid dispatching system in the context of the rapid development of new energy sources, so as to achieve the safe and stable operation of the entire power grid system in China. Only by fully understanding the development status of intelligent power network control and dispatching systems can we better develop and develop new types of future power grid dispatching technologies.

References

[1] Lin Rong. Some key technical issues of intelligent dispatching of large power grids. Global Market Information Review, 2016 (48): 125-125.
[2] Li Bijun, Zhou Xiaoning, Liu Qiang, et al. On-line prevention and control of grid operation security risks based on smart grid dispatching technology support system. East China Electric Power, 2014, 42 (6): 1057-1063.
[3] Chen Zizhao. Some key technical issues of intelligent dispatching in large power grids. Shandong Industrial Technology, 2014 (15): 10-16.
[4] Zhang Di, Miao Yan. Analysis on the key technologies of network security in smart grid. Communications World, 2019 (6): 211-212.
[5] Deng Zhiguang, He Xiqi, Yu Jie. Design and implementation of key technologies for power grid standby dispatch automation system. Electric Application, 2014 (23): 28-31.
[6] Lin Jinghuai, Mi Weimin, Li Zeke, et al. Remote operation safety error prevention technology for smart grid dispatching control system. Automation of Electric Power Systems, 2015 (1): 60-64.
[7] Zhou Guoliang, Lu Yijie, Wang Guilan. Key technologies for real-time analysis of big data panorama. Telecommunications Science, 2016 (4): 159-168.
[8] Wang Xingzhi, Bi Xiaoliang, Ge Chaoqiang. Discussion on Deepening Application of Smart Grid Dispatching Technology Support System. East China Electric Power, 2014, 42 (12): 2831-2834.

[9] Yin Chao. Some key technical issues of intelligent dispatching in large power grids. Heilongjiang Science and Technology Information, 2016 (7): 64-64.
[10] Zhou Zhihuan, He Chaolin, Liang Chao, et al. Construction and application of mobile platform for power grid dispatching. Electric Power Information and Communication Technology, 2014, 12 (12): 91-96.
[11] Chen Guang, Zhou Chao, Yang Hongshuang. Research on key technologies of emergency dispatching for distribution network. China Electric Power Education, 2014 (18): 121-122.
[12] Peng Yuanxiong. Research on key technologies for intelligent dispatching operation of distribution network. Smart City, 2016 (1): 70-71.
[13] Fu Z, Gatica L F, Sayas F J. Algorithm 949: MATLAB tools for HDG in three dimensions[J]. Acm Transactions on Mathematical Software, 2015, 41(3):1-21.
[14] Cockburn B, Fu Z, Hungria A, et al. Stormer-Numerov HDG Methods for Acoustic Waves[J]. Journal of Scientific Computing, 2017.
[15] Fu Z, Heuer N, Sayas F J . A non-symmetric coupling of Boundary Elements with the Hybridizable Discontinuous Galerkin method[J]. Computers & Mathematics with Applications, 2017: S0898122117305278.
[16] Wang Jing, Niu Rui. Key technologies for smart grid dispatch operation. China Science and Technology Information, 2015 (2): 180-181.
[17] Fan Hua. Situation-aware power grid automatic intelligent dispatching architecture and key technologies. Modern Manufacturing, 2016 (36): 97-98.

IOP Conf. Series: Materials Science and Engineering **799** (2020) 012026 doi:10.1088/1757-899X/799/1/012026

Research on Application of Digital Signal Processing Technology in Communication

Huang Lu, Yuan Xiaoyu, Wang Haodong, Li Jin, Ma Xuejiao, Zhang Caihong*

Qinghai Normal University, Xi'ning, Qinghai, 810000, China

*Corresponding author's e-mail: zhangcaihong2017@foxmail.com

Abstract. With the continuous development of modern information technology and the continuous improvement of people's living standards, electronic information technology has been widely used in people's lives. People's lives are inseparable from it. With its continuous development, its application range is getting wider and wider. This article first gives a brief overview of digital signal processing technology, analyzes the advantages and disadvantages of digital signal processing technology in the communication field, and the specific applications of speech compression coding and software radio in communication are analyzed in order to better promote the application of digital signal processing technology in the communication field.

1. Introduction

Digital signal processing technology, as an important product of the information age, exists in all aspects of the communication field and is an indispensable technology in the communication field. In recent years, China has increased its support for chips. Therefore, it has promoted the development and promotion of DSP chips, making it a control chip for many mainstream software communication products. In the field of communications, the equipment that applies digital signal processing mainly includes systems such as telephone communications, video voice, and voice signals, which effectively promote information exchange and information sharing. In the field of communication, digital signal processing technology is mainly applied to speech compression coding and software radio. Of course, there are still problems such as signal quality and signal transmission speed that need to be resolved. Therefore, in order to solve these problems, it is necessary to continue in-depth research on digital signal technology based on the current research results and promote the healthy development of the communication industry.

2. Digital signal processing technology

2.1. Digital Signal Processing Basic Theory

The core and symbol of digital signal processing technology is digital signal processor. Digital signal processing is a subject that uses computers or special digital processing equipment to process signals using numerical methods. It includes data acquisition, signal transformation, analysis, synthesis, filtering, evaluation and identification, etc., in order to extract information. Compared with the traditional analog processing method, digital processing has incomparable advantages. Digital signal processing systems can process digital signals as well as analog signals. Of course, the analog signal must be converted into a digital signal before it can be processed by a digital signal processing system. A typical digital signal processing flow is shown in Figure 1.

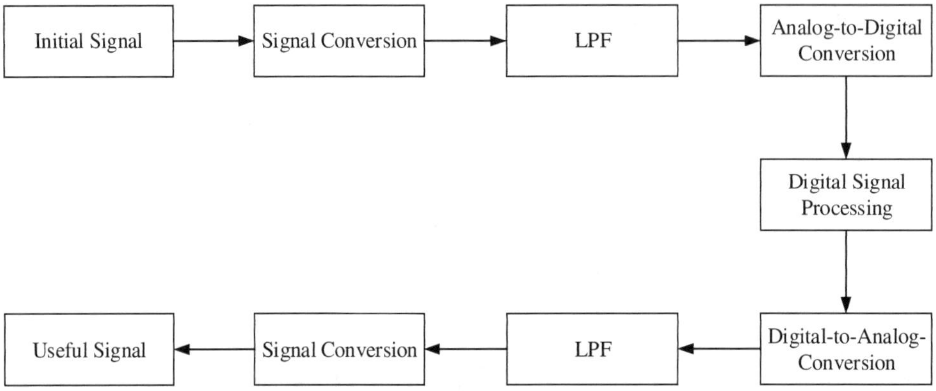

Figure 1. Digital Signal Processing Flow

The basic theory of digital signal processing includes the following:

- Pre-processing of analog signals: filtering unwanted frequency components and noise in the input analog signals to avoid spectral aliasing distortion after sampling;
- Time domain sampling and recovery of analog signals: analog-to-digital conversion technology, sampling theorem, quantization error analysis;
- Analysis of time-domain discrete signals and systems: representation and operation of signals, various transformations, description and analysis of time-domain and frequency domains of time-domain discrete signals and systems;
- Fast algorithms in digital signal processing: fast Fourier transform, fast convolution, etc.;
- Design and implementation of analog filters and digital filters;
- Multi-sampling-rate signal processing technology: The basic principle of the sampling rate conversion system and its efficient implementation method.

The quasi-signal processing system can only perform some conventional simple processing on signals, while digital signal processing uses numerical operations to implement signal processing, and many complex processing can be performed using computers. Therefore, digital signal processing applications will be more extensive.

2.2. Advantages of digital signal processing technology

Compared to analog signal processing, digital signal processing has many advantages, which are summarized as follows:

- Good flexibility. Digital signals are suitable for computer processing, and can also be realized by programmable devices. It is easy to change the parameters of digital signal processing systems through programming, so that the system can realize various processing functions.
- Stable and reliable, there is no impedance matching problem. As long as the design is correct, you can ensure the stable operation of the digital system, and the characteristics of the digital system are not easy to change with the changes in the use conditions. And because the digital systems at all levels are coupled through data, there is no impedance matching problem in analog circuits.
- High processing accuracy. The internal noise of the analog circuit and the external environment will affect the processing accuracy, while the digital system works in a binary state, so it is basically not affected by internal noise.
- Easy to encrypt and decrypt. With the increasing requirements of information security, encryption and decryption algorithms are becoming more and more complex, and only digital processing can solve this problem.
- Facilitate large-scale integration and miniaturization. Because digital circuits have low requirements for the consistency of circuit parameters, the basic units and basic modules that make up a digital system have a high degree of consistency, so it is convenient for large-scale integration and large-scale production.

- Easy to automate and multifunctional. Digital systems can easily perform corresponding operations based on various states, and a system can implement multiple functions.

2.3. Lack of digital signal processing technology

Although digital signal processing technology is more and more applied in reality, the problems still need to be further improved.

- Deal with speed. The speed of digital signal processing is a relatively common problem. It has been around since the birth of digital signal processing technology. However, due to factors such as equipment, environment, and technology, this problem is gradually being solved to facilitate communication.
- Quality problem. Information quality problems such as bad signal and unclear picture transmission also exist all the time. Digital signal processing technology needs to be better improved, and solutions such as multi-core processing are used to solve it.
- Anti-interference problem. Although the anti-interference ability has been significantly improved in digital signal processing technology, the anti-interference problem of the analog signal from the RF antenna needs to be improved. Improve all aspects of information transmission, ensure signal quality and effects, and provide more convenient services for communication.

3. Application research in communication

3.1. Speech compression coding

The main purpose of speech compression coding is to obtain clear and high-quality speech through corresponding equipment to convert information. Therefore, its transmission signal must have strong anti-interference ability and be able to transmit information in a narrow bandwidth spectrum in order to achieve reception. And fully restore the transmitted voice information to facilitate information communication. At the beginning of the speech coding system, the coding method mainly adopted large waveform coding, fully following the sampling guidelines of discrete Fourier transform, focusing on the use with the external environment, and ensuring the high-quality state of the speech signal. It is relatively fast, and it is easy to appear a series of problems in use, which affects the final voice signal quality. Another common form of speech coding is parameter coding. The difference is that the key parameters of the speech signal are used as the coding, but the coding rate is relatively low, which affects the speech effect.

A typical compression-sensing-based speech compression coding and reconstruction framework is shown in Figures 2 and 3.

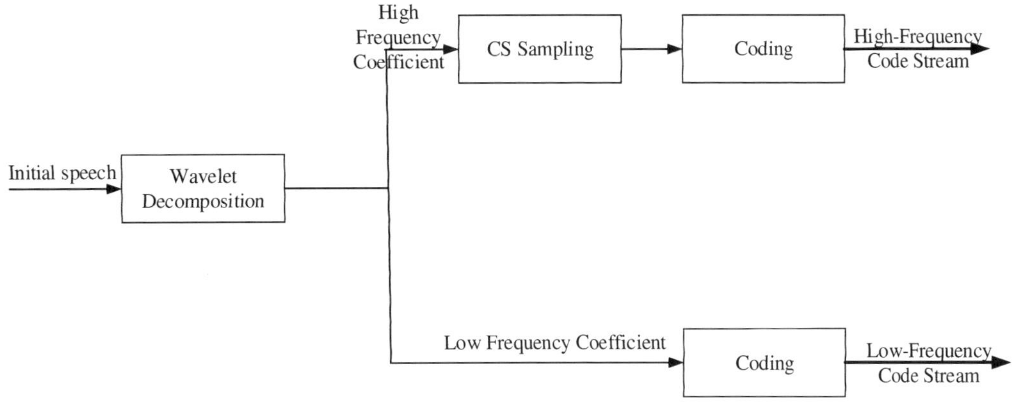

Figure 2. Sending voice compression and encoding

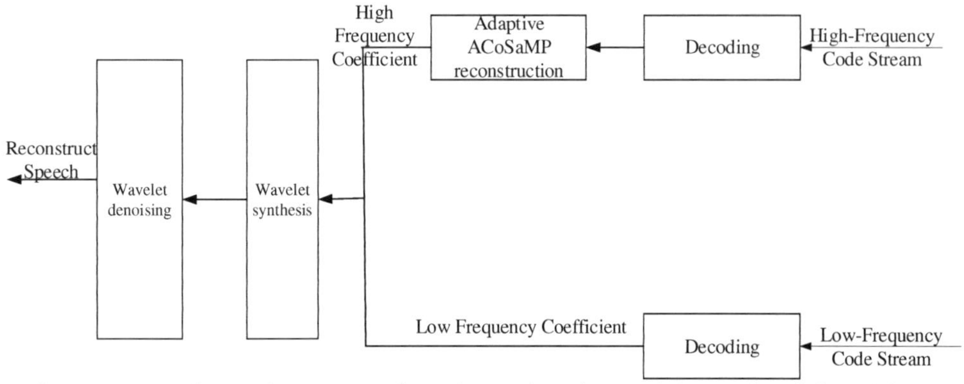

Figure 3. Decoding and reconstruction of speech at the receiving end and de-noising

In a speech compression system, it is mainly composed of a speech encoder, a digital storage medium, and a decoder. The speech input is dominated by the speech encoder, and the speech output is led by the speech decoder. The information transmission process is mainly the speech input of the speech encoder → the digital storage medium → the speech output of the speech decoder, which is the simplified speech compression coding. With the development of the times, the voice compression system mainly consists of a voice input module, a DSP module, an A / D conversion module, a D / A conversion module, and a voice output module. The key technology is the DSP module, which is the voice. Compression and decompression need to be implemented with the help of special algorithms and techniques to ensure that the voice signal is not damaged. The DSP chip has simple data processing characteristics and can implement multiplication and addition operations in a single instruction cycle. It is more suitable for voice compression systems. In addition, various hybrid encodings have also gradually emerged. For example, linear predictive encoding, as a new type of speech encoding technology, has a processing speed of 4 to 16 kilobytes per second, but it needs to match higher algorithms and digital signal processors. The digital signal processor can solve these problems, effectively improve the problem of voice transmission, and improve the stability and reliability of the voice compression system.

3.2. Software radio

Software-defined radio is a solution to the coexistence of multiple systems in the field of radio communications and the inability to formulate a unified standard between different systems. Because software radio implements various functions based on software programming, its main characteristics are manifested in flexibility and openness. Software-defined radio is mainly a communication software to rely on. This software can complete functions such as wireless calls and video surveillance with software programming. It has features such as functionalization and modularity. As long as the signal frequency band that its hardware system can process, the communication function in the corresponding frequency band can be increased by software. Digital signal processing technology has the characteristics of stable signal, fast transmission and strong anti-interference. It can be seamlessly connected with software radio and promote the application and development of software radio.

Software radio is a multi-band radio. It has a wideband antenna, a radio frequency front end, and an analog-to-digital / digital-to-analog conversion. It can support multiple air interfaces and protocols. In an ideal state, all aspects (including the physical air interface) can pass Software to define. The ideal software radio uses wideband analog-to-digital converters and digital-to-analog converters to perform A / D, D / A conversion at the antenna port, and digital signal processing at the back end is implemented by software of programmable devices. The structure of the radio is shown in Figure 4.

It is a characteristic of software-defined radio architecture to place the analog-to-digital converter and digital-to-analog converter as close to the RF front-end as possible. The status of the analog-to-digital converter and digital-to-analog converter in software-defined radio systems is also critical, which will directly affect the performance indicators of the entire system are excellent.

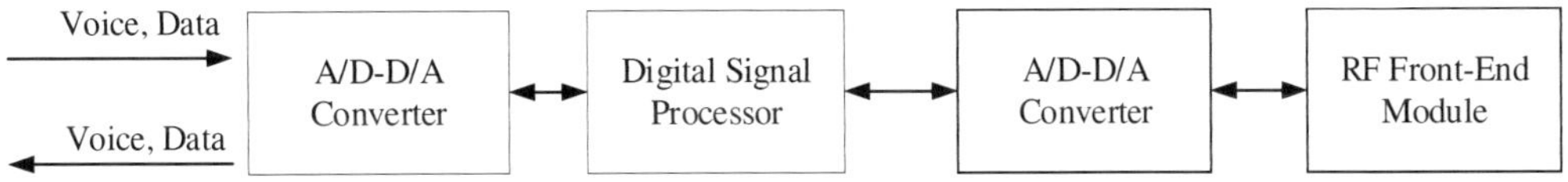

Figure 4. Ideal Software Radio Architecture

At present, the hardware implementation of digital processing mainly depends on two types of high-speed DSP chips and FPGAs. The respective characteristics of DSP and FPGA are relatively obvious, which are suitable for different occasions. At present, a development direction that is relatively well-recognized by everyone is the structure of DSP and FPGA. The purpose is to make the most of the advantages of these two processors and complement each other to achieve the parallel processing of signals in the entire system. When the clock is limited, the processing power of the system is improved as much as possible.

4. Conclusion

In the field of communication, the application of digital signal processing technology is mainly manifested in speech compression coding and software radio. Although its advantages are relatively obvious and it is a development trend in the future, there are still certain shortcomings, such as signal quality and signal transmission rate. Therefore, it is necessary to continue the research on digital signal processing technology in communication in the future to ensure more convenient and reliable communication. For example, high-speed digital processing technology and multi-core digital processing technology are under research.

References

[1] Vinay K. Ingle, "Digital Signal Processing and Its Implementation of MATLBA", Xi'an Jiaotong University Press, 1998.
[2] Meng Wenhan. Research on the application of digital signal processing technology in the communication field[J]. Electronic World, 2018, 552 (18): 124-127.
[3] Zhang Ze. Application analysis of modern digital signal processing technology in optical access network [J]. Electronic Test, 2015 (09): 102-104 + 107.
[4] Sun Bing. Research on the application and development of digital signal processing technology[J]. Information and Communication, 2015 (7): 177-177.
[5] Song Yuanqing. Development of Digital Signal Processing Technology [J]. Shandong Industrial Technology, 2016 (24): 130-130.
[6] Zhou Hongou. Design of Audio Signal Acquisition and Processing System Based on DSP [J]. Journal of Southwest University for Nationalities (Natural Science Edition), 2011,37 (S1): 100-103.
[7] Wang Zibo, Huang Shilong. Signal and Information Processing Technology Analysis of Emergency Communication System [J]. Electronic Test, 2018 (09): 123 + 119.
[8] Deng Caibao. Analysis of the Application of Intelligent Information Processing Technology [J]. Information System Engineering, 2018 (11): 42-43.

IOP Conf. Series: Materials Science and Engineering **799** (2020) 012027 doi:10.1088/1757-899X/799/1/012027

Research on the Main Problems and Countermeasures in the Construction of University Library Informatization

Weifeng Wang[1], Luyang Meng[2], Lei Wu[3], Jianfei Hu[4]

Corresponding Author: Weifeng Wang

[1,2,3,4]Library, University of Jinan, Jinan, Shandong Province, 250022, China

Abstract: University library informationization plays an important role in developing reading habits and knowledge literacy of teachers and students in universities. However, most college libraries in China still have problems such as not paying enough attention to informatization construction and low informatization construction. This paper adheres to the problem-oriented, and on the basis of analyzing the problems, puts forward countermeasures and suggestions to improve the level of informationization construction of university libraries.

1. Introduction

As an important foundation and guarantee for the development of a modern society, the construction of the library's informatization can enable the library to develop steadily in the fierce market competition. Library informatization can change the way people receive information, and can help readers obtain effective information resources. It is necessary for the library to carry out informatization construction, which can provide readers with more convenient and better services, and urge the library to gradually transform from the former document center to an information service center. At present, with the development and popularization of network information technology, China has entered the trend of informatization development and has shown a good development trend. As an important place for information collection, the library is an important base for people to master information and increase knowledge. The traditional library development model has been unable to meet its requirements in the new century. It is necessary to actively adopt informationization methods to further expand the library's capacity. Promote the development of libraries. How to build a library information development model under the new situation is an important subject. Based on this, this article conducts research on the informatization construction of university libraries, focuses on analyzing the existing problems, and proposes research countermeasures.

2. The Construction and Significance of University Library Office Informatization

2.1. Status of Informatization Construction in University Libraries

The university library is the school's document and information resource center. It plays an irreplaceable and important role in the construction of school informatization and shoulders the responsibility of teaching and research services. With the development of information technology and digital libraries, college libraries have entered a new era of digital construction. In the entire process of library management informationization in China, colleges and universities are at the forefront of development. Many universities have gradually used relevant computer technologies to manage libraries. China also attaches great importance to the development of library management informatization construction projects. The informatization construction of college libraries has become a major trend, and most library informatization construction has Began to take shape. Most colleges and universities in China have their own books and information related websites, which have advantages in hardware facilities, software resources, and talent team construction. Some colleges and universities have also accelerated the development of personal databases. Not only is the degree of informationization high, but it also integrates informationization with campus concepts and professional characteristics, such as the informationization construction of libraries in Tsinghua University and Peking University. However, different universities attach different importance to the informationization of library management. There are still some universities that are hindered in promoting the construction of informationization in library management, and their role is very small.

2.2. Significance of University Library Information Construction

Library information collection, management and other work contents have been changed by the development of informatization. The traditional book management model has been unable to meet the needs of school teaching and students' information resources, technology and services of the library. The development of informatization has gradually changed the school's teaching philosophy and teachers' teaching methods and methods, creating a good teaching environment for college teachers and students. Therefore, the management, service and technology of the library should adapt to the development of the times, and the teaching facilities should be continuously improved.

Promoting the informatization construction of college libraries has the following significance.

First, the informatization construction of university libraries is the actual need to realize resource sharing. The library relies on network technology, takes resource sharing as the guiding ideology, achieves the sharing of library resources through rational allocation of resources and integration of resources, provides a good teaching platform for teachers and students, and rationally arranges limited resources. Fully meet the needs of college teachers and students for resource sharing.

Second, it helps to improve the quality of library services. The development of network information technology is in a state of rapid development, and teachers and students are increasingly eager for knowledge. The use of digital network platforms by libraries is an irresistible trend. Strengthening university library management informationization can provide teachers and students with a rich knowledge base, and teachers and students can better communicate and learn. The quality of library services can be improved by strengthening the informatization of university library management libraries. By constructing a comprehensive Internet digital service integration platform, the degree of library management informatization can be improved to the greatest extent to facilitate readers' learning and living.

Third, it helps to improve basic human services. By promoting personalized information services and intelligently managing libraries through artificial intelligence and information, the effect of resource and knowledge sharing can be achieved. The informationization of book management can be realized quickly due to the convenience of digital network platform services, so as to realize the sharing of multi-dimensional book resources, and fundamentally promote the progress of human social science and technology exchange and spiritual civilization construction exchange.

Fourth, it is necessary to ensure that our own development meets the needs of social development. Under the development of science and technology, the integration of library business and information technology can truly meet the needs of the development of the times. With the rapid development of China's social economy, the development direction of libraries has gradually changed from a traditional model to a digital development model. Libraries must meet the rapid social and economic development by strengthening information construction and updating service forms and methods. The library plays a certain role in readers' thinking development and people's thinking mode. Therefore, in the new form, the library needs to actively change the traditional management mode and strengthen the information construction in order to truly provide high-quality services and meet the needs of society and Variety.

IOP Conf. Series: Materials Science and Engineering **799** (2020) 012027 doi:10.1088/1757-899X/799/1/012027

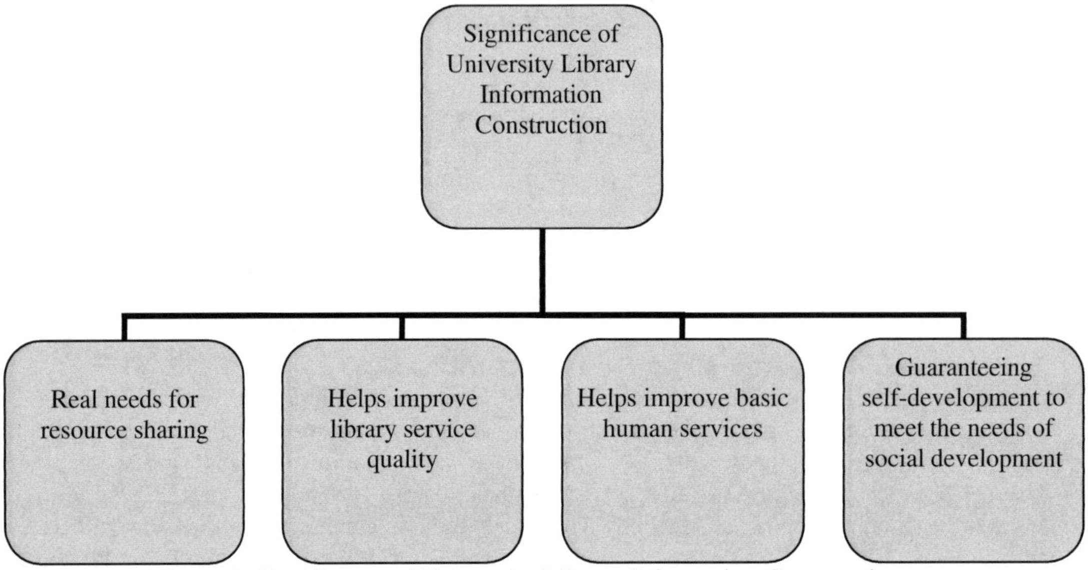

Figure 1. Significance of University Library Information Construction

3. Analysis of Factors Restricting the Construction of Office Informationization in University Libraries

3.1. The scientific development concept has not been formed. The informatization construction of university libraries requires the university management to establish a scientific development concept. At present, many colleges and universities still have insufficient awareness of the importance of informatization construction, lack of the concept of information resource construction, lack of long-term overall planning, neglecting the needs of teachers and students for information resources and the huge potential of academic resources, etc., affecting libraries The development direction of informatization. The development level of information resources is low, failing to build a data resource database with certain characteristics and scale, and failing to effectively integrate information resources. At the same time, the lack of plaques in information technology professionals, the uneven level of personnel and literacy, and the irrational team structure severely restrict the process of library information construction.

3.2. The degree of data resource sharing in library management is not high. In the process of library management informatization, the importance of data resource sharing was not recognized, and the awareness of information management was seriously insufficient. Many colleges and universities only provided services in accordance with the relevant requirements of the school passively, and did not take corresponding measures to better Promote the sharing of data resources. The readers' participation in the construction of library management and informationization is not high. Many readers do not realize their responsibilities. They believe that informationization construction is the responsibility of library managers. They only need to enjoy the convenient services brought by informationization. Readers have insufficient incentives for informatization construction. They do not understand the specific ways of informatization participation, so they cannot better participate in the informatization construction process.

3.3. Infrastructure construction is relatively weak. At this stage, many domestic college libraries have serious problems in terms of basic hardware facilities. Many computer software and hardware equipment lack the ability to update in a timely manner. Even some colleges and universities still stay in the manual management stage, which causes the actual efficiency of library management. low. Even in some college libraries where computer equipment has been installed, the construction process has

IOP Conf. Series: Materials Science and Engineering **799** (2020) 012027 doi:10.1088/1757-899X/799/1/012027

not really combined with their actual management situation. In addition, many college software developers have lower practical skills, which has led to the existence of many colleges for the development and maintenance of management systems. Many loopholes and shortcomings have not formed a coordinated situation.

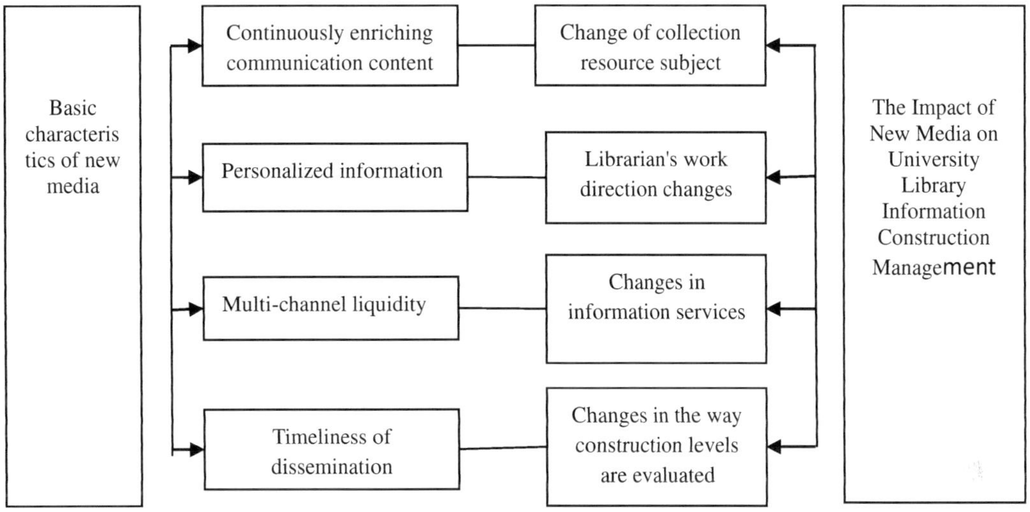

Figure 2. The impact of new media on university library information management

3.4. The level of personalized service is not high. At present, the digitalization of the collections of many college libraries is very low, and the network information resources that can be provided are very limited. At the same time, there are problems of underutilization of book management resources in many colleges. Although most libraries have dedicated electronic reading rooms for readers to read electronic information. However, in practice, the utilization rate of the library's electronic reading room is not high, resulting in a waste of resources. In addition, with the continuous rise of new media, it has had an important impact on the informatization construction of university libraries, but many universities have not yet recognized its importance in terms of management models and service consciousness, and lack diverse service channels.

4. Countermeasures to Accelerate University Library Informatization

In order to effectively solve the problems in library informatization construction and improve the level and quality of library informatization construction, it is necessary to strengthen the meaning of informatization, improve the infrastructure, strengthen the integration of resources, and attach importance to talent introduction and training.

4.1. Improve the awareness and construction level of information management. Attention should be paid to the construction of informatization, innovating the book management model, improving the book management information system, and comprehensively improving the service management level of colleges and universities. Facing the new problems of library information management, we must increase library infrastructure construction, pay attention to investment in book management systems, especially investment in information infrastructure construction, and ensure that capital investment can achieve maximum output. Actively create conditions to improve technical conditions and effectively improve personnel quality and management level. We must attach importance to the introduction and cultivation of high-quality information-based talents, be able to give corresponding guidance on information resource services in different disciplines, take necessary measures to strengthen the introduction and cultivation of talents, and comprehensively establish a sense of autonomous learning among staff.

4.2. Innovate the library service model. "Personalized service" is one of the important contents of the service concept reform of university libraries. Under the guarantee of the management quality of software and hardware facilities of university libraries, the service model, service concept, service content, and service methods in the library should be improved. We should analyze the reader's needs and adjust the literature resources through the specific requirements and personalized hobbies of the literature that the reader wants to know. It is necessary to improve the personalized service level of libraries, use the results of data analysis, and use data mining technology to rationally arrange the work of librarians and improve the utilization of resources. It is necessary to improve the integrated development of resources, innovate the development and use of information services, and improve the construction of library informatization in accordance with the actual situation of the school.

4.3. Effective integration of information resources. When libraries implement information construction, they need to make accurate positioning, make use of the characteristics of literature resources and network advantages, ensure that information resources can be used reasonably, establish distinctive information databases, and actively build new libraries to make them comprehensive. An important platform for sharing sexual information. A fast and accurate information system should be established, using computer technology, combined with the advantages of its own hardware conditions, to strengthen the management of materials, collections and documents in the library. Sorting and categorizing data by using scientific search engines. The accurate statistics of the database are used to arrange the literature resources, and to select and classify them, so that teachers and students can find the required literature materials accurately and quickly, and make full use of the library resources.

4.4. Improve the basic quality of library staff. The quality of library staff is related to the quality of library information construction. The quality of library staff determines the development of information construction. It is necessary to strengthen the education and training of library staff, and actively enhance the professional ethics of library staff. While improving their professional capabilities, they can also enhance the comprehensive literacy of library staff and truly realize library informationization. Innovative development. It is necessary to actively set up a reasonable internal organizational structure, improve the efficiency of library management, and improve the accuracy and effectiveness of information processing.

5. Conclusion

The library plays a very important role for students and teachers to obtain information resources. The advent of big data has made library information services more intelligent and personalized, and has improved library efficiency. University libraries should recognize the importance of information service innovation, actively expand the reader service work of university libraries, and let the majority of teachers and students feel the power of knowledge in a comfortable and warm environment. However, the main problems existing in the information management construction of libraries include insufficient information management awareness, relatively weak infrastructure construction, shortage of information construction talents, and lack of diversified service channels. To this end, we should increase the awareness of library information management services, attach importance to the construction of infrastructure, attach importance to the introduction and cultivation of high-quality information talents, and increase the intensity of integrated development of resources.

References

[1] Chen S F. On Informatization Construction of University Library[J]. Journal of Hengshui University, 2011.
[2] Jing Du, Daoliang Li, Hongwen Li. Application of Grey Prediction Model in Rural Informatization Construction[C]// 2010.
[3] Wei, Ying. The Construction of Colleges and Universities Library Unified Information System Based on Guangxi Regional Features[J]. Applied Mechanics & Materials,

543-547:3151-3155.
[4] Zhu Y. Position and Function of Library in Agricultural and Forestry Universities in the Construction of Rural Informatization[J]. 2008.
[5] Peng, Xu Fu, Shi, Shu Dong. The Research on University Librarys \"Four-in-One\" Cloud Computing Services Social System[J]. Advanced Materials Research, 791-793:1297-1300.
[6] CHEN Lin, Jiangsu Normal University. Research on the Characteristics of China Vocational Education Informatization Innovation[J]. Modern Educational Technology, 2014.
[7] Yin Feng. The Discussion on Information Resources Co-Construction and Sharing of Double-Type Society Building in Chang_Zhu_Tan City Cluster[J]. Library, 2009.
[8] WU Xiao-xia. Research on the Informatization Development Strategy of Yuncheng City in Eleventh-five-year Plan Period[J]. Sci-Tech Information Development & Economy, 2007.
[9] Xichuan Guo. "THE SCIENTIFIC AND TECHNICAL SERVICE FUNCTION OF AGRICULTURAL DIGITAL LIBRARY IN COUNTRYSIDE INFORMATIZATION DEVELOPMENT AND CONSTRUCTION UNDER NETWORK ENVIRONMENT." 2008.
[10] Chen Nenghua, Liu Kunxiong. A Study of the Transformation of Library Management Patterns in the Informatization Environment[J]. Journal of the Library Science in China, 2004.
[11] Atsuo Miyasawa. A historical overview and discussion of library functions forward the high-informatization society[J]. Journal of Information Processing & Management, 2008, 38(2):115-124.
[12] Chen H. Research on the Management of Digital Library Under Informatization Background[M]// Proceedings of the 9th International Symposium on Linear Drives for Industry Applications, Volume 2. 2014.
[13] Chen Nenghua, & Liu Kunxiong. (2004). A study of the transformation of library management patterns in the informatization environment. Journal of the Library Science in China.
[14] Duff, Alistair S. Four “e”pochs: the story of informatization[J]. Library Review, 52(2):58-64.
[15] Chen, S. F. (2011). On informatization construction of university library. Journal of Hengshui University.
[16] Weng Jia. On the Indexes Involved in Information Resource of the Measurement of Informatization[J]. Library & Information, 2007:21.
[17] L. Maosheng, F. Zhenjia, Z. Lili. The development, current state, and effects of community informatization in Mainland China[J]. Library Trends, 2013, 62(1):13-33.
[18] Soproni, Luminita, Horga, Ioan. Global communication as a result of globalization and informatization[J]. Mpra Paper, 2008.

IOP Conf. Series: Materials Science and Engineering **799** (2020) 012028 doi:10.1088/1757-899X/799/1/012028

Exploration of Teaching Reform of Computer Network Practice Course

Jiong Wang, Hua Zhang, Zhiqiang Ruan, Zhengyuan Wu, Ge Xu.

College of Computer and Control Engineering, Minjiang University, Fuzhou 350108, china

Fujian Provincial Key Laboratory of Information Processing and Intelligent Control (Minjiang University)

First author: Jiong Wang; Corresponding author: Ge Xu

Abstract: With the vigorous development of digital and network technology in China, computer and it's technology have been widely used in various industries. Therefore, information technology has become increasingly important in today's society. However, there are still many shortcomings in computer teaching in China, and reforms and improvements are urgently needed to give full play to the important role of computer technology. The research focuses on the following reforms of computer network practice course teaching reform.

1. Introduction

The computer in our country has been widely used in people's work and daily life. With the rapid development of computer technology, the traditional computer teaching is not effective enough to meet the needs of social development. Based on the social environment of computer development, the application of information technology in the computer public course teaching can replan the course content, improve the teaching mode and collect students' teaching evaluation to improve teaching effect. Computer course is an important public course in colleges and universities, which is directly related to the future employment of students. In view of the current situation of the development of computer public courses in China, there are still some problems, such as single content, low quality, lagging mode, slow information feedback and so on, which delay the course teaching reform.

2. Current situation analysis of computer teaching process

(1) The lack of students' practical ability

Computer courses are obviously comprehensive and involve a wide range of knowledge with a strong professionalism in electronics, communication, mathematics, etc. Students are prone to "feel confusion" in their studies due to the poor continuity between knowledge, and various problems with low levels of internalization, therefore, the learning results are not ideal.

Practical operation is an important part of computer teaching. Only by combining the theoretical knowledge in books with practical operation can students better understand and master. However, due to the influence of traditional exam-oriented education, the cultivation of students' practical ability is neglected [1]. In addition, the students themselves did not correctly understand the importance of computer practice courses, instead wasting time by playing games, which can not improve their own computer course learning level and quality, and is not conducive to the grasp of computer content and operating.

(2) The lack of advancement in textbooks

With the continuous deepening of information technology, the pace of updating computer software is also accelerating. However, the computer teaching textbooks used by most colleges and universities in China are still outdated, which is difficult to meet the needs of the development of the times. Therefore, under the old teaching version, it is difficult for students to apply what they have learned and keep up with the times, as well as highlighting the effectiveness of computer teaching[2].

Computer public course is a course closely connected with information technology. However, with more theoretical knowledge and less practical links, there is no obvious difference between the teaching of computer public course and other professional knowledge subjects. Teachers still adopt the traditional teaching mode, which seriously hinders the independent learning. In the one-way teaching mode, students' learning information is not received in a timely manner, which weakens the connection between teachers and students. Students are not motivated to study, so their interest is also weakening with poor classroom performance. Especially in the big data mode, more emphasis is placed on innovative teaching, information sharing and learning, as well as independent knowledge exploration. Obviously, backward teaching modes can no longer meet the needs of the development of new education.

(3) The individual differences of students are ignored.

Due to the individual's age, gender, personality, intellectual and non-intellectual factors, there are certain differences in students' grasp of computer knowledge and skills. However, students and teachers lack communication in classroom information, and the classroom teaching evaluation system lacks systematic and various features. Students cannot express their learning problems in public courses in a timely manner, and teachers cannot make targeted adjustments with insufficient feedback on teaching and learning based on student learning. Teachers do not pay attention to the individual differences of students may result in uneven computer skills of students[3].

3. The necessity of practice course teaching reform of computer network

With the vigorous development of digital and network technology, China has entered the era of Internet +. The theme of the era of information technology development promotes social science and technology, strengthens the sharing of information between Internet data. Based on the big data environment, the education industry has ushered in new development opportunities and changed the traditional teaching mode and enriched classroom teaching, as well as increasing inter-connectivity among students, teachers, and classroom information. As early as 2015, Premier Li Keqiang proposed to expand Internet business to organically integrate the Internet, cloud computing, big data, and various modern manufacturing industries. Therefore, the concept of Internet + should be "Internet Modern industries" [4]. Therefore, the economic development trend that China is facing should be the mode of Internet +.

With the development of the Internet, computer teaching has been enriched, and the teaching forms have also become diversified. For example, products from the Internet + era have been widely used in computer teaching as a new type of resources, such as micro-classes or MOOC classes. The limitations of traditional classroom time and place are broken. Students can use smart phones or other smart mobile devices anytime and anywhere to watch teaching videos on the Internet, which is conducive to the consolidation of students' knowledge. Under the teaching reform of computer network, countless gathered in the network makes every student can enjoy the common teaching resources. The reform can break through the geographical limitations of school age to realize the equality and dig their own potential and provide computer teaching results. The reform of computer network teaching has promoted the transformation of traditional instilled teaching into a mode that takes students as the mainstay and respects the development of students' individuality, and enables students to make full use of the Internet to improve their own level of practical operation, which is conducive to students' future development[6].

4. Analysis on the teaching reform method of practice course of computer network

(1) Improving students' practical operation

Computer teachers should change teaching idea, pay attention to students' practical ability in the teaching process. Taking the teaching content as the guiding theory, teachers should focus on students' ability in computer operation and the theory of knowledge and redistribute computer operation time. The theoretical knowledge time should account for a third of the total teaching hours. Students computer operation time should account for two-thirds to make students master more skilled operation. For example, in the basic operation of file processing, the teacher can prepare a copy first to make students create a new document and enter for training student's fingering and typing speed. This can make students follow to exercise both the practical ability and the initiative thinking as well as problem solving ability[7].

(2) Adopting diversified teaching methods and stimulating students' interest in learning

Einstein once said, "interest is the best teacher." Only computer teachers fully stimulate students' interest and enthusiasm, can students actively participate in teaching, which is helpful to improve the quality of classroom teaching. For example, when students enter the society, the interview brings about the problem of finding a job. Most companies or enterprises require students to show their resumes first[8].

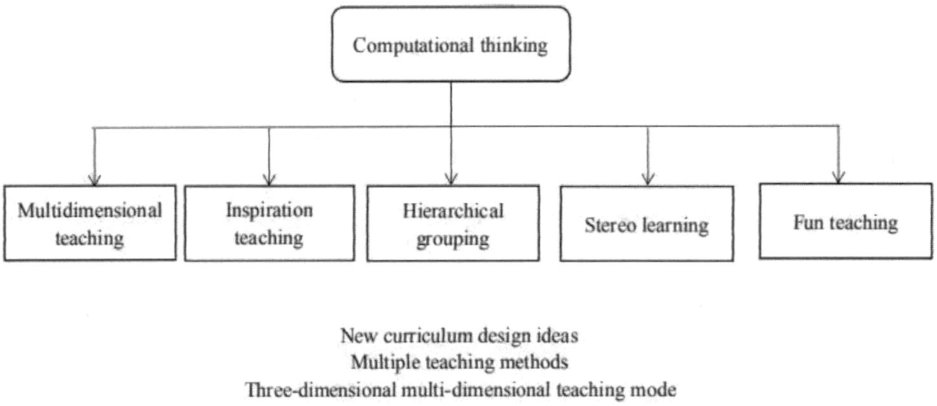

Fig. 1 Multiple teaching methods

Computer teaching based on "Internet +" should keep pace with the times to improve backward textbooks and increase advanced teaching content, as well as combining students' psychological characteristics to adapt to the needs of the times. For example, most students like to socialize online, and with the help of mobile smart devices, they can have online entertainment anytime and anywhere. Computer teachers can make full use of students' interest for students to select social software, such as QQ, WeChat, Weibo, Tieba. Teachers can teach students how to use and create a simple application or build a simple SaaS through WeChat public social networking sites. In this way, students cannot only satisfy their psychological curiosity, but also improve the application of knowledge and adapt to the needs of the times. In addition, teachers can also adopt the teaching form of "micro-class" to impart the key points and difficulties to students through micro-class and micro-video, so as to strengthen students' grasp of knowledge and improve the efficiency and quality of computer teaching, as shown in figure 1.

(3) Combining individual differences of students, using targeted teaching methods

If teachers need to refer to professional knowledge in the home teaching process, they can connect the professional courses with the main line to achieve a seamless connection of professional knowledge and weaken the students' sense of separation of professional knowledge. In order to improve the effectiveness of computer classroom teaching, computer teachers should pay full attention to the individual differences of students and adopt different teaching methods. For example, based on the psychological characteristics of students who like to play and joke, teachers can teach students to use "Photoshop" to make interesting pictures. At the same time, the method of layered teaching is used to carry out targeted teaching for students.

Teachers divide students into the following different levels. The students with strong cognitive ability and high level of computer operation are high-level students; students with intermediate cognitive skills and computer skills are at a medium level. Students with poor cognitive and computer skills are on average. In the teaching process of "Photoshop", teachers should pay attention to the operation and use of software tools for ordinary students. When students can master the various functions of the software, teachers should encourage them. For middle-level students, teachers not only need software tools, mastery and task requirements, they can also use classroom teaching software to count the number of times students answer questions and collect students feedback. In this process, students should be involved to meet the learning needs of students, scientifically construct and reform computer classrooms, check student learning progress, achieve teaching progress and evaluation, and let teachers and students think and work together.

5. Conclusion

In short, with the widespread popularity of computers and the rapid development of the Internet, it is imperative to improve the effectiveness of computer teaching. Based on this, computer teachers should change their teaching concepts, combine students' psychological characteristics and social development needs, and adopt a variety of teaching methods to fully stimulate students' learning interest, improve the quality and level of teaching, and strengthen students' practical ability to promote the development and progress of computer teaching in China.

Acknowledgement

This work is supported by the research Project of Education and Teaching Reform of Undergraduate Universities in Fujian Province under Grant number FBJG20190252, the Teaching Reform Research of Minjiang University under Grant number MJU2019B026, the Minjiang College undergraduate core curriculum construction project under Grant number MJW201833241, the fujian Province Young and Middle-aged Teacher Education Research Project (Class B) under Grant Nos. JB13176 and JB12159, and the National Nature Science Foundation of China under Grant 61871204.

References

[1] Peng Bing. 2019. Exploration on the Practical Teaching Reform of Computer Network Course in Higher Vocational Colleges [J]. Journal of Scientific Education, (25): 133-134.
[2] Zhu Lijin. 2019. Computer Network Technology Course Modular Practice Teaching Course Reform and Research [J]. Computer Knowledge and Technology, 15 (21): 36-37.
[3] Wang Yanping. 2019. Mixed Teaching Reform Practice of Computer Network Technology Course Based on Vocational Education Cloud Platform [J]. Modern Information Technology, 3 (16): 84-86.
[4] Liang Fuwei. 2019. Exploration and Practice of Teaching Reform of Network Operating System Courses [J]. Software Guide, Educational Technology, 18 (7): 59-62.
[5] Wan Qingda. 2019. Practice and Reflection on Teaching Reform of Computer Network Technology Course in Secondary Vocational Colleges [J]. Charming China, (12): 165.
[6] Tan Ming. Practice and Reflection on Teaching Reform of Computer Network Technology Course in Secondary Vocational Colleges [J]. Tiangong, 2019, (2): 46.
[7] Zhao Sijia, Yin Ting. 2019. Computer Network Course Teaching Reform Practice and Questionnaire Analysis [J]. All Circles, (2): 122, 126.
[8] Lu Zhifeng. 2019. Reform Practice of Computer Network Course Teaching Based on SPOC [J]. Heilongjiang Science, 10 (13): 62-63.

IOP Conf. Series: Materials Science and Engineering 799 (2020) 012029 doi:10.1088/1757-899X/799/1/012029

Research on Network APT Attack Intrusion Detection Technology Based on Machine Learning Algorithm

Qingchuan Meng[1,2*], YangYang[3], Fengzhi Wu[4], Xiang Chen[5], Xiaoming Chen[2]

[1]HeiLongJiang Province Cyberspace Research Center, Harbin, HeiLongJiang, 150090, China

[2]College of Computer Science And Technology, Harbin Engineering University, Harbin, HeiLongJiang, 150090, China

[3]University of Shanghai for Science and Technology, Shanghai, Shanghai, 200093, China

[4]Department of Mechanical Engineering, Massachusetts Institute of Technology, Cambridge, Massachusetts, 02139, US

[5]Shenyang Aerospace University, Shenyang, Liaoning, 110136, China

[*]Corresponding author's e-mail: mqchappy365@163.com

Abstract. The attack frequency of network advanced persistent threat (APT) is more and more higher, which seriously endangers the network security. In order to obtain high accuracy of network APT attack intrusion detection results, aiming at the limitations of current network APT attack intrusion detection model, a network APT attack intrusion detection model based on machine learning algorithm is proposed. A "one-to-one" network APT attack intrusion detection classifier is built through a neutral and excellent support vector mechanism of machine learning algorithm, and the current standard network APT attack intrusion detection database is adopted to verify the validity of the model. The accuracy of network APT attack intrusion detection is over 95%, and the detection error is far lower than the actual application range. It can be used in the actual network security management.

1. Introduction

With the continuous promotion of network applications, the issue of network security has attracted much attention from people. Traditional network security defense technologies mainly include data encryption and anti-virus software. These are passive defense methods, which cannot resist external intrusions, so network security cannot be effectively protected. Active network security prevention technology is mainly intrusion detection, which can detect the network security status in real time and find some illegal intrusion behaviors. Network intrusion has become the focus of current research. Among many network attacks, advanced persistent threat (APT) attack has become the most important threat to network security. It is difficult for traditional network security technologies such as firewalls to effectively protect it. Therefore, it is very important to research the APT attack intrusion prevention detection system[1-3].

In order to obtain network APT attack intrusion detection results with high accuracy, in view of the limitations of the current network APT attack intrusion detection model, a network intrusion detection model based on ant colony algorithm to determine the parameters of support vector machines was

IOP Conf. Series: Materials Science and Engineering **799** (2020) 012029 doi:10.1088/1757-899X/799/1/012029

proposed. We build a "one-to-many" network intrusion detection classifier, introduce the ant colony algorithm to determine the optimal parameters, and test the effectiveness of the model by using the current standard network APT attack intrusion detection database. The accuracy rate of network APT attack intrusion detection is over 95% , detection error is far lower than the actual application range.

2. Related theories

2.1. Support Vector Machine

Support vector machine is a kind of machine learning algorithm with excellent performance, specially for small samples, proposed by Vapnik and others. It works differently from neural network. Find an optimal plane and divide all training samples into two categories: one above the plane and the other below the plane. At the same time, keep the samples as far away from the optimal plane as possible. The samples on the optimal plane are called support vectors[4-5].

For a set of n samples $\{(x_1, y_1),\cdots,(x_i, y_i),(x_n, y_n)\}$, use the function $\varphi(x)$ to map the samples, and then classify the samples in the mapping space:

$$f(x) = \mathrm{sgn}(w \cdot \varphi(x) + b) \tag{1}$$

Where: w is the weight, b is the threshold.

To find the optimal classification plane, we must find the optimal w and b values, and it is very difficult to obtain the optimal w and b values by directly solving equation (1). Based on the principle of structural risk minimization, the following constraints are set:

$$y_1 \cdot (w \cdot \varphi(x_i) + b) \geq 1 \tag{2}$$

In order to speed up the modeling speed, a relaxation variable ξ_i is used to compromise the classification accuracy and classification error, so that the optimal classification plane can be transformed into the following form:

$$\min \frac{1}{2} w \cdot w + C\sum_{i=1}^{n} \xi_i \tag{3}$$

The corresponding constraints is:

$$y_i(w \cdot x_i + b) \geq 1 - \xi_i,\ \xi_i \geq 0,\ i = 1, 2, \cdots, n \tag{4}$$

Where C represents the degree of punishment for errors.

The Lagrange multiplier $\alpha_i > 0$ is introduced to obtain the dual form of equation (4):

$$\min \frac{1}{2} \sum_{i,j=1}^{n} \alpha_i \alpha_j y_i y_j (\varphi(x_i) \cdot \varphi(x_j)) + \sum_{i=1}^{n} \alpha_i \tag{5}$$

And has the following constraint:

$$\sum_{i=1}^{n} \alpha_i y_i = 0,\ 0 \leq \alpha_i \leq C \tag{6}$$

For the nonlinear classification problem, introducing the kernel function $k(x_i, x_j)$, we can get:

$$\min \frac{1}{2} \sum_{i,j=1}^{n} \alpha_i \alpha_j y_i y_j k(x_i, x_j) + \sum_{i=1}^{n} \alpha_i \tag{7}$$

Where $k(x_i, x_j) = \varphi(x_i) \cdot \varphi(x_j)$.

The optimal classification plane for support vector machines is:

$$f(x) = \mathrm{sgn}(\sum_{i=1}^{n} \alpha_i y_i k(x_i, x) + b) \tag{8}$$

Select the radial basis function, which is:

$$k(x, x_j) = \exp(-\frac{\|x - x_j\|}{2\sigma^2}) \tag{9}$$

Where σ represents the kernel width parameter.

2.2. Impact of parameters on network APT attack intrusion detection

An analysis of the working principle of support vector machines shows that the influence of parameters C and σ on their learning performance is very important. Select a training sample and analyze the correct rate of network intrusion detection under different parameters. The results are shown in Table 1. Analysis on Table 1 shows that even if the environment and data are the same, the intrusion detection accuracy rate of different parameters is still very different, so the optimal values of parameters C and σ need to be selected.

Table 1. Effect of parameters on learning performance of support vector machines

C	σ	Network APT intrusion detection accuracy rate /%
10	0.01	62.74
50	0.1	98.53
100	1	72.67
500	10	78.20
1000	100	95.74
5000	1000	67.49
10000	2000	77.40

2.3. Ant colony algorithm

Ant colony algorithm is a more commonly used search optimization algorithm. In the process of foraging, ants leave pheromones on the path. Other ants use pheromone to identify crawling paths. The higher the pheromone concentration, the more ants pass through the path, the higher probability that other ants choose the path[6].

If the number of ants is m, the following formula can be obtained:

$$m = \sum_{i=1}^{n} b_i(t) \tag{10}$$

Where $b_i(t)$ represents the number of ants on node i.

At time t, the path (i, j) residual pheromone concentrations of node i and j are:

$$\Gamma = \{\tau_{ij} \mid c_i, c_j \subset C\} \tag{11}$$

Where τ_{ij} is the pheromone concentration of (i, j).

In the initial stage of ant colony algorithm, $\tau_{ij}(0) = 0$, the transition probability $p_{ij}^k(t)$ of ant selecting the next node is:

$$p_{ij}^k(t) = \begin{cases} \dfrac{\tau_{ij}^\alpha \eta_{ij}^\beta}{k\tau_{is}^\alpha \eta_{is}^\beta(t)}, & s \in allowed_k, j \in allowed_k \\ 0 \quad , & \text{otherwise} \end{cases} \tag{12}$$

Where: η_{ij} is the local heuristic information transferred from node i to j; $allowed_k$ is the set of nodes not accessed; α and β are the weight parameters.

After a period of time, the ant colony completes a path search and needs to update the pheromone on the path:

$$\tau_{ij}(t + n) = (1 - \rho) \times \tau_{ij}(t) + \Delta\tau_{ij}(t) \tag{13}$$

$$\Delta\tau_{ij}(t)=\sum_{k=1}^{m}\Delta\tau_{ij}^{k}(t) \tag{14}$$

In the formula: ρ is the volatility of pheromone; $\Delta\tau_{ij}(t)$ is the pheromone increment on path (i, j); $\Delta\tau_{ij}^{k}(t)$ is the sum of pheromone, and its expression is:

$$\tau_{ij}^{k}=\begin{cases}\dfrac{Q}{L_k}, & \text{Ant } k \text{ passes through } (i,\ j) \text{ in this cycle}\\ 0\ , & \text{other}\end{cases} \tag{15}$$

3. Construction of network intrusion model

In network APT attack intrusion detection, the optimization problem of LSSVM parameters can be expressed by the following formula:

$$\max P(C,\sigma)\ \text{s.t.}\begin{cases}C\in[C_{\min},C_{\max}]\\ \sigma\in[\sigma_{\min},\sigma_{\max}]\end{cases} \tag{16}$$

The steps for network APT attack intrusion detection are as follows:

Step1: Collect network status information, extract the characteristics of network APT attack intrusion detection, and process the characteristics as follows:

$$x_1=(x-x_{\min})/(x_{\max}-x_{\min}) \tag{17}$$

Where $x_{\max}$ and $x_{\min}$ are the maximum and minimum, respectively.

Step2: Consider the support vector machine parameters (C, σ) as a path for ant colony crawling, and model the network APT attack intrusion detection training samples according to each set of parameters to get different detection accuracy rates.

Step3: The ant colony's pheromone update operation and node transfer are used to implement path crawling. Finally, the optimal parameter (C, σ) combination is found through path optimization.

Step4: Establish the optimal network APT attack intrusion detection model according to the optimal parameter (C, σ) combination.

Due to the classification of support vector machines for two categories, there are many types of network APT attacks, including mobile device attacks (MDA), malicious email attacks (MEA), and system vulnerability attacks (SVA). This article uses a "one-to-one" approach to build multiple classifiers, as shown in Figure 1.

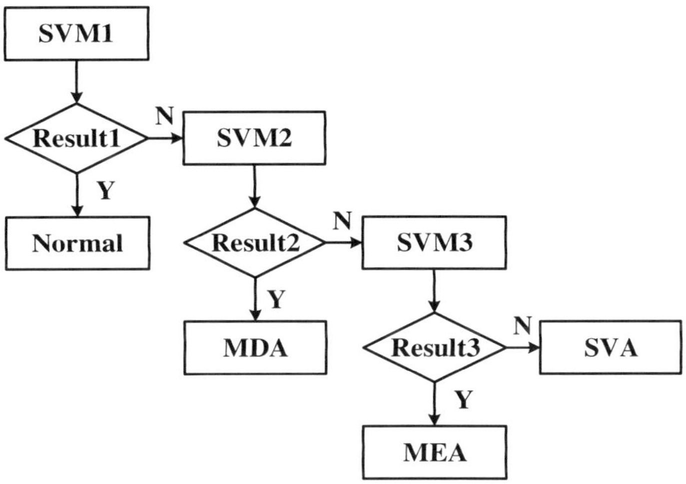

Figure 1. Classifier structure for network APT attack intrusion detection

IOP Conf. Series: Materials Science and Engineering **799** (2020) 012029 doi:10.1088/1757-899X/799/1/012029

4. Experimental results and analysis

The MIT KDD Cup's network APT attack intrusion detection dataset was selected as the test object, including three types of network APT attack intrusion behaviors: MDA, MEA, and SVA. Because the size of the data set is very large, 10% of the data is randomly selected for specific experiments. In order to make the experimental results of the proposed model (ACO-SVM) convincing, a BP neural network (BPNN) genetic algorithm optimized SVM (GA-SVM) network intrusion detection model is used as a comparative model, and the following indicators are used as the evaluation criteria of the experimental results:

$$\text{Accuracy rate} = \frac{\text{Correct number of samples detected}}{\text{Total number of samples}} \times 100\% \quad (18)$$

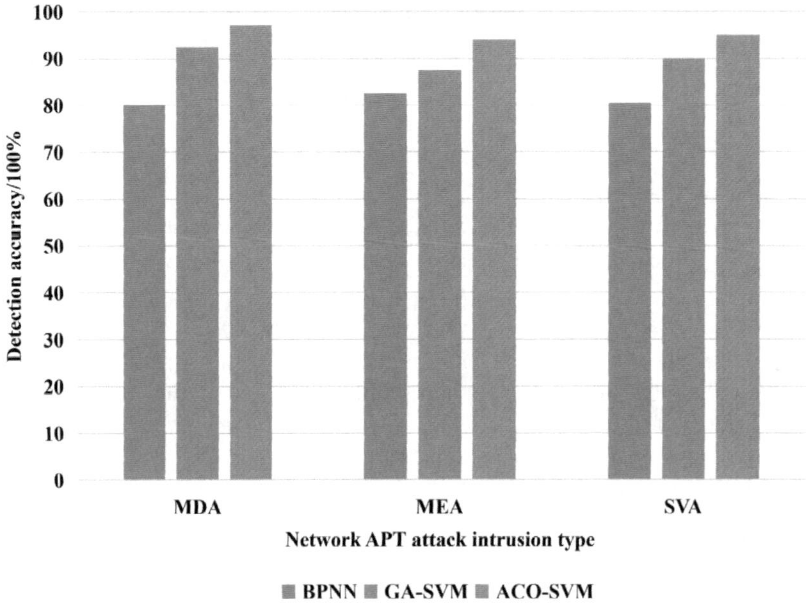

(a) Detection accuracy

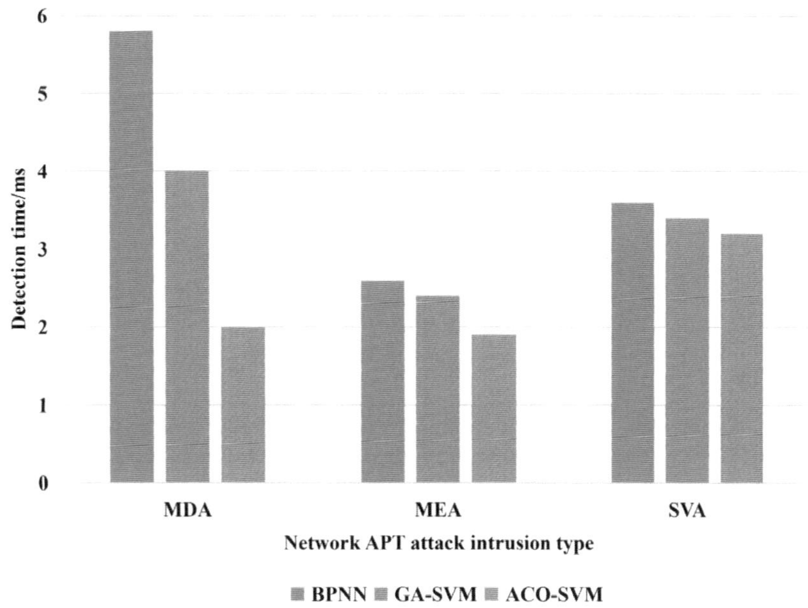

(b) Detection time

Figure 2. Comparison of network APT attack intrusion detection results

The simulation results are shown in Figure 2. As can be seen from Figure 2 (a), ACO-SVM has the highest accuracy rate of network APT attack intrusion detection of all models, followed by GA-SVM, the lowest rate of network intrusion detection is BPNN, and the false alarm rate is the lowest It shows that ACO-SVM can accurately recognize the intrusion behavior of APT attacks on the network and obtain ideal detection results. At the same time, it can be seen from Figure 2 (b) that the ACO-SVM network APT attack intrusion detection takes the least time and can meet the efficiency requirements of network APT attack intrusion detection. The superiority of the network APT attack intrusion detection result is obvious.

5. conclusions

The modeling of network APT attack intrusion detection is an important network security defense technology. The current intrusion detection model of network APT attack can not accurately describe the intrusion behavior of APT attack, resulting in the unsatisfactory intrusion detection results of network apt attack. Therefore, this paper proposes a machine learning algorithm based on network APT attack intrusion detection model. The support vector machine is used to fit the mapping relationship between network APT attack intrusion detection characteristics and network APT attack intrusion behavior, and a network APT attack intrusion detection model is established based on the mapping relationship. The experimental results show that the model can not only accurately identify the intrusion behavior of network APT attacks, but also the detection speed is quite fast. It obtains better intrusion detection results of network APT attacks than other models, and has a wide application prospect.

References

[1] Tang, J.C., Ming, X., Fu, S.J. (2018) A scheduling optimization technique based on reuse in spark to defend against apt attack. Tsinghua Science and Technology., 23: 550-560.

[2] Ma, Z., Li, Q. (2019) Discovering Suspicious APT Families Through a Large-Scale Domain Graph in Information-Centric IoT. IEEE Access., 7: 13917-13926.

[3] Alshamrani, A., Myneni, S., Chowdhary, A., et al. (2019) A Survey on Advanced Persistent Threats: Techniques, Solutions, Challenges, and Research Opportunities. IEEE Communications Surveys & Tutorials., 21: 1851-1877.

[4] Yu, S., Li, X.F., Zhang, X.L., et al. (2019) The OCS-SVM: An Objective-Cost-Sensitive SVM With Sample-Based Misclassification Cost Invariance. IEEE Access., 7: 118931-118942.

[5] Xu, X.H., Zuo, W.M., Lin, L., et al. (2018) F-SVM: Combination of Feature Transformation and SVM Learning via Convex Relaxation. IEEE Transactions on Neural Networks and Learning Systems., 29: 5185-5199.

[6] JayaBrindha, G., Gopi Subbu, E.S. (2018) Ant Colony Technique for Optimizing the Order of Cascaded SVM Classifier for Sunflower Seed Classification. IEEE Transactions on Emerging Topics in Computational Intelligence., 2: 78-88.

IOP Conf. Series: Materials Science and Engineering **799** (2020) 012030 doi:10.1088/1757-899X/799/1/012030

Research on WT-SVD Bilayer Filter Denoising for Downhole Signal

Yuyan Yang[1], Aiqing Huo[2] and Feng Qi[3]

[1]School of Electronic Engineering, Xi'an Shiyou University, Xi'an, Shaanxi, 710000, China

[2]School of Electronic Engineering, Xi'an Shiyou University, Xi'an, Shaanxi, 710000, China

[3]School of Electronic Engineering, Xi'an Shiyou University, Xi'an, Shaanxi, 710000, China

[1]2682125672@qq.com

[2]huoaiqing@sina.com

[3]562593663@qq.com

Abstract. Aiming at the problem of singularity and mutation points after wavelet denoising with improved layered threshold when denoising the downlink signal received in the downhole in rotary steering drilling, a bilayer filtering denoising method based on singular value decomposition is proposed. This method combines wavelet denoising and singular value decomposition denoising, and adopts the WT-SVD bilayer filtering denoising method to construct a matrix of the wavelet improved layered threshold denoising signal, and then performs singular value decomposition and recovery on the matrix signal. The number of selected singular values and the structure of the reconstruction matrix were determined through experiments, and the signal waveforms, signal-to-noise ratio, root mean square error and correlation coefficients of three different denoising methods including singular value decomposition, wavelet improved layered threshold and WT-SVD bilayer filtering are compared and studied. The final experimental results show that the WT-SVD bilayer filtering denoising method is effective.

1. Introduction

In the rotary steering drilling system, the surface monitoring device uses a variable drilling fluid displacement method to send a control instruction to the downhole steering tool via a downward channel. The downhole signals constitute the control instructions in the form of "three drops and three rises" pulse width coding [1]. The downlink communication receiving device in the downhole steering tool obtains the corresponding change in the drilling fluid displacement by detecting and collecting the output voltage of the downhole mud motor. After identifying and interpreting the generation time and sequence of the pulse edges of the downlink signal and the pulse widths of the five instruction codes, an accurate downlink control command word is obtained [2]. Noise is generated due to the environmental interference during the transmission of the transmitted signal. Therefore, denoising the received downhole signal has become a key technology for obtaining control instructions.

The wavelet threshold denoising method proposed by Donoho proved that there is an optimal global threshold in the wavelet transform process, but the signal after denoising has delay and oscillation [3-4]. In addition, many research on the reasonable selection of thresholds and threshold functions shows that different thresholds and threshold functions have different effects on the denoising process [5-6]. The improved layered threshold denoising method is a more effective and flexible denoising method after improving the threshold function. Although this method basically removes the noise, it still fails to solve the problem of singularity and mutation points in the downhole received and transmitted signals, and these singular points and abrupt points will seriously affect the recognition of the pulse width signal [7].

2. WT-SVD bilayer filtering denoising method

WT-SVD (Wavelet Transform-Singular Value Decomposition) bilayer filtering denoising is a wavelet denoising method based on the singular value decomposition, which combines wavelet transform (WT) denoising and singular value decomposition (SVD) denoising to achieve the signal's denoising requirements. The principle of denoising is as follows: wavelet denoising is performed on the downhole signal, and then the singular value decomposition method is used to further process the denoised signal.

2.1. Wavelet layered threshold denoising

Compared with the global threshold, the layered threshold denoising method is based on the correlation of the wavelet coefficients to determine the appropriate threshold denoising of each layer. It will not seriously "kill" the wavelet coefficients and the high-frequency part of useful signal, and can ensure the integrity of the signal [8]. Therefore, in view of the shortcomings of the soft threshold, hard threshold, and global threshold methods, and the differences in noise coefficients at different scales after wavelet decomposition, this paper uses an improved layered threshold denoising method in the first process of filtering and denoising to ensure smoothness after the threshold is applied transition [9]. The expression of the improved threshold function used in this method is shown in equation (1).

$$\hat{w}_{(j,k)} = \begin{cases} \operatorname{sgn}\left(w_{(j,k)}\right)\left(\left|w_{(j,k)}\right| - \dfrac{\lambda^2}{\left|w_{(j,k)}\right| + \lambda e^{\sqrt{\left|w_{(j,k)}\right|-\lambda}}}\right) & \left|w_{(j,k)}\right| \geq \lambda \\ 0 & \left|w_{(j,k)}\right| < \lambda \end{cases} \tag{1}$$

When the absolute value of the wavelet coefficient is less than the critical threshold, the wavelet coefficient is directly set to 0 as a conventional threshold function. When the absolute value of the wavelet coefficient is greater than or equal to the critical threshold, a new expression is used to shrink the wavelet coefficient. Figure 1 shows a comparison of soft threshold, hard threshold, and improved threshold functions.

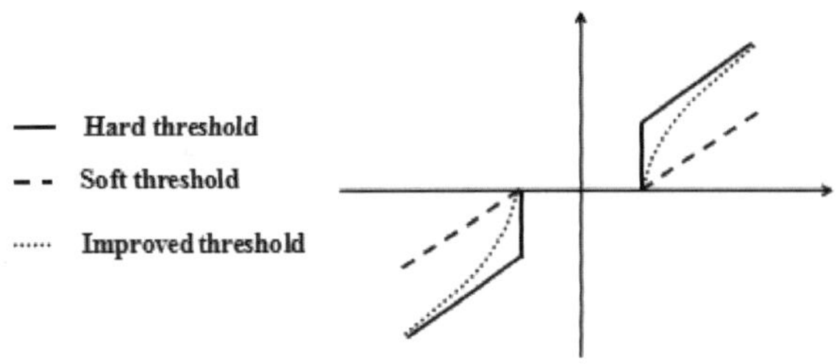

Figure 1. Comparison of soft, hard and improved threshold functions.

The improved threshold function is between soft and hard threshold functions, and has continuity and flexibility in the wavelet domain, overcomes the discontinuity and reconstruction-induced oscillations of the hard threshold function, and the constant deviation of the soft threshold function [10-11].

2.2. Singular value decomposition denoising

The singular value decomposition denoising method mainly reflects the correspondence between the information components of the signal and the singular values of the reconstruction matrix. Smaller singular values reflect components with less information in the signal, while larger singular values reflect components with larger amount of information [12]. Therefore, it is necessary to reasonably remove signal components that reflect small singular values and use large singular values to recover the original signal. It can not only filter out the noise interference, but also keep the characteristic information of the original signal from being destroyed, and finally obtain the denoised signal more accurately. The basic steps of singular value decomposition and denoising are:

(1) Constructing a matrix by using the downlink signal $S=(s(1),s(2),s(3),...,s(N))$;

(2) Perform singular value decomposition on the matrix E to obtain $U=R^{m\times m}$, $\Sigma\in R^{m\times n}$ and $V\in R^{n\times n}$, then $E=U\cdot\Sigma\cdot V^T$;

(3) Select a larger singular value for the singular value matrix, and obtain a processed matrix Σ^*;

(4) Use the matrices $U=R^{m\times m}$, $V\in R^{n\times n}$ and Σ^* to reconstruct the matrix to obtain $E^*=U\cdot\Sigma^*\cdot V^T$;

(5) A new data sequence S^* is constructed from all the elements in the first row of the matrix E^* and the elements in the second row from the n-th column to the m-th row and the n-th column, that is the signal after denoising.

The wavelet denoising method using only the improved layered threshold can basically remove the noise, but still cannot solve the problem of singularity and mutation points in the downhole received and transmitted signal. Therefore, this paper combines improved layered threshold denoising and singular value decomposition denoising, and performs bilayer filtering denoising to remove the remaining noise in the signal to achieve better denoising effect.

3. Key points of bilayer filtering denoising

The key steps in applying bilayer filtering denoising to signal processing to reduce signal noise include two aspects: selecting the number of non-zero singular values and determining the structure of the reconstruction matrix.

3.1. Selection of the number of singular values

Although the singular value decomposition method has certain denoising advantages, there are still some difficulties in selecting the number of singular values. If the number of selected singular values is large, there will still be a lot of noise in the denoised signal; on the contrary, when the number is small, the original signal will be destroyed, and cause signal distortion[13].To this end, different schemes are used to select the number of singular values, such as: ① the characteristic mean method; ② the differential spectrum method; ③ the singular median method. Through a large number of experiments, it is concluded that when $L\le n$, $p=L$; when $L>n$, there is always $p=2n$. Among them, L represents the number of rows selected by the reconstruction matrix, n represents the number of dominant frequencies after one-dimensional signal fast Fourier transform, and p represents the number of singular values selected by the reconstruction matrix after singular value decomposition.

3.2. Determination of the reconstruction matrix structure

When selecting the number of matrix rows, the general method is to try it out based on specific signals, observe and analyze the characteristics of each set of singular value matrices, and expect to obtain a satisfactory set of singular value matrix components [14]. This trial-and-error method requires a lot of calculations, and it is demanding on the signal analysis experience of researchers.

For the matrix $E = U \cdot \Sigma \cdot V^T$ constructed by the downhole signals, the amount of information δ_i contained in each component Σ_i signal obtained by the separation is determined by the corresponding singular value σ_i. The smaller σ_i, the smaller the corresponding information amount Σ_i, where the information amount Σ_i contained in δ_i can be calculated by equation (2).

$$\delta_i = \frac{\sigma_i}{\sigma_1 + \sigma_2 + \cdots + \sigma_i}, \quad i = 1,2,\ldots,r \tag{2}$$

Signal components with relatively small amounts of information have little effect in practice, so the number of rows and columns of a matrix can be appropriately selected according to this measure.

The specific method of selecting the number of rows and columns of the matrix is as follows: Select different rows in turn to construct the matrix, use the singular value of the corresponding matrix to obtain the components of each information amount, and then analyze the connections and changes between the components. If it starts with a certain information component δ_i, and the subsequent information components gradually approach zero in order, the remaining components after the i-th information component have little effect, so the number of rows $m = i$ of the matrix can be determined. The advantage of this method is that it does not need to calculate and analyze the amount of information of each component, and only uses singular values to judge. It greatly reduces the amount of calculation based on a comprehensive inspection of the information contained in all singular values. Finally, for the calculation of the number of columns, the continuous truncation method is $n = \mathrm{int}(N/m)$, and the Hankel matrix method is $n = N - m + 1$.

4. Experiment and result analysis

4.1. Determination of the number of singular values

Figure 2 is a group of signals received downhole, including a signal sequence of 512 sampling points, that is, $N = 512$. Perform fast Fourier transform on the signal sequence to obtain the number and distribution of the main frequencies. The spectrum of the signal is shown in Figure 3. The number of main frequencies of the signal can be obtained. The main frequency is the number of points indicated by the arrows in the figure, that is, the number of points with large amplitude.

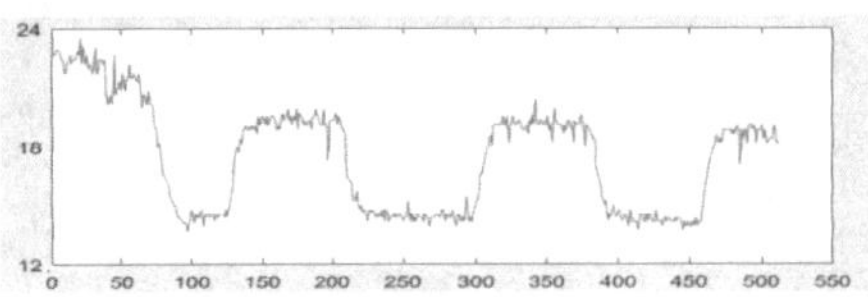

Figure 2. Downlink signal received downhole.

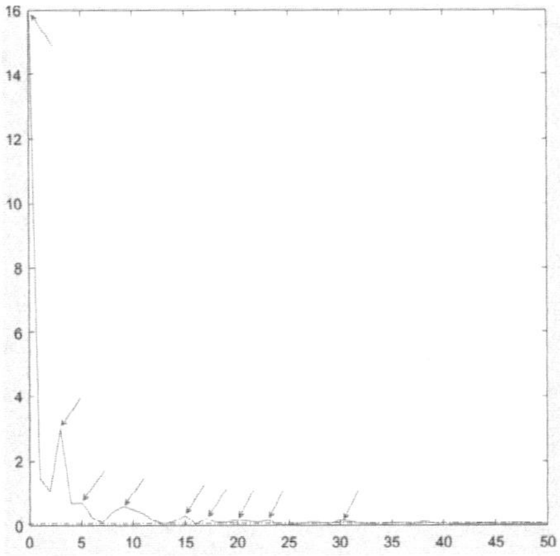

Figure 3. Signal spectrum.

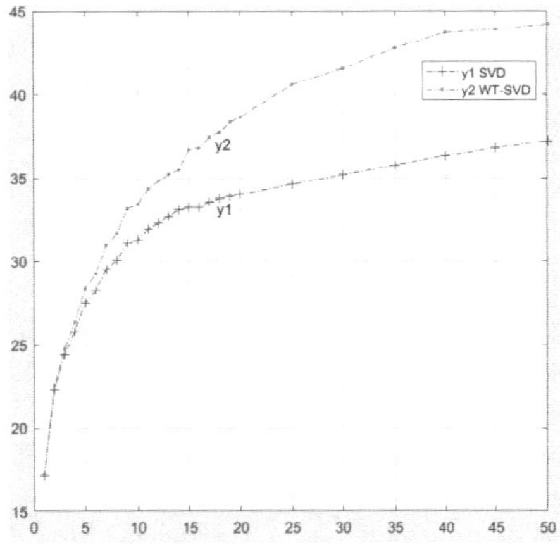

Figure 4. Signal-to-noise ratio curve at different singular values.

The biggest advantage of singular value decomposition is not only the ability to remove noise points, but the more important point is to simplify the data. Based on the singular value matrix data results obtained by performing singular value decomposition on the transmitted signal, combining the characteristics of singular value decomposition, it is concluded that the singular values decay rapidly from large to small, and 256 singular values are obtained after decomposition. According to formula (2), it can be obtained that the sum of the singular values of the first 10% and even the first 1% accounts for more than 95% of the sum of the singular values. These singular values basically include the information contained in the signal.

Figure 4 is the signal-to-noise ratio (SNR) curve when the WT-SVD reconstruction matrix takes different singular values. It shows the trend of the SNR of the two noise reduction methods of singular value decomposition (SVD) and wavelet transform-singular value decomposition (WT-SVD). It can be seen that the signal-to-noise ratio curve is increasing with the number of singular values. When the number of singular values reaches 18, the upward trend is gradually slower. However, we cannot judge the denoising effect based on this criterion alone. Therefore, by selecting four representative singular values of 10, 18, 25, and 40, a waveform comparison experiment is performed on the denoised signal.

Figure 5 shows the comparison of signal waveforms after denoising with different singular values. It can be seen that the signal waveform is smoothest when the number of singular values is 18, and the singular points are basically removed. Combining Figures 4 and 5, it can be concluded that the denoising effect is best when the number of singular values is 18.

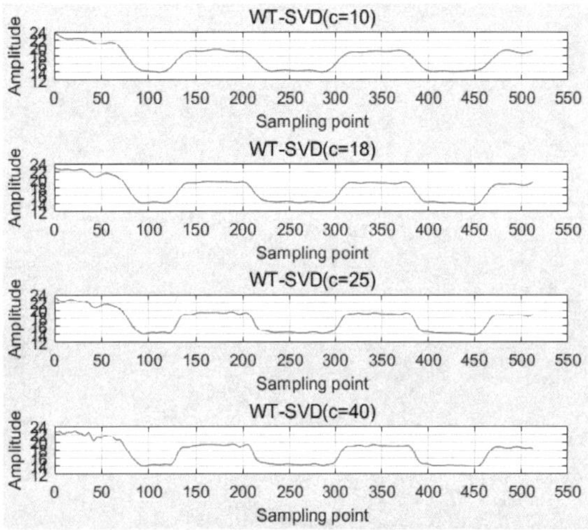

Figure 5. Signal waveforms after denoising with different singular values.

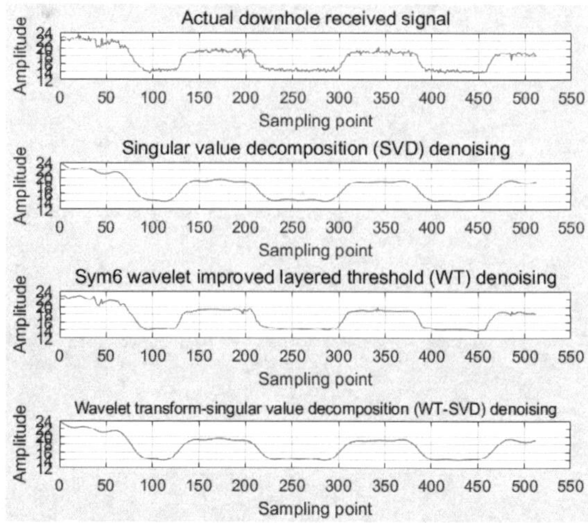

Figure 6. Denoised signal waves with three different denoising methods.

4.2. Determination of the reconstruction matrix structure

In the rotary steering drilling system, the number of sampling points $N = 512$ of the downstream signal is sampled. When using the Hankel matrix method to construct the matrix, the optimal number of rows should be 256 rows and the number of columns should be determined as 257 columns when considering each sampling point. The reason is that when the number of lines is $L = 2,3,\ldots,255$ in sequence, the subsequent sampling points in the signal will not be obtained, and its number is $512 - 2L$, which will eventually cause some useful information to be removed. When the number of lines is $L = 257,258,\ldots,511$, the number of signal sampling points will be insufficient. There are a total of $2L - 512$ sampling points missing, and repeated acquisition of the signal sampling point values will result in information redundancy. Therefore, the denoising effect is best when the number of rows of the reconstruction matrix is half of the number of sampling points, and it is finally determined that the number of rows of the reconstruction matrix is $L = 256$.

4.3. Comparative experiment of multiple denoising methods

In order to verify the denoising effect of the WT-SVD bilayer filtering denoising method, the singular value decomposition (SVD) denoising method, the Sym6 wavelet improved layered threshold (WT) denoising method, and the wavelet transform-singular value decomposition (WT-SVD) denoising method have been compared with simulation experiments. Figure 6 shows the signal waveforms of the three methods after denoising.

From the signal waveform in Figure 6, it can be seen that the original signal is greatly affected by noise. Pure singular value decomposition and denoising cannot effectively remove noise. After Sym6 wavelet improved layered threshold denoising, most noise has been filtered out. However, there are still some singularities, indicating that the noise has not been completely eliminated. The singular value decomposition denoising method is used to denoise the signal after the improved layered threshold denoising, which basically eliminates all the noise contained in the signal, and the signal waveform is smoother and the distortion is smaller.

Table 1. SNR, RMSE and correlation coefficient of different noise reduction methods.

	Signal-to-noise ratio(SNR)	Root mean square error(RMSE)	Correlation coefficient ρ
SVD	33.7227	0.2601	0.9910

WT	35.8822	0.2028	0.9946
WT-SVD	37.7109	0.1643	0.9964

From the data in Table 1, it can be seen that the signal-to-noise ratio of the signal is 37.7109, the root mean square error is 0.1643, and the correlation coefficient is 0.9964 after WT-SVD bilayer filtering denoising. These three indicators have been improved compared to the previous two denoising methods, indicating that the WT-SVD bilayer filtering denoising method is suitable for the analysis and processing of the received signal downhole, and can achieve better denoising effects. The signals denoised by bilayer filtering can be more conducive to accurate identification of the three-drop and three-liter pulse widths.

5. Conclusions

This paper proposes a bilayer filtering denoising method based on singular value decomposition. By combining improved layered threshold denoising and singular value decomposition denoising, a WT-SVD bilayer filtering denoising method is obtained. After applying it to the actual downhole received signal, the improved layered threshold denoising is applied to the downlink signal, and then the matrix is constructed using Hankel matrix method. After determining the number of singular values and the structure of reconstruction matrix, the matrix is subjected to singular value decomposition and reconstruction to recover the signal. The experiment compares the WT-SVD bilayer filtering denoising method with other methods, and verifies that the method can effectively remove singularity and mutation points in noise. It has a better denoising effect and provides a guarantee for the coding identification of the downhole received signal.

Acknowledgments

The author is very grateful to the mentor, Ms. Huo Aiqing, for her meticulous guidance, colleague Qi Feng's careful help, and the support of colleagues in the laboratory.

References

[1] Tang Nan, Huo Aiqing, Wang Yuelong, Cheng Weibin. (2010) Development of Downlink Communication Receive Function of Rotary Steering Drilling System[J]. Acta Petrolei Sinica, 31(1): 157-160.

[2] Li Qi, Peng Yuanchao, Zhang Shaohuai, Liu Zhikun. (2007) Research on Downhole Transmission Technology of Rotary Steering Drilling Signals[J]. Acta Petrolei Sinica, 28(04): 108-111.

[3] Donoho D L, Johnstone I M. (1994) Ideal spatial adaptation via wavelet shrinkage[J]. Biometrika, 81(12): 425-455.

[4] Donoho D L. (1995) De-noising by soft-thresholding. IEEE Transactions on Information Theory, 41(3): 613-627.

[5] JOY J, PETER S, JOHN N. (2013) Denoising using soft thresholding[J]. International Journal of Advanced Research in Electrical, Electronics and Instrumentation Engineering, 2(3): 1027-1032.

[6] Bruno C.Z. Honório, Rodrigo D. Drummond, Alexandre C. Vidal, Alexandre C. Sanchetta, Emilson P. Leite. (2012) Well log denoising and geological enhancement based on discrete wavelet transform and hybrid thresholding[J]. Energy, Exploration & Exploitation. 30(3): 417-434.

[7] Huo Aiqing, Qi Feng, and Zhu Bing. (2019) Improved segmented threshold noise reduction method for downhole received signals[J]. Logging Technology, 43(01): 92-96.

[8] Mortezanejad R, Gholami A. (2016) Optimization of wavelet- and curvelet-based denoising algorithms by multivariate SURE and GCV[J]. Journal of Geophysics and Engineering, 13(3): 378-390.

[9] Zhang Junjie, Feng Yu. (2018) Research on wavelet denoising of partial discharge signal of power cable with layered threshold[J]. Electronic Measurement Technology, 41(05): 68-72.
[10] Zhao Hongtu, Liu Fang, Hou Shouming. (2017) Frequency wavelet power signal denoising based on improved threshold function[J]. Measurement & Control Technology, 36(08): 20-25.
[11] Cui Yan, Chen Shijun, Qu Meng, He Shanhong. (2017) Research and Application of New Threshold De-noising Algorithm for Monitoring Data Analysis in Nuclear Power Plant[J]. Journal of Shanghai Jiaotong University (Science), 22(3): 355-360.
[12] Qian Zhengwen, Cheng Li, Li Yinghong. (2011) Signal Denoising Method Using Singular Value Decomposition[J]. Vibration.Test and Diagnosis, 31(04): 459-463 + 534-535.
[13] Yan Qiangqiang, Sheng Shouzhao, Zhou Jun, Lin Qunzhen. (2018) Research on an Improved Singular Value Decomposition Noise Reduction Method[J]. Electro-Optics and Control, 25(09): 22-25 + 41.
[14] Xu Yankai, Shuang Kai. (2014) Transient Weak Signal Detection Based on Wavelet Singular Value Decomposition[J]. Journal of China University of Petroleum (Edition of Natural Science), 38(03): 181-185.

IOP Conf. Series: Materials Science and Engineering **799** (2020) 012031 doi:10.1088/1757-899X/799/1/012031

Research on Cow Gesture Recognition Based on Multi-label Chain GBDT and LR Fusion

ZhiQiang Zhao, Zheng Zhu and ShunAn Ye

Automation College, Chongqing University of Posts and Telecommunications, Chongqing 400065, China

1050176902@qq.com

Abstract. In this paper, by wearing a nine-axis motion sensor on the neck of a cow, the three-axis acceleration, three-axis angular velocity and three-axis angular motion data of the cow are collected, and the cow's sports behavior is divided into six postures: standing, resting, walking, and so on, which are used as secondary marks. After the data is balanced and sampled, the six types of gesture labels are divided into three categories as the first level labels, namely, static, small motion and large motion. Gradient Boosting Decison Tree and logistic regression fusion algorithm are used to learn and predict the first-level label, and the prediction result is used as the new feature to learn and predict the second-level label, so as to achieve the purpose of classification of cow gesture recognition. The experimental results show that the average accuracy of the model is 93.6%, and the AUC value is 96.1%. This study can provide an effective basis for dairy cow on estrus and health monitoring.

1. Introduction

Beef and milk are two important nutrient supplies on the Chinese table, so how to effectively reduce the cost of dairy farming and increase the milk production is very important. Studies have shown that cows with disease or estrus, the amount of behavioral exercise will have corresponding fluctuations[1]. Cows in estrus are often more irritated than unoccupied cows[2], and cows in estrus are twice as likely to travel each day. In addition, more than 90% of cows in estrus will have the behavior of climbing across other cows[3-5].

2. Materials and Method

The nine-axis data acquisition node designed in this paper is shown in Fig.1. The controller uses the low-power STM32F103ZET6 minimum system module, which has one SDIO interface and up to five USART interfaces. The SDIO interface is used for an external SD card module, and the two USART interfaces are respectively connected to the JY901 nine-axis attitude sensor module and the ESP8266WIFI module. The JY901 nine-axis attitude sensor integrates an attitude solver with a dynamic Kalman filter algorithm to output 3D acceleration, 3D angular velocity and 3D angular data. The ESP8266WIFI module is used for node network calibration to enable nodes to obtain precise time points. The data acquisition node collects the nine-axis raw data of the cow at a sampling rate of 10hz, and saves the real-time network time to the SD card, which is used to establish and predict the machine learning model.

Fig.1 Nine-axis data acquisition node

This article divides the six types of sports behaviors of cows into six categories and labels them. Since the amount of data on the standing and resting posture of the cow is too large, and the amount of data for the crawling and running posture is too small, the raw data needs to be balancedly sampled. The balanced data is added to the first-level label, and the six cows' sports postures are roughly divided into three postures, as shown in Fig.2. The GBDT and LR fusion algorithm is used to learn and predict the first-level label, and the prediction result is added as a new feature to the feature vector group. Finally, the second-level label is learn and predicted.

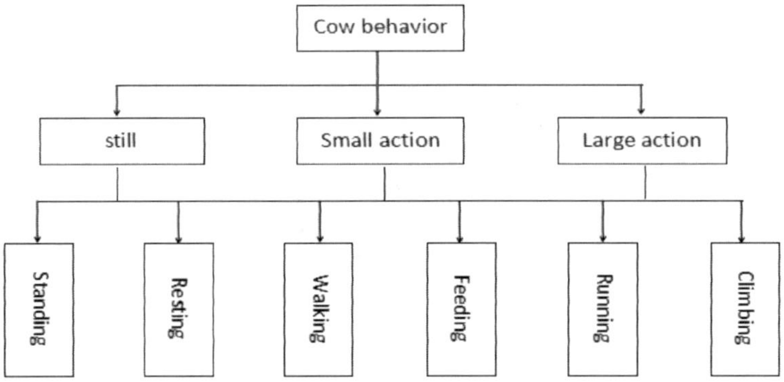

Fig.2 Classification of cow behavior

3. Data Preprocessing and Model Building

3.1. Data preprocessing

Take the nine-axis raw data from the SD card and put the six types of behavior tags on the original data according to the timestamp:

$$Y_2 = [S\tan ding:1, \mathrm{Re}\,sting:2, Walking:3, Feeding:4, Running:5, C\mathrm{lim}bing:6] \tag{1}$$

The triaxial acceleration (a_1, a_2, a_3) raw data is then derived as a 19-dimensional feature vector:

$$[\overline{a}_1, \overline{a}_2, \overline{a}_3, a_{1\max}, a_{2\max}, a_{3\max}, a_{1\min}, a_{2\min}, a_{3\min}, \Delta_{a1}, \Delta_{a2}, \Delta_{a3}, |a_1|_{\max}, |a_2|_{\max}, |a_3|_{\max}, \gamma_{a1a2}, \gamma_{a1a3}, \gamma_{a2a3}, \sum(\overline{a}_1, \overline{a}_2, \overline{a}_3)] \tag{2}$$

Among them, $[\overline{a}_1, \overline{a}_2, \overline{a}_3]$ represents the average of the three directions, reflecting the trend of cow movement in the three-axis direction; $[a_{1\max}, a_{2\max}, a_{3\max}]$ represents the maximum value of the three directions, reflecting the instantaneous acceleration change of the cow's strenuous exercise; $[a_{1\min}, a_{2\min}, a_{3\min}]$ represents the minimum of the three directions; $[\sum(\overline{a}_1, \overline{a}_2, \overline{a}_3)]$ represents the sum of the average values of the three directions, comprehensively reflects the magnitude of the change in the three-axis direction; $[\Delta_{a1}, \Delta_{a2}, \Delta_{a3}]$ represents the standard deviation of the three directions, reflecting the maximum instantaneous trend of cow movement; $[\gamma_{a1a2}, \gamma_{a1a3}, \gamma_{a2a3}]$

IOP Conf. Series: Materials Science and Engineering **799** (2020) 012031 doi:10.1088/1757-899X/799/1/012031

indicates the correlation of the three directions The coefficient reflects the degree of correlation between the three axes of the cow's movement; $[|a_1|_{max}, |a_2|_{max}, |a_3|_{max}]$ indicates the largest absolute value of the three directions, reflecting the maximum change in the triaxial direction of the cow's movement.

Similarly, the triaxial angular velocity (w_1, w_2, w_3) raw data is derived as a 19-dimensional eigenvector:

$$[\overline{w}_1, \overline{w}_2, \overline{w}_3, w_{1\max}, w_{2\max}, w_{3\max}, w_{1\min}, w_{2\min}, w_{3\min}, \Delta_{w1}, \Delta_{w2}, \Delta_{w3}, |w_1|_{\max}, |w_2|_{\max}, |w_3|_{\max}, \gamma_{w1w2}, \gamma_{w1w3}, \gamma_{w2w3}, \Sigma(\overline{w}_1, \overline{w}_2, \overline{w}_3)] \tag{3}$$

The triaxial angle $[\theta_1, \theta_2, \theta_3]$ raw data is derived as a 19-dimensional eigenvector:

$$[\overline{\theta}_1, \overline{\theta}_2, \overline{\theta}_3, \theta_{1\max}, \theta_{2\max}, \theta_{3\max}, \theta_{1\min}, \theta_{2\min}, \theta_{3\min}, \Delta_{\theta 1}, \Delta_{\theta 2}, \Delta_{\theta 3}, |\theta_1|_{\max}, |\theta_2|_{\max}, |\theta_3|_{\max}, \gamma_{\theta 1\theta 2}, \gamma_{\theta 1\theta 3}, \gamma_{\theta 2\theta 3}, \Sigma(\overline{\theta}_1, \overline{\theta}_2, \overline{\theta}_3)] \tag{4}$$

During the data collection process, the standing and resting postures of the cows often appear, and the two poses of running and climbing are less likely to occur. Therefore, the experimental data set is extremely unbalanced. In order to balance the data set, the following steps are required:

S1. according to the Y_2 label, the original sample is divided into a majority class Y_{2s} and a minority class Y_{2l};

S2. Calculate N neighbors of each sample, and Δ is the number of N neighbors that are inconsistent with their own categories;

S3. For any sample Y_i, if $Y_i \in Y_{2s}$ and $\Delta > N/2$, delete sample Y_i .if $Y_i \in Y_{2l}$, increase the number of minority Y_{2l};

S4. Synthesizing a new sample increases the number of a few types of Y_{2l}, and the number of new samples added is:

$$m = (\max Y_{2s} - \min Y_{2l}) \bullet \frac{1}{4} \tag{5}$$

S5. generating m new samples according to the samples in the minority class Y_{2l}, determining the number of new samples of the i-th sample in the minority class Y_{2l}, denoted as $n = m \cdot k / \sum_{i=l}^{Y_{2l}} k_i$, synthesizing the new sample according to the i-th sample is expressed as: $\hat{x} = x_i + (\overline{x}_i - x_i)\lambda$;

Where $\max Y_{2s}$ represents the maximum number of samples in the majority class Y_{2s}, $\min Y_{2l}$ represents the number of samples with the least number of samples in the minority class Y_{2l}, $\overline{x}_i$ represents any point in the nearest neighbor of the sample point x_i, λ is the random factor value (0, 1), k_i is an intermediate parameter, denoted as $k_i = \frac{\Delta}{N}$.

Then add the balanced data set to the first level label $Y_1 = [still:1, Small\ \ action:2, Large\ \ action:3]$, where Y_1 is determined as follows:

$$\begin{cases} Y_1 = 1 \ , \ Y_2 = [1,2] \\ Y_1 = 2 \ , \ Y_2 = [3,4] \\ Y_1 = 3 \ , \ Y_2 = [5,6] \end{cases} \tag{6}$$

Second, the maximum and minimum normalization of the data set;

Finally, a recursive feature elimination method [6] is used for the label A to select a 9-dimensional feature from the 57-dimensional features. Using the SVM support vector machine model as the evaluation model, one feature is deleted from the 57-dimensional feature to train a model, then the feature is returned, and the next feature is deleted... until 57 SVM support vector machine models are obtained. The obtained 57 model scores were sorted in order, and the 9 models with the smallest scores were selected. The nine features corresponding to the nine models are selected from the five-dimensional features.

Summary The above data preprocessing process is:

(1) The 9-dimensional raw data set is derived into 57 dimensions and labeled Y_2;

(2) Balanced sampling processing of 57-dimensional data sets;

(3) Add a first-level label Y_1 and normalize it;

(4) Select the optimal 9-dimensional feature by using the recursive feature elimination method.

3.2. Model establishment

The preprocessed data set is divided into a first training set and a first test set, a second training set and a second test set.

The first training set and its corresponding label Y_1 are trained to obtain a first Gradient Boosting Decison Tree. The first test set is predicted using the first Gradient Boosting Decison Tree to obtain a first prediction result. The first test set and the first prediction result are combined as a first chained data set. According to the first chain data set and the corresponding label Y_2, the second Gradient Boosting Decison Tree is trained.

The process of training the first Gradient Boosting Decison Tree and the second Gradient Boosting Decison Tree includes:

Gradient Boosting Decison Tree is:

$$f(x) = f_0(x) + \sum_{t=l}^{T} \sum_{j=1}^{J} c_{tj} I(x \in R_{tj}) \tag{7}$$

The loss function of the Gradient Boosting Decison Tree is:

$$L(y, f(x)) = -\sum_{k=1}^{K} y_k \log p_k(x) \tag{8}$$

In the t-th training process, the negative gradient error of the corresponding category l of the i-th sample is calculated, expressed as:

$$r_{til} = y_{il} - p_{l,t-1}(x_i) \tag{9}$$

Using $(x_i, r_{ti})(i = 1,2...m)$ to fit the CART regression tree, the t-th regression tree is obtained, and the corresponding leaf node region $R_{tj}, j = 1,2..., J$, for each sample in the leaf node, seeks to minimize the loss function, that is, fit the best output value C_{tj} of the leaf node, expressed as:

$$c_{tj} = \frac{K-1}{K} \frac{\sum_{x_i \in R_{tjl}} r_{til}}{\sum_{x_i \in R_{tjl}} |r_{til}|(1 - r_{til})} \tag{10}$$

The second training set is input to the first Gradient Boosting Decison Tree to obtain a second prediction result. The second training set and the second prediction result are combined as a second chained data set. The second chain data set is input to the second Gradient Boosting Decison Tree to obtain a leaf node corresponding to the second Gradient Boosting Decison Tree. The leaf nodes are individually heat-coded to obtain a sequence of leaf nodes.

The first logistic regression model is trained by using the 40% leaf node and the Y_1 tag corresponding to the leaf node as a training set. Using the first logistic regression model to predict 60% of the leaf nodes, the prediction results are obtained. Combine 60% of the leaf nodes and their

IOP Conf. Series: Materials Science and Engineering **799** (2020) 012031 doi:10.1088/1757-899X/799/1/012031

predictions as a third chained data set. The second logistic regression model is trained with the third chain data set and its corresponding Y_2 label.

The second test set was used to test the validity of the entire model. Identify and classify the six behavioral postures of cows.

The multivariate logistic regression is solved as follows:

$$\begin{cases} P(y=k \mid x,\theta) = e^{x\theta k} / 1 + \sum_{t=1}^{K-1} e^{x\theta k} & k = 1,2,...K-1 \\ P(y=k \mid x,\theta) = 1/1 + \sum_{t=1}^{K-1} e^{x\theta k} & k = 1,2,...K-1 \end{cases} \tag{11}$$

4. Conclusion

On November 1, 2019, a collection of experimental data was collected from a dairy farm in Tumut Zuoqi, Hohhot, Inner Mongolia. By comparing the video and sensor timestamps, the selected time sliding window is 2s, and a total of 34,315 experimental data are cropped. Because the amount of experimental data for each category varies widely, the experimental data needs to be balanced. The amount of standing and restroom data is too large for downsampling. Upsampling is performed with a small amount of data for feeding and crawling. The results are shown in Table 1.

Table 1 Data Balanced Sampling Results

Behavior classification	Standing	Resting	Walking	Feeding	Running	Climbing
Before data balance	**12988**	**10389**	4758	**1830**	3870	**480**
After data balance	**8000**	**6000**	4758	**3000**	3870	**900**

After the original data is balanced and normalized, the SVM is used to evaluate the model, and the recursive feature elimination method is used to reduce the 57-dimensional features to 9 dimensions, and the optimal 9-dimensional feature vector $[\bar{a}_2, \bar{a}_3, \theta_{3\max}, a_{2\min}, \Delta_{a3}, \Delta_{w1}, |a_2|_{\max}, |w_1|_{\max}, \gamma_{w1w3}]$ is obtained. For the GBDT parameters in the model: the weak learner CART regression decision tree number is 100, the random factor is 12; logistic regression parameters: Cs=np. logspace(-4,1,50), cv=3,penalty='l2', Solver='lbfgs', tol=0.01, multi_class='multinomial', random factor is 12.

In this experiment, the average accuracy of the classification and classification of the six behavioral postures of dairy cows was 91.8%. The confusion matrix for each behavioral gesture classification result is Table 2. The ROC curve is shown in Fig.3.

Table 2 Dairy cow behavior posture classification results

Observed \ Prediction	Standing	Resting	Walking	Feeding	Running	Climbing
Standing	**3175**	19	7	10	0	0
Resting	6	**2368**	11	11	0	0
Walking	0	9	**1597**	272	18	0
Feeding	0	0	77	**1116**	7	6
Running	0	11	24	18	**1371**	113
Climbing	0	0	0	20	46	**340**

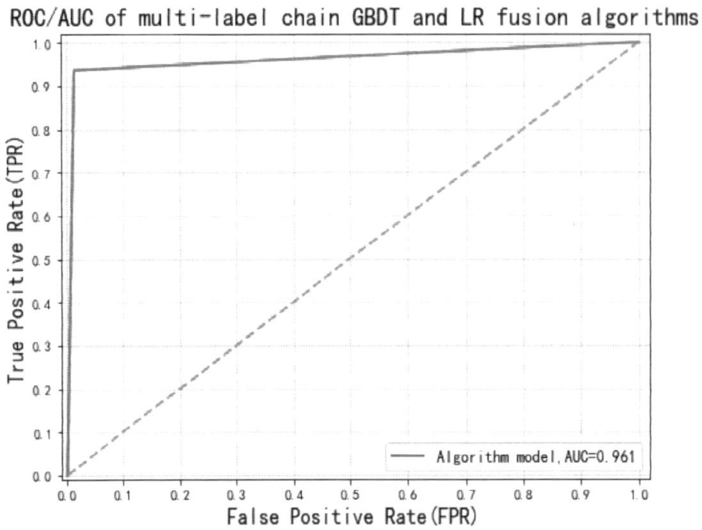

Fig.3 Multi-label chain GBDT and LR fusion algorithm ROC/AUC

In this study, a cow behavior recognition model based on multi-label chain GBDT and LR fusion was used to classify six behavioral characteristics of cow standing, lying, walking, eating, running and climbing. The accuracy rate for standing, lying, eating and running is as high as 90%. The accuracy rates for distinguishing between slow walking and climbing are 84.2% and 83.7%, respectively. The reasons for the analysis may be as follows: On the one hand, the amount of data is too small. Because the cow's crawling behavior is rare, it is difficult to record the first time. On the other hand, the cow's movement posture will appear complex, such as the cow may still be walking while eating. However, the reasons for these two aspects can be solved by long-term cows and long-term collection and testing, which will further improve the accuracy of the model.

References

[1] Li Xiaoshan, Yang Fengli. Effects of Foot and Soul Diseases on Reproductive Performance of Dairy Cows[J]. China Animal Husbandry and Veterinary Medicine,2014,41(05):248-251.

[2] Chapinal, N., et al. "Technical Note: Validation of a System for Monitoring Individual Feeding and Drinking Behavior and Intake in Group-Housed Cattle." Journal of Dairy Science 90.12(2007):0-5736.

[3] Roelofs, J. B., et al. "Various behavioral signs of estrous and their relationship with time of ovulation in dairy cattle." Theriogenology 63.5(2005):0-1377.

[4] Solano, J., et al. "Relationships between social behaviour and mounting activity of Zebu cattle (Bos indicus)." Applied Animal Behaviour Science 94.3-4(2005):197-203.

[5] Diskin, Michael G., and J. M. Sreenan. "Expression and detection of oestrus in cattle." Reproduction Nutrition Development (EDP Sciences) 40.5(2000):481-491.

[6] Guyon, Isabelle, et al. "Gene Selection for Cancer Classification using Support Vector Machines." Machine Learning 46.1-3(2002):389-422.

IOP Conf. Series: Materials Science and Engineering **799** (2020) 012032 doi:10.1088/1757-899X/799/1/012032

Design and Implementation of Unmanned Agricultural Machinery

Ru Jang [1], Farkhod Kasimov [1], Dong Zhang [1*] and Kulyash Kaliyeva[2]

[1]Institute of Automation, Shandong Academy of Sciences, Jinan, 250000, China

[2]Kazakh National Research Technical University, Almaty, 050000, Kazakhstan

[*]Corresponding author's e-mail: jnzd156@163.com

Abstract. The unmanned agricultural machinery can greatly improve operating efficiency and can cover a large area of farmland operations with very little manpower, which saves agricultural investment costs. In this paper, the hardware and software design of unmanned agricultural machinery are introduced, the unmanned agricultural machine is developed, the autonomous driving experiments in different paths are carried out, and the accuracy of the operation is tested. The results of the experiments show that the unmanned agricultural machinery has high tracking accuracy and can meet the needs of field operations.

1. Introduction

Despite the rapid development of China's agricultural machinery industry in recent years, no large-scale agricultural machinery manufacturing enterprise with large scale and strong international influence has been formed, and the overall level of agricultural machinery technology in China has lagged behind, especially in the application of unmanned control technology. The demand for the unmanned of agricultural machinery technology is becoming more and more urgent [1-2].

The Thirteenth Five-Year Plan National Science and Technology Innovation Plan proposed that agricultural transformation and upgrading should be the main target, specifically focus on the development of key technologies of unmanned agricultural equipment [3].Promoting the application of information technology on agricultural machinery is not only in line with the world's development trends, but also in accordance with China's major needs to improve the level of agricultural intelligence, and has great significance for accelerating the development of smart agriculture[4].

In this article, we developed the unmanned agricultural machinery which can realize field information collection and processing, generate electronic maps, plan the optimal driving path, and drive autonomously according to the planned path.

2. System hardware design

The overall design of unmanned agricultural machinery is shown as Figure 1. The system is mainly composed of industrial computer, GPS navigation system, steer and control system [5]. GPS navigation system is mainly composed of GPS receiving device and supporting antenna. By processing the position and attitude data received from the GPS receiver, the boundary of the plot is displayed on the touch screen. The path generation software plans a reasonable operation path based on the situation of the plot and the operation requirements. When the agricultural machinery is operating, it displays the position of the agricultural machinery in real time by dynamic tracking data and realizes automatic navigation.

IOP Conf. Series: Materials Science and Engineering **799** (2020) 012032 doi:10.1088/1757-899X/799/1/012032

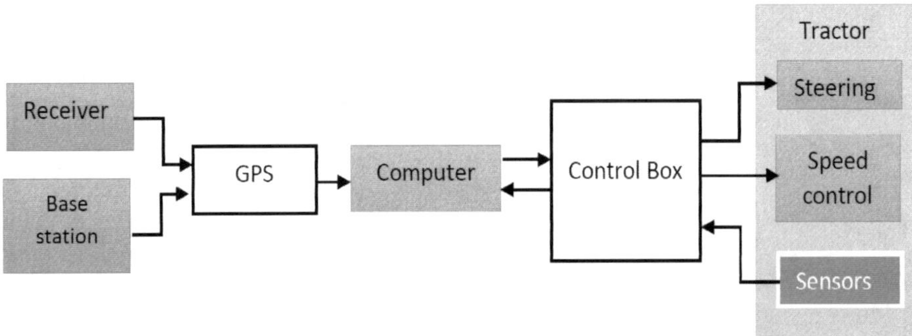

Figure 1. Overall design of unmanned agricultural machinery.

We chose Advantech ARK-3500 industrial computer as the system control centre. This type of industrial computer has rich I / O interfaces, including 4 USB 3.0 and 8 COM ports. The differential positioning accuracy of the selected GPS device can be within 3 cm, the data update rate can reach 50 Hz, and the power consumption of the GPS device is low, which is suitable for field navigation. The ATmega328p chip is selected as the main chip of the steering control system. It has six PWM channels, eight 10-bit ADCs, and a programmable serial USART, the operating speed can reach 20MHz [6]. The circuit design of the steering controller is shown as Figure 2.

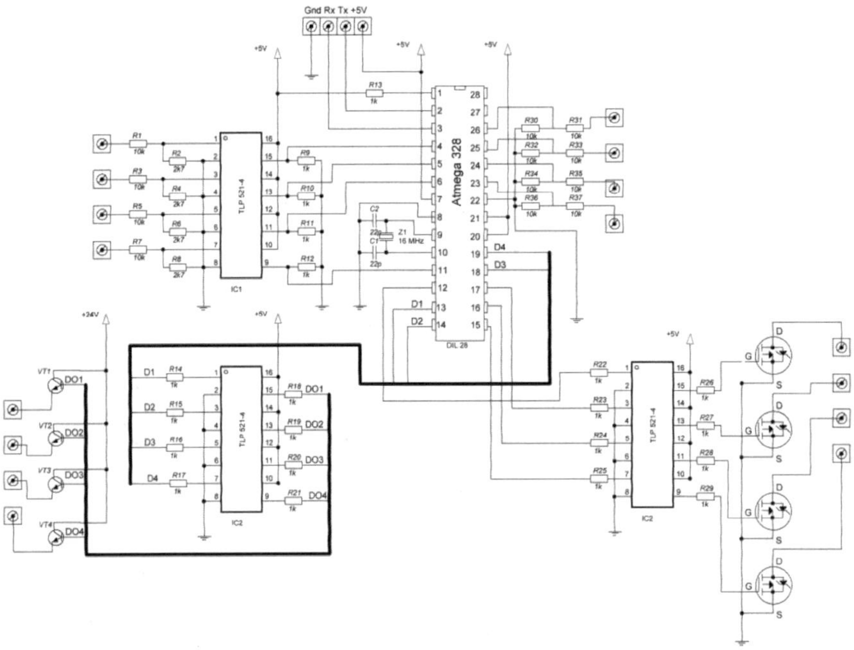

Figure 2. Circuit design of the steering controller.

3. System software design

In the unmanned agricultural machinery system, the industrial computer communicates with the navigation system and the steering control system. The touch screen displays and processes status information of each system, farmland outline maps, and man-machine dialogue information. As for the heading tracking algorithm, we choose to use the incremental PID calculation method. The sum of the heading at the current moment and the estimated heading is used as the system feedback. The input deviation of the controller is the difference between the expected heading and the feedback. The logic block diagram of the heading tracking algorithm is shown as Figure 3.

The unmanned agricultural machinery software is written in C++ language, and the software interface is shown as Figure 4. The software can visually show the deviation between the planned path and the real-time position of the agricultural machinery, and monitor the status of the GPS system, steering system, communication system and other parts.

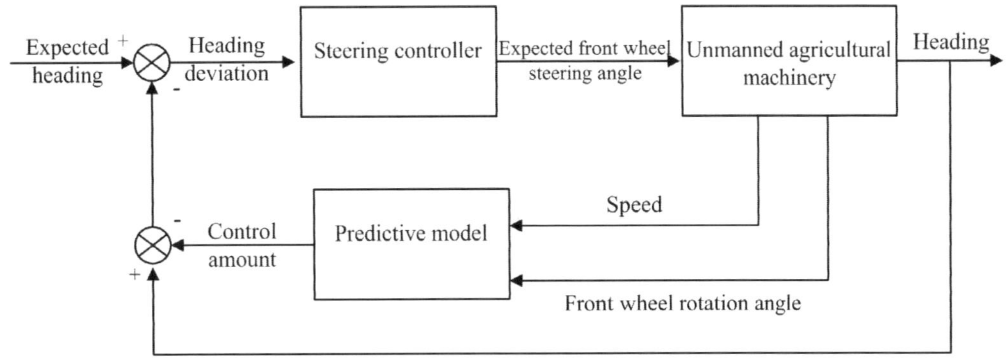

Figure 3. Logic block diagram of the heading tracking algorithm.

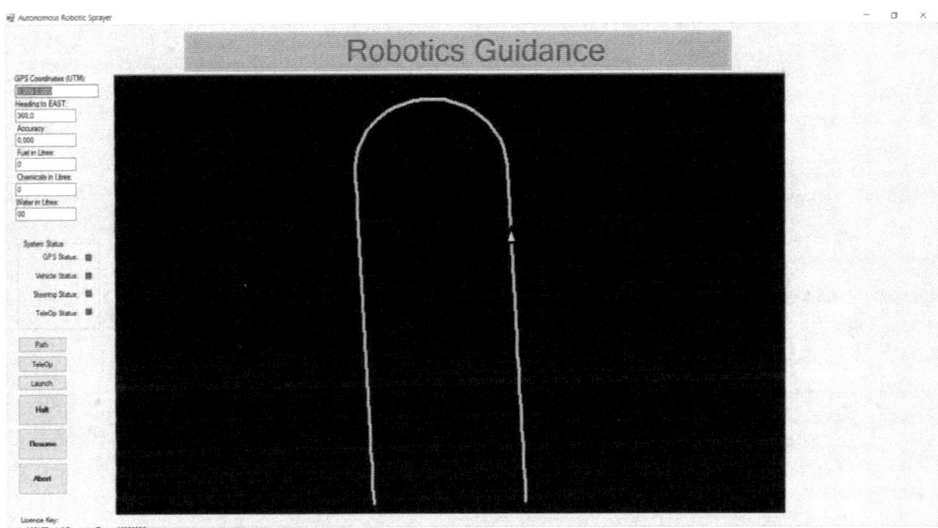

Figure 4. Unmanned agricultural machinery software interface.

4. Experiments

We have modified the 3WP-3000 pesticide spraying agricultural machine, and the completed experimental prototype is shown as Figure 5. After starting, the agricultural machine tracks the target path at a speed of about 2m/s, and the host computer records the position, speed, and heading information of the agricultural machine to evaluate the unmanned performance of the agricultural machine.

IOP Conf. Series: Materials Science and Engineering **799** (2020) 012032 doi:10.1088/1757-899X/799/1/012032

Figure 5. The unmanned agricultural machine prototype.

We tested the tracking of straight path and curved path of the unmanned agricultural machinery, compared the real driving path and the target path. The comparison diagrams of straight path tracking are shown in Figure 6(a), the comparison diagrams of curved path tracking are shown in Figure 6(b). The lateral tracking error of the straight path is shown in Figure 7(a), and the lateral tracking error of the curved path is shown in Figure 7(b).

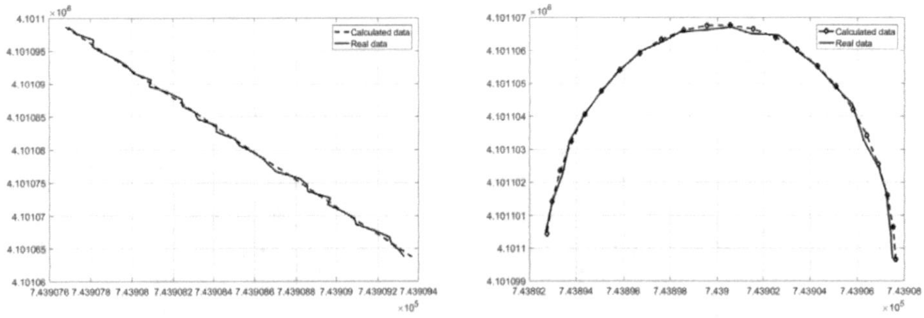

(a) (b)

Figure 6. Comparison of real path and target path.

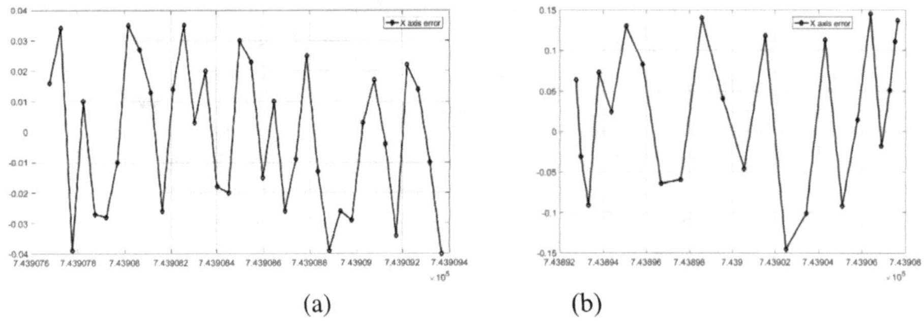

(a) (b)

Figure 7. The lateral tracking error.

The experimental results show that the unmanned agricultural machinery can complete the task of tracking planned path well. The maximum lateral deviation of the straight path tracking is about 4cm, and the maximum lateral deviation of the curved path tracking is about 15cm.

5. Conclusion

The hardware and software design of the unmanned agricultural machinery is introduced in details in the paper. The experiments show that the accuracy of the agricultural machinery tracking is high, the overall tracking error can be kept within a very small range and the unmanned agricultural machinery can effectively complete unmanned tasks.

Acknowledgments
This work was financially supported by the Shandong Academy of Sciences International Science and Technology Cooperation Project (2019GHPY22), and Shandong Academy of Sciences-Lanshan District Industry-University-Research Collaborative Innovation Fund Project (2017CXY-5).

References

[1] Kitchen N R, Snyder C J, Franzen D W, Wiebole W J. (2002) Educational needs of precision agriculture. Precision Agriculture, 3(2), 341-351.
[2] Naiqian Zhang, Maohua Wang, Ning Wang. (2002) Precision agriculture-a worldwide overview. Computers and Electronics in Agriculture, 36(2), 113-132.
[3] Sun Jie. (2017) "13th five-year national science and technology innovation plan": Optimize the innovation of agricultural and rural areas. Chinese Rural Technology, 12, 10.
[4] Li Jun, Ma Rong. (2011) Tractor automatic navigation technology based on multi-sensor fusion Agricultural Mechanization Research, 33(12), 58-60.
[5] Dong Sheng, Yuan Chaohui, Gu Chao. (2017) A survey of research on unmanned agricultural machinery control platform based on multidisciplinary technology fusion. Journal of Agricultural Engineering, 33(8),1-11.
[6] Thomas Noulis. (2015) Mixed-Signal Circuits. Taylor and Francis, England.

IOP Conf. Series: Materials Science and Engineering **799** (2020) 012033 doi:10.1088/1757-899X/799/1/012033

Digital Signal Processing Optimization Algorithm for Terahertz Communication Based on Short Convolution Iterative Algorithm and Adaptive Blind Equalization Algorithm

Dingsheng Deng[1]

[1]Sichuan University Nationalities, KangDing, SiChuan,626001, China

[*]Corresponding author's e-mail: dds0904@scun.edu.cn

Abstract. This paper systematically studies the crosstalk problem caused by the low parallelization rate and tail-to-back tailing in the high-speed digital signal parallel processing in terahertz communication. First, in order to improve the parallelization rate in system application, a high-speed digital signal in the network is filtered in parallel based on a short convolution iterative algorithm for the input signal. Second, an adaptive blind equalization algorithm is proposed to solve the problem crosstalk is used to restore the original signal. Finally, the effectiveness of the equalization algorithm is verified by MATLAB, and the designed algorithm is implemented by Labview simulation software. The effectiveness of the designed algorithm is verified by simulation analysis. The high-speed signal after parallel filtering and equalization can restore the information of the original signal well.

1. Introduction

With the progress of science and technology and the acceleration of the pace of life, people's demand for network communication speed is increasing day by day. Since the mid-1990s, wireless network applications began to develop rapidly in the world, and the transmission rate of megabits has been unable to meet the current actual communication needs. However, the high-speed communication of T-bit has been successfully realized in the laboratory with terahertz technology. The traditional hardware system is difficult to support such a high communication rate. Serial communication is the internal communication mode of the equipment, even if the auxiliary equipment such as antenna is added by external means, the internal communication is still the same. Increasing the sampling rate, bandwidth and modulation signal order is the main way to improve the communication rate in serial communication. Today, the demand is expanding rapidly, and the actual effect is not enough. In view of this, the industry urgently needs to find another way to achieve this goal from other perspectives[1-2].

In recent years, parallel signal processing technology has been widely used in various fields because of its advantages over the traditional serial method. Without changing the single channel capacity, the communication speed has been significantly improved. The United States and other developed countries started this research earlier, and China also paid enough attention to this technology. As early as the mid-1990s, NASA's JPL laboratory has proposed a concept of parallel receiving device; until 2012, the high-speed parallel principle prototype based on high-frequency terahertz unit with communication rate up to 2 Gbps developed by China Academy of engineering physics finally came out[3-4].

IOP Conf. Series: Materials Science and Engineering **799** (2020) 012033 doi:10.1088/1757-899X/799/1/012033

2. Significance of developing terahertz communication

The electromagnetic wave whose frequency is between 100 GHz and 10 THz is called terahertz wave. Because of the particularity of this frequency band, the research on it is still in the primary stage in academic and engineering circles.

The development of new wireless communication frequency band has been more and more focused on solving the contradiction of spectrum resources. With the rapid development of wireless communication, there is an obvious trend of scarcity. The key to solve the problem is the terahertz frequency band, which is suitable for future wireless communication. Although the choice of technical route is diverse, wireless communication system is usually realized by solid-state electronic equipment, and the system can be integrated on chip in the future, which is very important for the commercialization of terahertz wireless communication system. The most significant advantage of using terahertz frequency for wireless communication is that terahertz frequency band has abundant bandwidth resources. On the ground, it is connected to the mobile communication base station through terahertz wireless connection, high-speed access to dense area, reverse data transmission, and finally connected to remote users to solve the "last kilometer" problem; in space, for high-capacity and high-speed wireless communication between satellites, it is called the best choice because of the characteristics of terahertz wave that is not easy to decay in vacuum. First of all, terahertz wave has unique advantages for carrying confidential information: because of its length and directionality, it is easier to achieve ultra-high communication bandwidth in this frequency band; secondly, in the battlefield and other harsh environments, the attenuation of terahertz wave is less than that of light wave attenuation, which is convenient for high-speed transmission. Therefore, terahertz communication technology not only has high academic value, but also has a good prospect in practical application

3. Overview of mainstream parallel algorithms

3.1. Parallel filtering algorithm

In practical application, if the communication rate is not very high, then the parallel algorithm can be realized by adding sharing. It can run from 2 Gbps to 5 Gpbs and consume the least hardware resources. However, with the increase of communication speed and the number of parallel channels, the pure addition sharing method has lost its practical application value in some cases due to the factors of too high data dimension, too much computation overhead and low fault tolerance. As for parallel processing, fast finite impulse response algorithm (FFA) and iterative short convolution algorithm (ISCA) are two parallel filtering algorithms used in high-speed environment.

The essence of FFA algorithm is a finite impulse response filter. It inputs the original serial sequence into parallel input sequence after serial parallel conversion, and then decomposes the filter coefficients into similar parallel sequences similarly. After being decomposed into parallel sequences, parallelization is expressed in matrix form. This algorithm does not need to use the general overlap technique in the cyclic or aperiodic convolution based method. Can significantly reduce hardware performance requirements. Because of its excellent linear phase characteristics and unconditional stability, FFA algorithm is widely used in many fields such as video, image processing and wireless communication. In some applications, such as high-speed remote sensing satellite receiver, 4G communication system, their throughput requirements for FIR filter are also increased due to their high data transmission rate, such as cellular data communication and portable medical applications. On the other hand, in the field of equipment, there are strict requirements for the power consumption of FIR filter. Parallel technology can be used to improve the throughput of FIR filter and reduce the power consumption of FIR filter, which is its main advantage. But generally, the number of parallel structures increases significantly with the increase of parallel series, making it difficult to be applied in the system with high complexity 。 How to solve the complexity of parallel FIR filter has been an important topic in the past decade. However, in most designs of FIR filter, the parallel structure can not tolerate the hardware overhead. In addition to the FFA algorithm, there is still a more suitable algorithm - ISCA algorithm, which can obtain better hardware resource utilization efficiency than FFA algorithm, so as to improve the FFA algorithm is difficult to adapt to the problem of high complexity system. In this algorithm, parallel filtering is realized by fast short convolution. The fast short convolution algorithm was first proposed by achaji in 1989, and then improved by Cheng C and Cheng C, parhi K. Especially, when the length of FIR filter is very large, this kind of linear convolution structure based on ISC can save a lot of hardware cost.

3.2. Parallel equalization algorithm

With the increasing demand for high-speed data transmission, the existing digital modulation and demodulation technology has been stretched out, which is difficult to meet the high demand of the public. At this time, terahertz communication is in the public eye. At present, the research of terahertz communication mainly focuses on terahertz high-speed signal processing, but ignores the accuracy of signal transmission. At present, most of the academic literature does not consider equalizer, but in the actual communication, the problem of code crosstalk is widespread. There are many reasons for crosstalk, such as finite length truncation and sampling period deviation. In order to recover the original signal, equalization technology is needed to reduce or even eliminate the adverse effects of these inter code interference. However, due to the non-linear characteristics of the actual channel and devices, it is impossible to accurately observe the impact caused by them. At the same time, considering the magnitude time-varying characteristics of the channel itself, the equalization algorithm must have self adaptability. In this paper, the parallel adaptive blind equalization algorithm is selected.

4. Basic principle of algorithm

4.1. Iterative short convolution algorithm

In parallel filtering algorithm, the iterative short convolution algorithm (ISCA) is more efficient than FFA in hardware resource utilization. In this algorithm, some multiplication units are replaced by addition units and delay units, while fast convolution is used to reduce the number of sub filters, and the coefficient symmetry of sub filters is guaranteed as much as possible. The number of multipliers needs to be optimized by relying on coefficient symmetry. ISCA algorithm equation is as follows:

$$S_{2M-1} = A_{M\_mn}(Q_m \otimes Q_n)H_{M\_nm}(P_m \otimes P_n)X_M \tag{1}$$

On the basis of formula (1), the formula of multi-level iteration can be further derived, which is suitable for the case of high parallel series. The formula of multi-level iteration is as follows:

$$S_{2M-1} = A_{M\_mk}(A_{M\_mn} \otimes I_{(2k-1)\times(2k-1)})(Q_m \otimes (Q_n \otimes Q_k))H_{M\_nmk}(P_m \otimes (P_n \otimes P_k))X_N \tag{2}$$

In formula (2):

$$\begin{cases} N = Mk = mnk \\ H_{N\_mnk} = diag[(P_m \otimes (P_n \otimes P_k))(h_0, h_1, \cdots, h_{N-1})^T] \end{cases} \tag{3}$$

4.2. Parallel adaptive blind equalization algorithm

Set the transmitted information symbol as s(n)=a(n)+jb(n), the equalizer length as *N*, and the signal vector received by the equalizer as follows:

$$X(n) = [x_n, x_{n-1}, \cdots, x_{n-N+1}] \tag{4}$$

The equalizer coefficient is:

$$W(n) = [w_0(n), w_1(n), \cdots, w_{N-1}(n)]^T \tag{5}$$

The renewal equation of adaptive coefficient based on stochastic gradient method can be expressed as follows:

$$W(n+1) = W(n) + \mu e(n) X^*(n) \tag{6}$$

Where μ is the parameter of adaptive updating step size, and its error function is.

$$e(n) = y(n)(R_C^2 - |y(n)|^2) \quad (7)$$

The formula of adaptive blind equalization algorithm is as follows:

$$y(Lk+i) = W^T(Lk+i)X(LK+i) \quad (8)$$

If the above formula is transformed in advance and the equilibrium coefficient at the current time is replaced by that at the next time, it can be obtained:

$$y(Lk+i) \approx W^T(Lk)X(Lk+i) \quad (9)$$

This transformation method can simplify the calculation of the iterative process, and will not have a significant impact on the performance of the algorithm. Where k is the time index corresponding to n under L-channel parallelism. By expanding the above formula (10), we can get:

$$W[L(k+1)] \approx W(Lk) + \mu \frac{L}{LA} \sum_{m=0}^{LA-1} e(Lk+m)X^*(Lk+m) \quad (10)$$

Pipeline delay is unavoidable in the hardware implementation phase of the algorithm, so it must be considered in the modeling process. In this system, the pipeline delay of the error feedback loop is defined as $D1$. Since the converged coefficients are stable and slow in the adaptive updating, formula (10) can be approximately expressed as follows:

$$W[L(k+1)] \approx W(Lk) + \mu \frac{L}{LA} \sum_{m=0}^{LA-1} e(Lk+m-D1)X^*(Lk+m-D1) \quad (11)$$

Combining short convolution to update coefficients adaptively can realize parallelization, and the hardware cost is low. Take L=8, LA=2. As shown in Figure 1, the overall implementation structure is obtained, and the flow of parallel adaptive blind equalization algorithm based on short convolution is described.

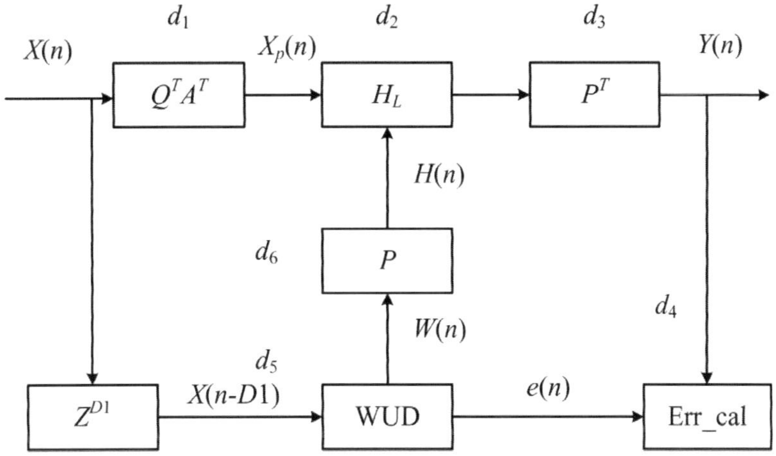

Figure 1. Overall implementation structure of adaptive blind equalization

Through the matrix Q^T and matrix A^T preprocessing, d_1-d_6 is the pipeline delay, through $X_q(n)= X_p(n)H(n)$ filtering and the matrix P^T data combination processing can get the final equalization results. Here we need to limit the pipeline delay conditions:

$$D1 = d_1 + d_2 + d_3 + d_4 \quad (12)$$

5. Algorithm simulation

The effectiveness and performance of the designed algorithm need to be verified by simulation. For the verification platform of the adaptive blind equalization algorithm, MATLAB is selected. The simulation results of the algorithm are shown in Figure 2. As can be seen from the figure, the result of the equalization is ideal, and the virtual devices used in the simulation process are low-performance hardware, and can still achieve fast equalization.

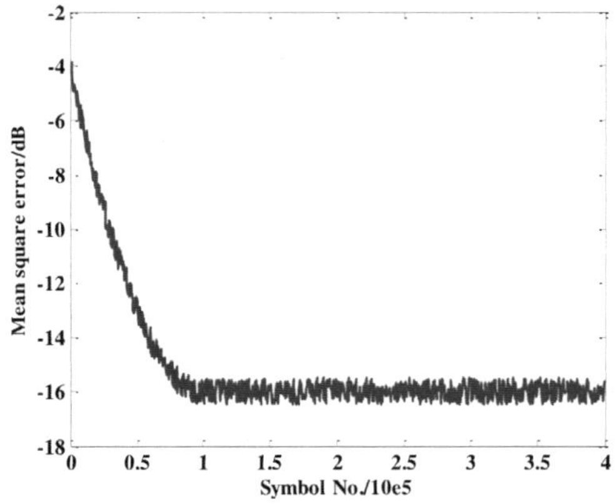

Figure 2. Simulation of Parallel Adaptive Blind Equalization Algorithm

We adopt Labview simulates the entire process of transmitting signal filtering and equalization algorithms, and uses 5Gbps 16QAM signals to simulate input signals. Select the modulation phase filter for parallelization and select 32-channel parallel simulation. First, the modulation data is parallelized on four channels through polyphase decomposition. Each of the four channels uses the ISCA algorithm for eight channels in parallel, so that the total number of parallel channels eventually becomes 32. Signal reduction is performed by a parallel equalization algorithm. Figure 3 shows the frequency spectrum of the output data. It can be seen from the parallel output data spectrum in Figure 3 that the parallel signal spectrum is highly similar to the general serial transmission spectrum. And the signal bandwidth does not change after parallelization. The transmission speed of the parallel signals meets the requirements, and the amplitude within the bandwidth is stable, indicating that the data equalization processing effect is better.

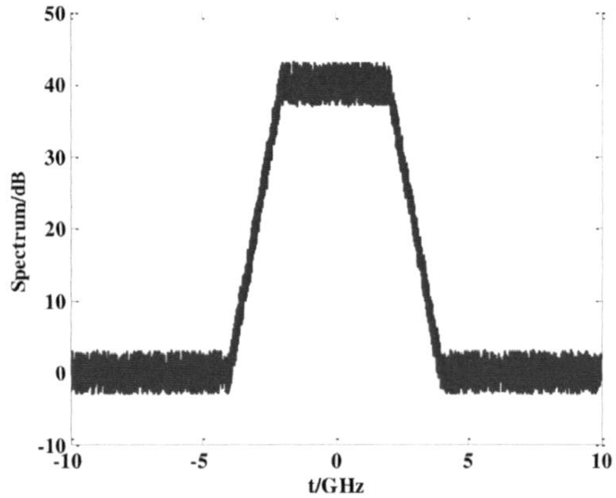

Figure 3. Parallel output spectrum

In summary, the effectiveness of the filtering and equalization algorithms described in this article can be verified by simulation. Provide a solution for parallel processing of high-speed digital signals in terahertz communication in hardware systems in the future.

6. Conclusion

Intelligent mobile devices, big data, the Internet of Things, and artificial intelligence services continue to penetrate all aspects of human life, so terahertz communication needs sufficient rates to support high-speed wireless communications. Terahertz science and technology has achieved success in many key areas over the past two decades: spectroscopy, communications, radar, astronomy, meteorology, petroleum, chemistry, military security, and breakthrough breakthroughs, especially in the fields of defense and aerospace. Research on terahertz science and technology still has a long way to go, and the development of terahertz communication is highly expected. Improving communication speed, from component integration to system integration, and short-range high-speed terahertz secure wireless communication are the major development trends of terahertz communication research.

At the same time, with the advent of the 5G era and the introduction of efficient and secure new communication technologies, the digital media industry will also face new technological innovations. With the gradual popularization of 5G, the communication industry and the digital media industry will also have explosive growth, which will greatly reduce the arrival time of the next technology bottleneck period. Terahertz technology is leading the frontier of today's communication neighborhood with its superior communication speed and security. The research of high-speed signal processing technology in terahertz communication can not only serve the public at the moment, but also serve as a technological precipitation, which is more conducive to the research of next-generation technology. With the vigorous development of terahertz technology today, the focus on hardware development has been recognized by the industry, but good software algorithms cannot be ignored. After the hardware system is determined, the algorithm needs to be continuously improved to improve performance and system stability. In the future work, we will continue to conduct in-depth research on signal processing algorithms in terahertz communication.

References

[1] Dai, J.M., Clough, B., Ho, I.C, et al. (2011) Recent Progresses in Terahertz Wave Air Photonics. J. IEEE Transactions on Terahertz Science and Technology, 1: 274–281.
[2] Yu, Q., Gu, J.Q., Yang, Q.L, et al. (2017) All-Dielectric Meta-lens Designed for Photoconductive Terahertz Antennas. J. IEEE Photonics Journal, 9: DOI:10.1109/JPHOT.2017.2721503.
[3] Tu, H., Meng, X.X., Sun, Z.Y., et al. (2019) A Fast 220-GHz Real Aperture 3-D Personnel Screening System With a Novel-Shaped Mirror Design. J. IEEE Transactions on Terahertz Science and Technology, 9: 273–261.
[4] Zhao, H., Zhang, L.L., Huang, S.X., et al. (2018) Terahertz Wave Generation From Noble Gas Plasmas Induced by a Wavelength-Tunable Femtosecond Laser. J. IEEE Transactions on Terahertz Science and Technology, 8: 299–304.
[5] Zhao, J., Zhang, L.L., Wu, T., et al. (2016) Wavelength Scaling of Terahertz Wave Absorption via Preformed Air Plasma. J. IEEE Transactions on Terahertz Science and Technology, 6: 846–850.

IOP Conf. Series: Materials Science and Engineering **799** (2020) 012034 doi:10.1088/1757-899X/799/1/012034

Research on characteristic parameters of acoustic actuator radiation plate

Zhouli Li[1], Cailing Qian[1] and Kaikai Shi[1*]

[1]Instrument Engineering, Xi'an Shiyou University, Xi'an, Shan Xi, 710065, China

* Corresponding author's e-mail: shikk_1995@163.com

Abstract. As the final output element of the magnetostrictive actuator, the material, size and shape selected by the radiant plate will affect the output efficiency of the actuator directly. Based on the theoretical analysis on natural frequency of thin circular plate and thick circular plate with peripheral fixing-supported, the result shows that the ratio of the thickness of circular plate to the square of its radius can be determined by material and vibration mode of the circular plate ,and in portion to its natural frequency . when frequency of circular plate and circular plate vibration mode shapes are selected, the ratio of the thickness to the area of the circular plate is constant, which provides the basis for acoustic actuator radiation plate geometric design.

1. Introduction

Magnetostrictive material is a new type of functional material, which has a strain value ten to hundreds times higher than piezoelectric material, and it has many excellent characteristics, such as high output power, microsecond response speed, wide operating band and so on. It has important applications in many fields, especially in the application of micro displacement, which shows its unique advantage. In the perspective of engineering application, the electromagnetic mechanical coupling process of magnetostrictive actuator can be simulated by a relatively simple mathematical model. The radiation plate is simplified as the vibration model of the forced excitation force of the peripheral fixed center as shown in Figure 1 [1]. In this paper, the circular plate vibration model is used to analyze the vibration law of the radiation plate, which provides a reference for the dynamic design of the transducer.

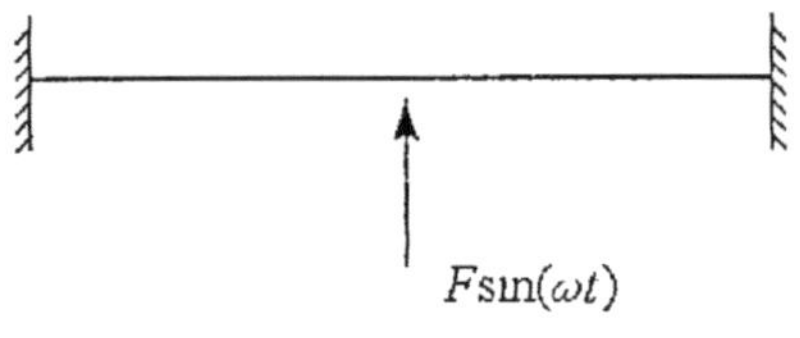

Figure1. forced vibration model of radiation plate

2. Vibration equation of elastic circular plate Circular

Thin plate is also a common thin plate element in Engineering, Polar coordinates are the most convenient way to analyze the transverse vibration of circular thin plates [2-4]. Polar coordinates are shown in Figure 2.

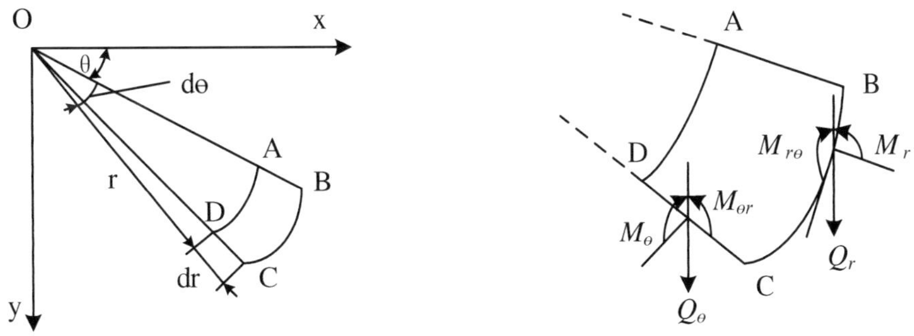

(a) **(b)**

Figure 2. The value of circular thin plate in polar coordinates

The relationship between polar coordinates and rectangular coordinates is

$$r^2 = x^2 + y^2, \theta = \arctan\frac{y}{x} \tag{1}$$

then

$$\begin{cases} \dfrac{\partial r}{\partial x} = \dfrac{x}{r} = \cos\theta \\ \dfrac{\partial r}{\partial y} = \dfrac{y}{r} = \sin\theta \\ \dfrac{\partial \theta}{\partial x} = -\dfrac{y}{r^2} = -\dfrac{\sin\theta}{r} \\ \dfrac{\partial \theta}{\partial y} = \dfrac{x}{r^2} = \dfrac{\cos\theta}{r} \end{cases} \tag{2}$$

using the above relationship, we can get

$$\begin{cases} \dfrac{\partial \omega}{\partial x} = \dfrac{\partial \omega}{\partial r}\dfrac{\partial r}{\partial x} + \dfrac{\partial \omega}{\partial \theta}\dfrac{\partial \theta}{\partial x} = \cos\theta\dfrac{\partial \omega}{\partial r} - \dfrac{\sin\theta}{r}\dfrac{\partial \omega}{\partial \theta} \\ \dfrac{\partial \omega}{\partial y} = \dfrac{\partial \omega}{\partial y}\dfrac{\partial r}{\partial y} + \dfrac{\partial \omega}{\partial \theta}\dfrac{\partial \theta}{\partial y} = \sin\theta\dfrac{\partial \omega}{\partial r} + \dfrac{\cos\theta}{r}\dfrac{\partial \omega}{\partial \theta} \end{cases} \tag{3}$$

then

$$\nabla^2\omega = \frac{\partial^2 \omega}{\partial r^2} + \frac{1}{r}\frac{\partial \omega}{\partial r} + \frac{1}{r^2}\frac{\partial^2 \omega}{\partial \theta^2} \tag{4}$$

From the vibration differential equation of thin plate

$$D_0\nabla^4\omega + \rho h\frac{\partial^2 \omega}{\partial t^2} = p(x, y, t) \tag{5}$$

The vibration differential equation of thin plate in polar coordinate system is obtained

$$D_0\nabla^4\omega + \rho h\frac{\partial^2 \omega}{\partial t^2} = p(r, \theta, t) \tag{6}$$

Among the equation, ∇^4 is Biharmonic operator, defined in polar coordinates as

$$\nabla^4 = \nabla^2\nabla^2 = \left(\frac{\partial^2}{\partial r^2} + \frac{1}{r}\frac{\partial}{\partial r} + \frac{1}{r^2}\frac{\partial^2}{\partial \theta^2}\right)\left(\frac{\partial^2}{\partial r^2} + \frac{1}{r}\frac{\partial}{\partial r} + \frac{1}{r^2}\frac{\partial^2}{\partial \theta^2}\right) \tag{7}$$

From Fig. 2 (a), the micro element $ABCD$ is formed by the intersection of two radii with included angle dθ and two arcs with radius r , r + dr , On the boundary of the micro element, there are moments M_r and M_θ , Torque $M_{r\theta}$ and shear Q_r , Q_θ , As shown in Figure 2 (b). If the x-axis direction is cut to coincide with the radius r direction, then M_r, M_θ, $M_{r\theta}$ and Q_r, Q_θ are the same as M_x, M_y, M_{xy} and Q_x, Q_y at the same point, so get

$$\begin{cases} M_r = -D_0\left(\frac{\partial^2\omega}{\partial x^2}+\mu\frac{\partial^2\omega}{\partial y^2}\right)_{\theta=0} = -D_0\left[\frac{\partial^2\omega}{\partial r^2}+\mu\left(\frac{1}{r}\frac{\partial\omega}{\partial r}+\frac{1}{r^2}\frac{\partial^2\omega}{\partial\theta^2}\right)\right] \\ M_\theta = -D_0\left(\frac{\partial^2\omega}{\partial y^2}+\mu\frac{\partial^2\omega}{\partial x^2}\right)_{\theta=0} = -D_0\left(\frac{1}{r}\frac{\partial\omega}{\partial r}+\frac{1}{r^2}\frac{\partial^2}{\partial\theta^2}+\mu\frac{\partial^2\omega}{\partial r^2}\right) \\ M_{r\theta} = -D_0(1-\mu)\left(\frac{1}{r}\frac{\partial^2\omega}{\partial\theta\partial r}-\frac{1}{r^2}\frac{\partial\omega}{\partial\theta}\right) \\ Q_\theta = -D_0\left[\frac{\partial}{\partial y}\left(\nabla^2\omega\right)\right]_{\theta=0} = -D_0\frac{\partial}{r\partial\theta}\left(\nabla^2\omega\right) \end{cases} \tag{8}$$

3. Solution of vibration equation

For circular thin plates, the origin of polar coordinate system should be based on the center of the circle, assuming that the radius of the circle is a, then the corresponding boundary conditions at $r=a$ are classified as follows

1) Fixed edge

$$\omega|_{r=a}=0, \frac{\partial\omega}{\partial r}|_{r=a}=0 \tag{9}$$

2) Simply supported edge

$$\omega|_{r=a}=0, M_r|_{r=a}=0 \tag{10}$$

3) Free edge

$$M_r|_{r=a}=0, \left(Q_r+\frac{1}{r}\frac{\partial M_{r\theta}}{\partial\theta}\right)|_{r=a}=0 \tag{11}$$

Let the main vibration of the circular plate be

$$\omega(r,\theta,t)=W(r,\theta)\sin(\omega t+\varphi) \tag{12}$$

The corresponding free vibration equation of equation (6) is

$$\nabla^4 W-\beta^4 W=0 \tag{13}$$

where

$$\beta^4=\frac{\rho h\omega^2}{D_0} \tag{14}$$

equation (12) can be rewritten as

$$\left(\nabla^2+\beta^2\right)\left(\nabla^2-\beta^2\right)W=0$$

So the solution of equation (12) is

$$\left(\frac{\partial^2}{\partial r^2}+\frac{1}{r}\frac{\partial}{\partial r}+\frac{1}{r^2}\frac{\partial^2}{\partial\theta^2}+\beta^2\right)W=0 \tag{15}$$

$$\left(\frac{\partial^2}{\partial r^2}+\frac{1}{r}\frac{\partial}{\partial r}+\frac{1}{r^2}\frac{\partial^2}{\partial\theta^2}-\beta^2\right)W=0 \tag{16}$$

Let the main mode $W(r,\theta)$ be

$$W(r,\theta)=R(r)\cos n\theta, n=1,2,\dots \tag{17}$$

Substituting equation (17) into equation (15) and equation (16), the following two equations are obtained

$$\frac{d^2R}{dr^2}+\frac{1}{r}\frac{dR}{dr}+\left(\beta^2-\frac{n^2}{r^2}\right)R=0 \tag{18}$$

$$\frac{d^2R}{dr^2}+\frac{1}{r}\frac{dR}{dr}-\left(\beta^2-\frac{n^2}{r^2}\right)R=0 \tag{19}$$

equation (18) is n-order Bessel equation, and its general solution is

$$R(r)=J_n(\beta r)+C_2N_n(\beta r) \tag{20}$$

where $J_n(\beta r)$ and $N_n(\beta r)$ are the first and second kind of Bessel functions of real variables respectively. Equation (19) is n-order modified Bessel equation, and its general solution is

$$R(r)=C_3I_n(\beta r)+C_4K_n(\beta r) \tag{21}$$

In the formula, $I_n(\beta r)$ and $K_n(\beta r)$ are the first type and the second type of Bessel functions of virtual quantum respectively. The general solution of equation (13) is

$$W(r,\theta)=[C_1N_n(\beta r)+C_2N_n(\beta r)+C_3I_n(\beta r)+C_4K_n(\beta r)]\cos n\theta \tag{22}$$

If the plate is solid, then $\beta r=0$ in the center of the circular plate, and $N_n(\beta r)$ and $K_n(\beta r)$ will tend to infinity. In fact, ω and $\frac{\partial\omega}{\partial r}$ in the center of the circular plate should be finite, so there must be $C_2=C_4=0$, so equation (22) is simplified as

$$W(r,\theta)=[C_1J_n(\beta r)+C_3I_n(\beta r)]\cos n\theta \tag{23}$$

Using the two boundary conditions of the outer edge of a circular plate, we can obtain the homogeneous equations of C_1 and C_3. Thus, the frequency equation and the main mode shape are derived. The boundary conditions at $r=a$ expressed by $R(r)$ can be obtained in this way, substituting equation (17) into equation (9) to equation (11), we can get

1) Fixed edge

$$R(a)=0,\frac{dR}{dr}|_{r=a}=0 \tag{24}$$

2) Simply supported edge

$$R(a)=0,\left[\frac{d^2R}{dr^2}+\mu\left(\frac{1}{r}\frac{dR}{dr}-\frac{r^2}{n^2}R\right)\right]|_{r=a}=0 \tag{25}$$

3) Free edge

$$\begin{cases}\left[\dfrac{d^2R}{dr^2}+\mu\left(\dfrac{1}{r}\dfrac{dR}{dr}-\dfrac{n^2}{r^2}R\right)\right]|_{r=a}=0\\ \left[\dfrac{d}{dr}\left(\dfrac{d^2R}{dr^2}+\dfrac{1}{r}\dfrac{dR}{dr}-\dfrac{n^2}{r^2}R\right)+\dfrac{(1-\mu)n^2}{r^2}\left(\dfrac{1}{r}R-\dfrac{dR}{dr}\right)\right]|_{r=a}=0\end{cases} \tag{26}$$

4. Frequency solution

Calculation of the first three natural frequencies of a solid circular plate with fixed outer boundary without radial nodes (pitch diameter). Let the radius of circular plate be a , Substituting $R(r)=C_1J_n(\beta r)+C_3I_n(\beta r)$ into equation (24), we can get

$$C_1J_n(\beta a)+C_3I_n(\beta a)=0 \tag{27}$$

$$\frac{d}{dr}\left[C_1J_n(\beta r)+C_3I_n(\beta r)\right]|_{r=a}=0 \tag{28}$$

Because

IOP Conf. Series: Materials Science and Engineering **799** (2020) 012034 doi:10.1088/1757-899X/799/1/012034

$$\frac{d}{dr}J_n(\beta a)=\beta\left[\frac{n}{\beta r}J_n(\beta r)-J_{n+1}(\beta r)\right]$$

$$\frac{d}{dr}I_n(\beta a)=\beta\left[\frac{n}{\beta r}I_n(\beta r)+I_{n+1}(\beta r)\right]$$

So formula (b) becomes

$$C_1\left[\frac{n}{\beta r}J_n(\beta a)-J_{n+1}(\beta a)\right]+C_3\left[\frac{n}{\beta r}I_n(\beta a)+I_{n+1}(\beta a)\right]=0 \tag{29}$$

The coefficient determinant of the system of equations composed of formula (27) and formula (29) must be zero, i.e

$$\begin{vmatrix} J_n(\beta a) & I_n(\beta a) \\ \frac{n}{\beta a}J_n(\beta a)-J_{n+1}(\beta a) & \frac{n}{\beta a}I_n(\beta a)+I_{n+1}(\beta a) \end{vmatrix}=0 \tag{30}$$

So the frequency equation is

$$J_n(\beta a)I_{n+1}(\beta a)+J_{n+1}(\beta a)I_n(\beta a)=0 \quad n=0,1,2,\ldots \tag{31}$$

When $n = 0$, there is no pitch diameter in the circular plate. The above formula becomes

$$J_n(\beta a)I_1(\beta a)+J_1(\beta a)I_0(\beta a)=0 \tag{32}$$

The roots of equation (32) are arranged as $\beta_{0,0}$, $\beta_{0,1}$, $\beta_{0,2}$ The first subscript is the number of pitch diameter, the second subscript is the number of pitch circle, and the first three roots are

$$\beta_{0,0}=3.196, \beta_{0,1}=6.306, \beta_{0,2}=9.440$$

Therefore, from equation (14), the first three natural frequencies without pitch diameter are

$$\omega_{0,0}=\frac{10.21}{a^2}\times\left(\frac{D_0}{\rho h}\right)^{\frac{1}{2}}, \omega_{0,1}=\frac{39.77}{a^2}\times\left(\frac{D_0}{\rho h}\right)^{\frac{1}{2}}, \omega_{0,2}=\frac{88.90}{a^2}\times\left(\frac{D_0}{\rho h}\right)^{\frac{1}{2}}$$

According to formula (23), the corresponding main mode shape is

$$W_{n,s}=C_{n,s}\left[J_0(\beta_{0,s},a)+b_{0,s}I_0(\beta_{0,s},r)\right] \quad s=0,1,2,\ldots$$

Where

$$b_{0,s}=-\frac{J_0(\beta_0,r)}{I_0(\beta_0,r)}$$

If $n = 1$ is taken in the frequency equation (31), the natural frequency of a circular plate with one pitch diameter can be calculated, and so on. And takes

$$W_{n,s}=C_{n,s}\left[J_n(\beta_{n,s}r)+b_{n,s}I_n(\beta_{n,s}r)\right]\cos n\theta \quad ,n、s=0,1,2,\ldots$$

where $b_{n,s}=-\frac{J_n(\beta_n,r)}{I_n(\beta_n,r)}, \beta_{n,s}^4=\frac{\beta h}{D_0}\omega_{n,s}^2$. The natural frequency of a circular plate is usually expressed as

$$\omega=\frac{k}{a^2}\left(\frac{D_0}{\rho h}\right)^{\frac{1}{2}} \tag{33}$$

Where ω is natural frequency, k is coefficient of frequency, a is radius of circular plate, ρ is material density h is the thickness of the circular plate, D_0 is Bending stiffness of sheet, and $D_0=\frac{Eh^3}{12(1-\mu^2)}$, μ is poisson's ratio of materials, E is Young's modulus of material. It can be obtained by simple reasoning according to equation (33), substitute D_0 into it,

$$\omega = \frac{k}{a^2}\left[\frac{E}{12\rho\left(1-\mu^2\right)}\right]^{\frac{1}{2}} h \tag{34}$$

The definition of radiant panel τ characteristic parameters, by reasoning

$$\tau = \frac{h}{\pi a^2} = \frac{\omega}{\pi k}\left[\frac{12\rho\left(1-\mu^2\right)}{E}\right]^{\frac{1}{2}} \tag{35}$$

Where E and x are the material parameters of the radiant plate, and k is the frequency coefficient, which is related to the vibration pattern of the radiant plate. So learn that tau only associated with radiant panel material, and the radiant panel vibration mode. When the radiating plate material and configuration are determined, its $\frac{1}{\pi k}\left[\frac{12\rho\left(1-\mu^2\right)}{E}\right]^{\frac{1}{2}}$, that is the constant of C , and τ equaling to $C\omega$. The $\frac{h}{a^2}$ is proportional to the natural frequency of the plate. After selecting the natural frequency of the circular plate, Therefore τ is a constant shown as (35). In other words, when the natural frequency and mode of the circular plate are selected, the ratio of the characteristic parameters of the designed radiant plate, the ratio of thickness to area of a circular plate, is constant.

5. Conclusion

Through the theoretical derivation of the formula, it is concluded that when the material, mode and natural frequency of the circular plate are selected, the characteristic parameters of the designed radiant plate and the ratio of thickness to area of a circular plate are constant. The characteristic parameters provide the basis for the geometric design of the radiant plate of acoustic actuator.

References

[1] Wang, JX. (2006) Approximately computed piston vibration of circular eradiation plate of columned magnetostricative actuator. Journal of Baotou University of Iron and Steel Technology., 2006(03):246-248.
[2] Xu, X. (2000) Elasticity Solution for Axisymmetric Vibration Problem of Thick Circular Plate. Chinese quarterly of Mechanics. 2000(01):59-65.
[3] Wang, JX.(2019) Acoustic field analysis in resonant cavity of ultra-high frequency thermoacoustic refrigerator. Cryogenics. 2019(04):38-41+47.
[4] Zhao, ZL. (2014) Vibration mechanics. National Defense Industry Press, Beijing.

IOP Conf. Series: Materials Science and Engineering **799** (2020) 012035 doi:10.1088/1757-899X/799/1/012035

ENSO prediction based on Long Short-Term Memory (LSTM)

Cao Xiaoqun[1], Guo Yanan[1,2*], Liu Bainian[1], Peng Kecheng[1,2], Wang Guangjie[3] Gao Mei[1,2]

[1]College of Meteorology and Oceanography, National University of Defense Technology, Changsha Hunan 410073, China

[2]College of Computer, National University of Defense Technology, Changsha Hunan 410073, China

[3]Chinese People's Liberation Army 61741, Beijing 100000, China

[*]Corresponding author's e-mail: guoyanan@nudt.edu.cn

Abstract. El Niño-Southern Oscillation (ENSO), as a global climate event with cyclical characteristics, often causes global climate anomalies and produces non-negligible economic and social impacts. Therefore, the prediction and research of ENSO events are important for understanding and solving global climate change issues. It has important scientific and practical significance. Previous research on ENSO events mainly used traditional statistical analysis and numerical simulation methods. This study explores the use of deep learning to improve the accuracy of El Niño-Southern Oscillation (ENSO) prediction. Based on long-term and short-term memory neural networks, the time series of meteorological and marine elements were analyzed. In the meantime, the sea surface temperatures (SST) and sea level pressure were predicted to calculate the Southern Oscillation Index (SOI) to reflect the ENSO phenomenon. Finally, this article takes the Niño3.4 regional data from the National Centers for Environmental Prediction (NCEP) dataset as an example, and uses the model proposed in this paper to compare with traditional statistical regression methods. The results show that the Long Short-Term Memory (LSTM) has a good effect in the prediction of ENSO events, and has certain scientific significance and practical value for the prediction of ENSO events.

1. Introduction

El Niño-Southern Oscillation (ENSO) is a global climate event. It has periodic characteristics and has a significant impact on global climate change [1]. The ENSO events is one of the most important factors affecting climate disasters such as summer droughts, floods, and low temperatures in the winter. It is also the strongest interannual change signal in the tropical Pacific air-sea coupling system [2]. At the same time, severe ENSO events often lead to global climate anomalies and have economic and social impacts that cannot be ignored. Therefore, prediction and research of ENSO events are also of great scientific and practical significance for understanding and solving global climate change problems [3]. The current research on ENSO events is mainly to analyse and study the causes and future development trends of ENSO events by defining indexes. However, due to the different physical mechanisms of different types of ENSO events, there is currently no unified index standard for the diagnosis and prediction of ENSO events. With the rapid development of artificial intelligence, machine learning methods can be used to select key indexes and make ENSO events predictions. In addition, related scholars use numerical models to simulate and study ENSO events, but long-term

IOP Conf. Series: Materials Science and Engineering **799** (2020) 012035 doi:10.1088/1757-899X/799/1/012035

model integration will cause a huge accumulation of errors and eventually lead to the failure of prediction.

Due to the nonlinear interaction of the air-sea system and its own chaotic characteristics, in practice, most numerical prediction models are difficult to simulate the annual average changes of elements such as SST and cannot accurately predict the occurrence and development of ENSO events. In addition, due to the randomness of the spatiotemporal evolution of different types of ENSO events, traditional statistical regression analysis methods also have great uncertainty in predicting ENSO. In recent years, with the continuous development of machine learning methods, many scholars have started to use machine learning or deep learning technology to forecast meteorological elements and short-term climate predictions, and have obtained relatively satisfactory results [4,5]. If the SST or index prediction problem is defined as a time series regression problem, a long-term short-term memory neural network (LSTM) can be used to predict the ENSO events.

2. LSTM-based ENSO events prediction model

2.1. Problem definition

El Niño and Southern Oscillation events occurred in the eastern Pacific region, which can be divided into uniform grids by using longitude and latitude information. Each grid point has multiple meteorological element information, including sea surface temperature (SST), sea surface wind field, and sea level pressure. At a specific time, these meteorological element information is stored in a two-dimensional matrix in a horizontal distribution. If the change information of meteorological elements is taken as a research object within a period of time, the information of these meteorological elements is stored as three-dimensional matrix data and processed.

In this study, the problem of forecasting ENSO events is considered as a regression forecasting problem. A neural network model is used to solve such a statistical problem. To predict the changes in meteorological elements such as sea temperature (SST) and sea level pressure based on historical meteorological information. Further, calculate the Southern Oscillation Index (SOI), the Oceanic Niño Index (ONI), etc. For example, according to the sea surface temperature information of M months and other relevant meteorological element information, the sea temperature changes from $M + 1$ to $M + t$ months are predicted, where t represents the forecast duration.

2.2. RNN model and LSTM introduction

Recurrent neural network (RNN) is developed from a feedforward neural network (FNN). There is a certain difference between them. FNN only use layers and connections between layers to achieve information transfer. The loop structure was introduced in the network, that is, the connection between the neuron and itself was established [6]. Through this connection, the RNN can store the input at the previous time point in the network and affect the network output in the next step. In contrast, FNN can only map the input to the output layer through the hidden layer, and RNN can map a complete piece of historical information to each output neuron, making full use of historical information. Therefore, compared with FNN, RNN has more advantages in the prediction problem where the input and output are time series data.

Long Short-Term Memory (LSTM) is a special form of RNN. It introduces the concept of gate and adds the function of long-term and short-term memory so that the information is no longer attenuated. It is solved by designing forget gate, input gate and output gate. It has solved the long-term dependence problem of RNN, and has excellent performance in language translation, machine control, document summary, and speech recognition [7]. For the long-term series of meteorological data, consider that the El Niño and Southern Oscillation events are the products of the interaction between the ocean and the atmosphere. From the perspective of the physical mechanism, SST, sea surface wind field, sea level pressure, and sea surface height are all related to ENSO events. So long-term short-term memory (LSTM) is considered for predictive research on multifactor-related ENSO events.

2.3. Experimental methods and steps

According to the characteristics of the ENSO events, LSTM is used to predict the ENSO events using the historical information of time series such as sea level pressure, sea surface temperature, and sea surface wind field. The data used in this research is the reanalysis data of the Climate Forecasting System of the National Centers for Environmental Prediction (NCEP). The data set includes sea temperature, sea surface wind speed, sea level pressure, and precipitation rate, etc. Data set time span from 1979 to 2018.

The training set and test set are obtained by sliding method. First, set the sliding window size to the sequence length, then use the moving average method to expand the selected data set, and finally slide the amplified data set through the sliding window. The samples required for model training are divided into a training set and a test set according to a 4: 1 ratio. In addition, this paper uses Root Mean Square Error (RMSE) as the loss function during training. The data set is processed, and the Southern Oscillation Index (SOI), Nion3, ONI Index, PNA Index, and average precipitation are calculated. Finally, a correlation analysis is performed on the obtained results to determine the variables input to the prediction model, thereby predicting the occurrence of ENSO events.

3. Analysis of results

This study uses the sea surface temperature (SST), sea surface wind field, sea surface pressure, geopotential height data and precipitation data to calculate SOI, Nion3, ONI, PNA, and average precipitation, etc. Calculate the correlation between them to determine the dependent variable of the input. The indexes and statistical results are shown in Figure 1. According to the results in the figure, it can be seen that each index has the characteristics of periodic changes, so it is feasible to predict the ENSO event through a deep learning model.

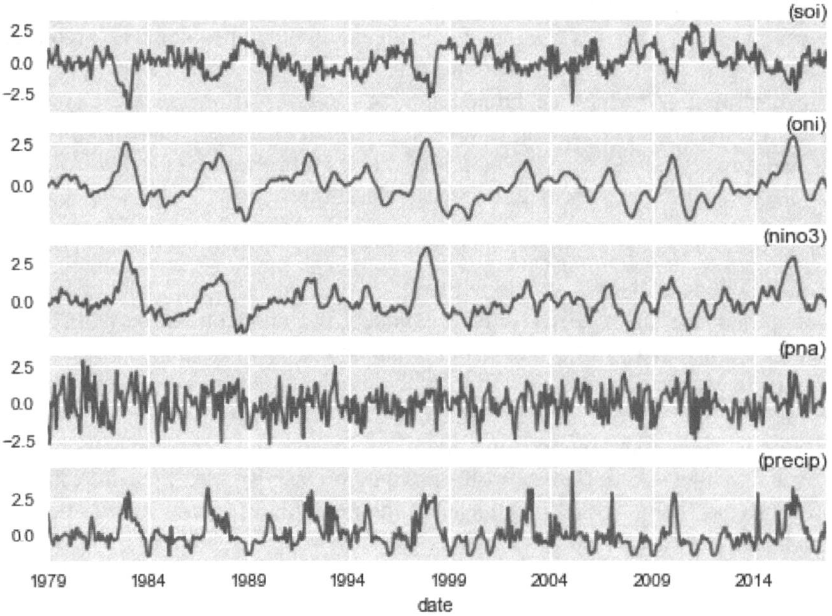

Figure 1 Inter-annual changes of different indices

The correlation degree of each index and statistical results is shown in Figure 2. According to the results in the figure, it can be concluded that the correlation between SOI and ONI, Nino3 and precipitation is the largest. Therefore, by entering relevant data into the deep learning model, SOI is calculated. The SOI is used as a prediction index to analyze and predict the occurrence of ENSO events.

IOP Conf. Series: Materials Science and Engineering **799** (2020) 012035 doi:10.1088/1757-899X/799/1/012035

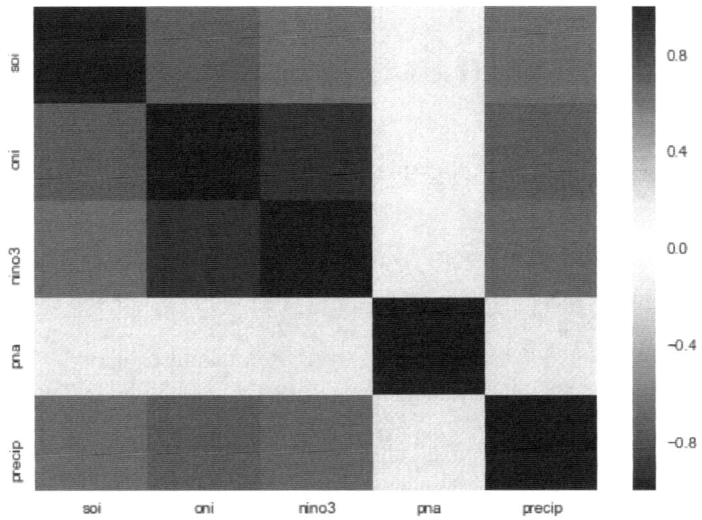

Figure 2 Calculation results of correlation coefficients between different indices

In order to test the reliability of the prediction results of the deep learning model, the Southern Oscillation Index (SOI) prediction results were compared with the statistical results of actual observation data. The comparison results are shown in Figure 3. The solid blue line is the observation result, and the solid red line is the prediction results based on LSTM. By comparison, it can be concluded that the prediction results based on the LSTM are basically consistent with the observations. Therefore, the prediction of ENSO events based on the LSTM has some scientific significance and applicable value.

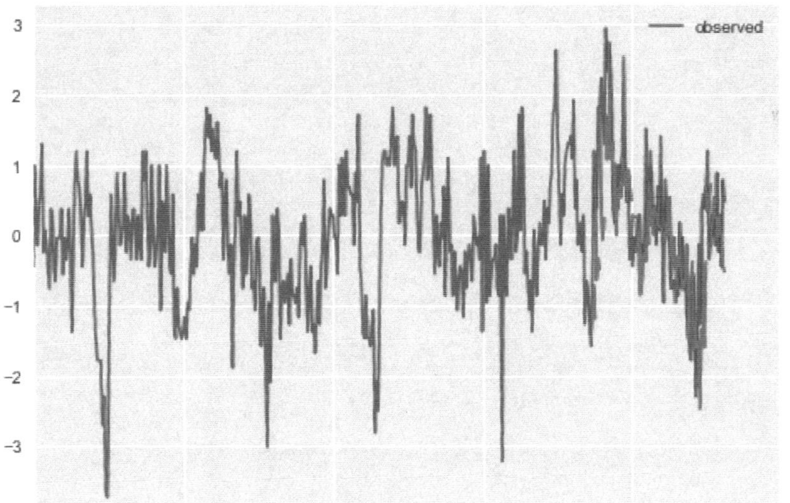

Figure 3 Comparison of SOI based on observations and predictions

4. Conclusion

This paper studies the prediction of ENSO events based on LSTM. By preprocessing and diagnosing meteorological time-series data, we select the variable with strong correlation with ENSO events to train a prediction model, and use the prediction results to calculate Southern Oscillation Index (SOI) which reflects the ENSO phenomenon and has achieved good results. The results of this study show that deep learning methods have great potential in the field of ENSO events prediction. It was also

IOP Conf. Series: Materials Science and Engineering **799** (2020) 012035 doi:10.1088/1757-899X/799/1/012035

found in the experiment that the construction of the prediction model depends heavily on the quality of the data set. When there is a large error in the input data, the prediction result is poor. In the future, a regularization method will be introduced to further study the problem.

Acknowledgement
National key research and development program of China (Grant of No. 2018YFC1506704)

References
[1] Larkin, N. K., & Harrison, D. E. (2002). Enso warm (el niño) and cold (la niña) event life cycles: ocean surface anomaly patterns, their symmetries, asymmetries, and implications. Journal of Climate, 15(10), 1118-1140.
[2] LI Tim. (2010). Impacts of atlantic sea surface temperature anomalies on indo-east asian summer monsoon-enso relationship. Science Bulletin (22), 158-168.
[3] Josephine Brown, Matthew Collins, Alexander W. Tudhope, & Thomas Toniazzo.(2008) . Modelling mid-holocene tropical climate and enso variability: towards constraining predictions of future change with palaeo-data. Climate Dynamics, 30(1), 19-36.
[4] Jones Nicola. (2017). How machine learning could help to improve climate forecasts[J]. Nature, 548(7668):379-380.
[5] Krasnopolsky V M, Fox-Rabinovitz M S. (2006). Complex hybrid models combining deterministic and machine learning components for numerical climate modeling and weather prediction[J]. Neural Networks, 2006, 19(2):122-134.
[6] Hang Li. (2018) Deep learning for natural language processing: advantages and challenges[J]. National Science Review, v.5(1):28-30.
[7] Fazle Karim, Somshubra Majumdar, Houshang Darabi, et al. (2017) LSTM Fully Convolutional Networks for Time Series Classification[J]. IEEE Access, PP (99):1662-1669.

IOP Conf. Series: Materials Science and Engineering **799** (2020) 012036 doi:10.1088/1757-899X/799/1/012036

Research on Design and Control of Dual-Energy Source System for Pure Electric Vehicle

Wei Min

School of Electronic Engineering, Xi'an Shiyou University, xian, shanxi, 710065, China

Abstract: Based on the analysis of the structure of dual-energy source storage system and the requirements of drive motor for the power of the battery and super capacitor under different operating conditions, and according to the motor's required power and the battery and super capacitor's SOC (State of Charge) and other parameters, a corresponding distribution fuzzy control strategy of dual-energy source power is established. A complete electric vehicle model of a dual-energy source pure electric vehicle driven by a DC (Direct Current) motor is constructed by Matlab. The system simulation results show that the battery-super capacitor dual-energy source system can simultaneously meet the dual requirements of energy and power of the pure electric vehicle. Both the economy and power have been improved. Compared with the single-battery power supply of pure electric vehicles, the electric motor drive system of pure electric vehicles with dual-energy sources is more efficient. And the mileage for a single continuous charge, the maximum speed, the shortest acceleration time of 50 km/h and maximum gradeability of 50km/h are improved. The mileage increases by 32.4%. At the same time, the charge and discharge current and temperature of the battery are significantly reduced, and the service life of the battery is effectively.

1. Introduction

The energy crisis and environmental pollution problems have promoted the rapid development of electric vehicles around the world[1]. Traditional pure electric vehicles use battery as the only energy source, which has the problems of long charging time, low specific power, and inability to meet the short-term power requirements of automobiles. The acceleration, climbing, braking performance and energy recovery efficiency of the motor are seriously affected and the motor's requirements for on-board power cannot be fully met[2-3]. In order to solve the above problems, the battery and super capacitor constitute a dual-energy source. The super capacitor has a high power density, a long service life, and the ability to withstand instantaneous high-current charge and discharge, which can effectively make up for the shortcomings of the battery, and improve the mileage of pure electric vehicles.

In addition, the drive motor and control technology are also the key technologies of electric vehicles. The application of brushless DC (Direct Current) motors with electric and electric generation functions in dual-energy sources pure electric vehicles can help solve this problem. In the operation of automobiles, the driving motor is switched and controlled between electric and power generation according to the control requirements. Energy utilization and recovery are performed between the large capacitor and the battery.

On energy distribution strategies for dual-energy sources electric vehicles, the current research mostly focuses on single energy management of dual-energy source [4-5] based on the charge and discharge characteristics of battery and capacitors. It is still not ideal in terms of energy brake recovery and driving mileage improvement of electric vehicles, etc. Therefore, combining the drive motor control technology and dual-energy source technology of electric vehicle, a fuzzy control strategy is designed, and an overall model of the dual-energy source control system for DC motors is built. the system performances are verified by simulation.

2. the structure of dual-energy source drive control system

The structure of the dual-energy source drive control system is shown in Figure 1. The battery is the vehicle's main power source, and the super capacitor is the vehicle's auxiliary power source. The battery and the super capacitor form a parallel structure, with a DC/DC converter adjusting the terminal voltage of the super capacitor, realizing power allocation according to a certain control strategy.

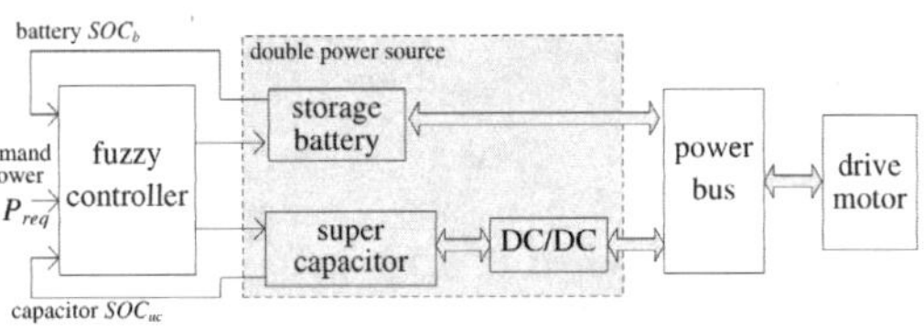

Figure 1 system structure diagram of motor drive with source

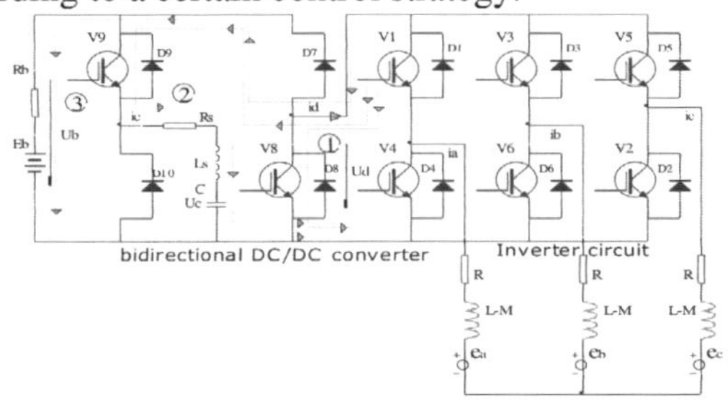

Figure 2 priority of the capacitor and the battery dual-energy source charging

3. The circuit design for motor drive and braking with dual-energy source

During the operation of a pure electric vehicle, the drive motor generally works in two operating states: electric and power generation. During the electric operation, the dual-energy source supplies power to the drive motor. And during the power generation operation, the capacitor is fully charged first, then the battery is charged by the drive motor. In this paper, we focus on studying this kind of complex power generation operation. The circuit is shown in Figure 2.

In this process, the six power tubes of the inverter bridge circuit are turned off by the control signal of the main board. When the power generation control preferentially charges the capacitor, v8 and v9 are respectively modulated by PWM (Pulse Width Modulation) [6-7]. In Figure 2, when V8 is turned on by the PWM modulation, the loop① is a BOOST boosting circuit that stores magnetic field energy for the winding coils of the brushless DC motor. When V8 is turned off by the PWM modulation and V9 is turned on, the motor winding coils store the magnetic field energy to charge the capacitor by D7, as shown in the loop②.

After the capacitor is fully charged, the battery is charged. As in the case of charging the capacitor, the magnetic field energy is stored in the DC motor winding coils through the loop① in Figure 2. When V8 is turned off, the magnetic field energy stored in the motor winding coils is used to charge the battery through D7, as shown in the loop③.

4. The design of fuzzy control strategy

The design of the system fuzzy control strategy is based on the premise of ensuring the dynamic performance of electric vehicles, making full use of the characteristics of high current charging and discharging of super capacitor, while reducing the impact on the battery of the frequent high current charging and discharging, extending the service life of the battery and increasing the energy recovered by the regeneration brake system. The driving mileage of the electric vehicle and the vehicle's dynamic performance are also increased [8-9].

4.1. Structure of fuzzy controller

The three-input and single-output fuzzy controllers have the input variables as the required power p_{req} of the motor drive system, the battery SOC_b(State of Charge), and Super capacitor SOC_{uc}. The output variable is the power factor $K_{uc}$of the super capacitor.

IOP Conf. Series: Materials Science and Engineering **799** (2020) 012036 doi:10.1088/1757-899X/799/1/012036

4.2. Design of fuzzy control rules

The number of fuzzy controller language values is related to the accuracy of the fuzzy control and the calculation workload. Considering the overall situation, the language value of each variable of the controller is shown in Table 1.

Table 1 Language values of input and output variables of fuzzy controller

mode	Input / output variable	Language value
Discharge mode (p_{req}> 0)	p_{req}	TS, S, M, B, TB
	SOC_b	L, M, H
	SOC_{uc}	L, M, H
	K_{uc}	TS, S, M, B, TB
Charging mode (p_{req} <0)	SOC_b	L, M, H
	SOC_{uc}	TL, L, M, H
	K_{uc}	TS, S, M, B

Combined with the dual-energy source working mode and various driving conditions of electric vehicles, the fuzzy control rules in the discharge mode and charge mode are analyzed.

When the required power of the driving motor is $p_{req} > 0$, the dual-energy source is in the discharge mode, and the pure electric vehicle is in the driving state. According to the different characteristics of the battery and the super capacitor, the power battery should be at low current and low power at this time. The super capacitor should provide large peak power to ensure the maximum efficiency of the dual-energy source system.

If $p_{req} > 0$ and the power increase $\Delta p_{req} > 0$is large, the motor is in an emergency acceleration state (starting, accelerating or overtaking), and it is likely to decelerate. At this time, K_{uc} ,at least, should be large. if $p_{req} > 0$ and the power increase$\Delta p_{req} > 0$ is small, it means that the motor is driving at a constant or low speed, and the required power is mainly provided by the battery, and the super capacitor is used as an auxiliary power source to supplement. At this time, K_{uc} should be smaller.

When the required power of the drive motor is $p_{req} < 0$, the dual-energy source is in charging mode, and the pure electric vehicle is in the brake feedback state.

If $p_{req} < 0$ and the remaining battery quantity is small, that is, SOC_b is small, the feedback energy of the drive motor is preferentially used to charge the battery. At this time, the power ratio K_{uc} of the super capacitor is smaller.

If $p_{req} < 0$ and a speed increase is required after braking, in order to provide the instantaneous power required for speed increase, the feedback energy of the drive motor is used to charge the super capacitor first. At this time, the power ratio K_{uc} of the super capacitor is higher.

Finally, overcharge phenomenon should be prevented. That is, if the battery's remaining power SOC_b is high, the battery's proportion of power is small, while K_{uc} is high. The fuzzy rules of the fuzzy controller for the discharge and charge mode are shown in Figures 3 and 4 respectively.

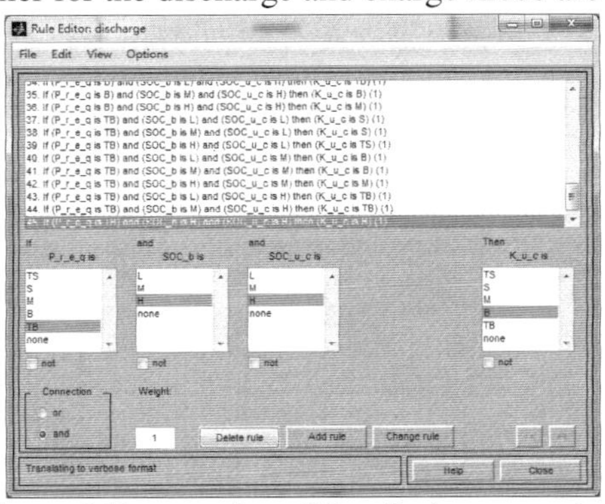

Figure 3 Fuzzy rules of the discharge mode

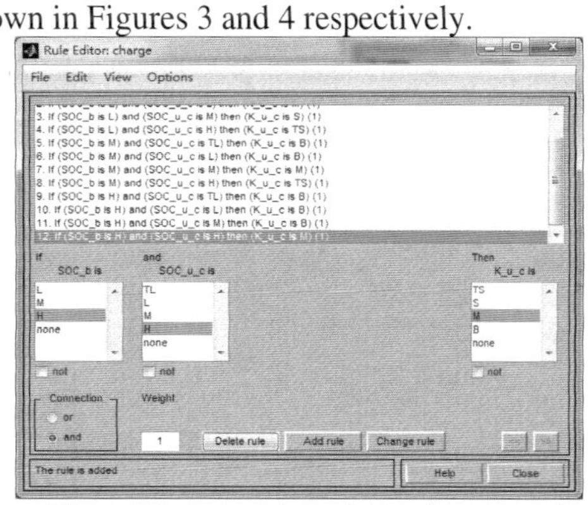

Figure4 Fuzzy rules of the charge mode

IOP Conf. Series: Materials Science and Engineering **799** (2020) 012036 doi:10.1088/1757-899X/799/1/012036

The structure of the dual-energy source control strategy model built on the Matlab/Simulink platform is shown in Figure 5.

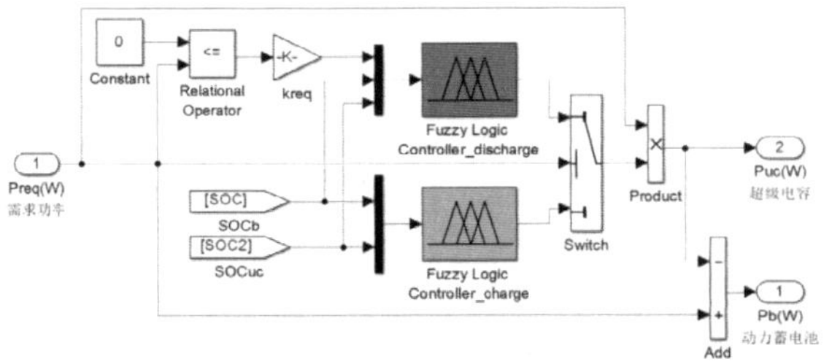

Figure 5 Structure diagram of dual-energy source control strategy model

5. System Modeling

Under the Matlab environment, the fuzzy controller, motor, composite-energy, DC/DC converter and other component models are constructed. Then they are Integrated into the advisor simulation software. The dual-energy source electric vehicle model interface is shown in Figure 6.

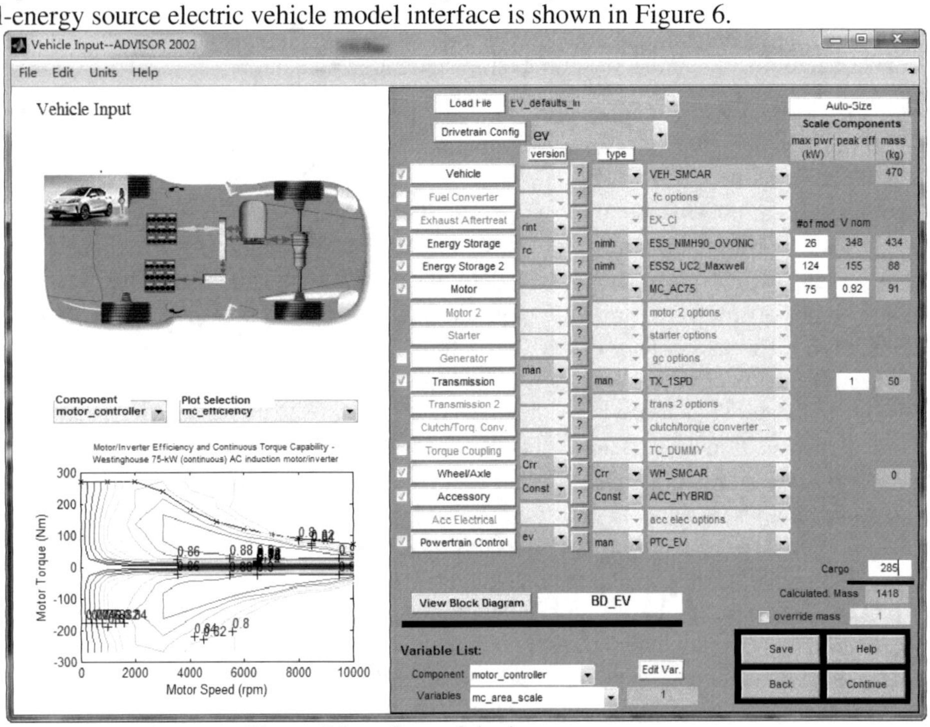

Figure 6 interface of dual-energy source

IOP Conf. Series: Materials Science and Engineering **799** (2020) 012036 doi:10.1088/1757-899X/799/1/012036

6. System Simulation and Analysis

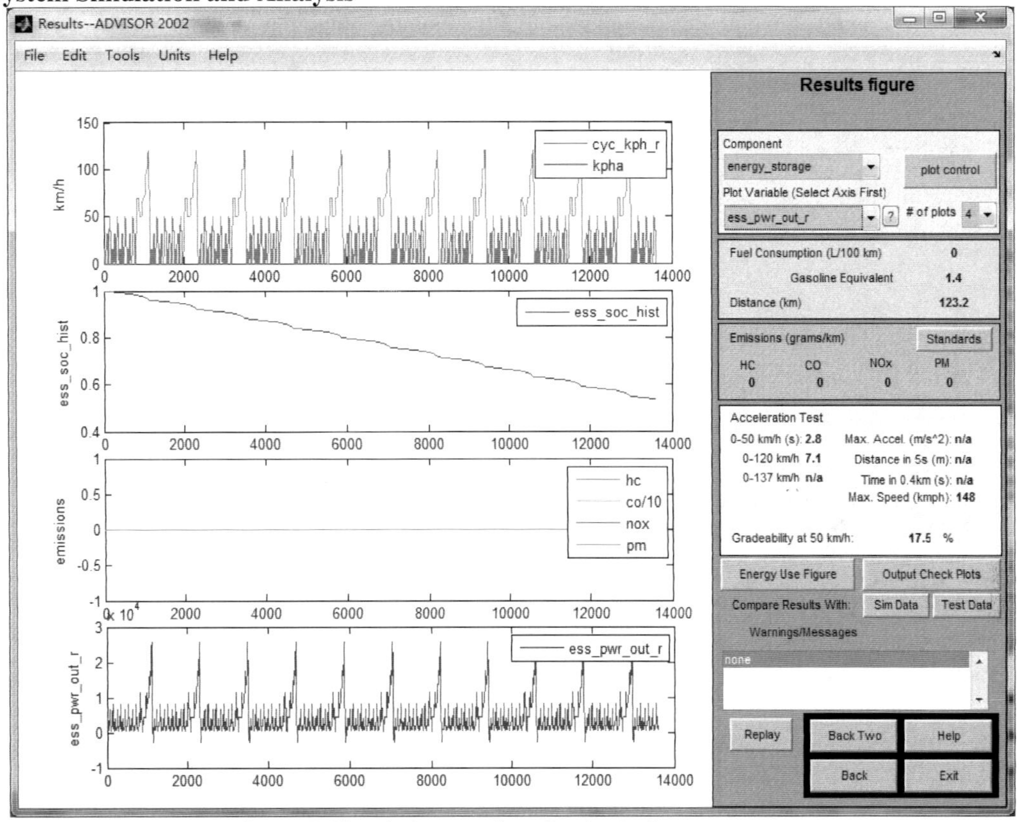

Figure 7 Simulation results of a single-battery source

The number of battery and super capacitors (only dual-energy sources), the peak power of the drive motor, the test load quality and so on, are set in the simulation system interface. While the type of cycle conditions, the number of cycles, the acceleration performance test, the climbing performance test and other functions are selected.

Simulation experiments were performed under NEDC typical operating conditions. The simulation results of a single-battery and a dual-energy source pure electric vehicle are shown in figures 7and 8respectively. The comparison results are shown in Table 2.

IOP Conf. Series: Materials Science and Engineering **799** (2020) 012036 doi:10.1088/1757-899X/799/1/012036

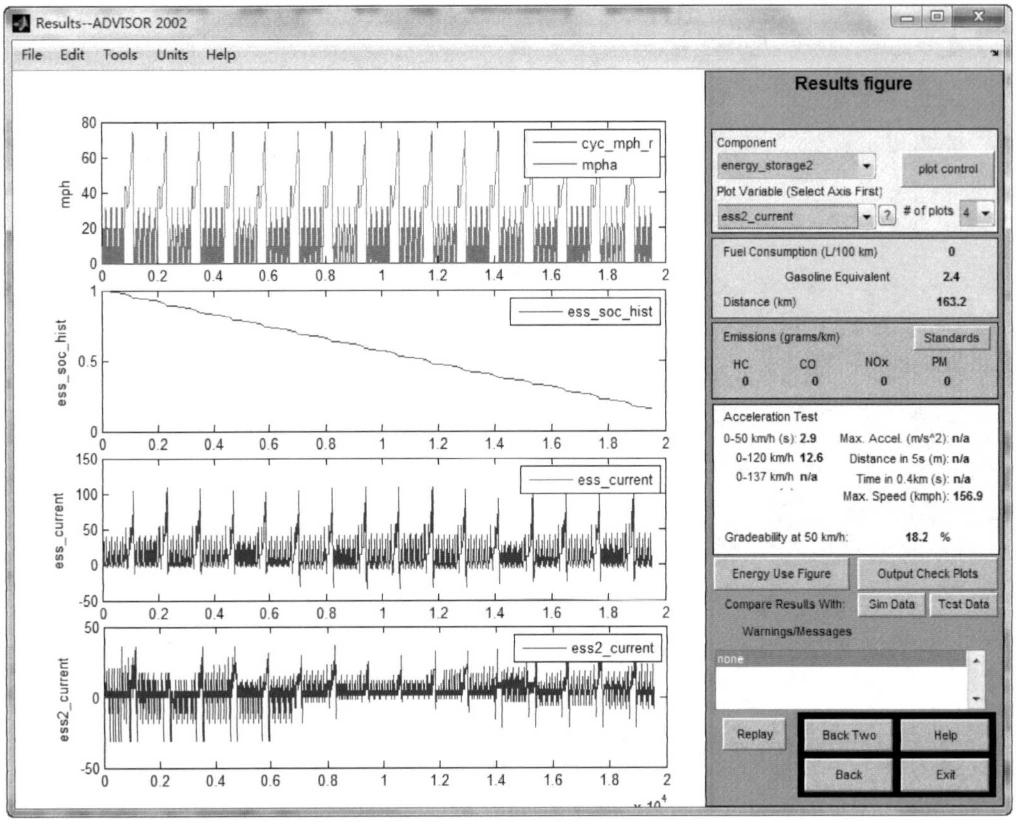

Figure8 Simulation results of dual-energy sources

Table 2 Comparison of dynamic performance results

	Single- power supply	Dual-energy source	Compared
Driving mileage	123.2Km	163.2Km	Up 32.47%
Maximum speed	148.0Km/h	156.9Km/h	Up 6.01%
Minimum acceleration time of 50Km / h	2.9 s	2.8 s	Up 3.57%
Maximum gradeability of 50Km / h	17.5%	18.2%	Up 4%

It can be seen from the simulation results that under the same operating conditions, the pure electric vehicles based on dual-energy sources are compared with those based on single-battery power. The continuous mileage has increased significantly by 32.47%, and the maximum vehicle speed, the shortest acceleration time of 50Km/h and the maximum gradeability of 50Km/h are also increased slightly. From the analysis of current changes, the battery current in the dual-energy source system is relatively stable, avoiding large current shocks. Analyzing the temperature changes, super capacitor ease the interior heat accumulation of the battery during driving, extending the service life of the battery.

7. Conclusion

Aiming at current problems of the low energy utilization and recovery efficiency of pure electric vehicles' composite-energy technology, combining the control technology of electric vehicle drive motor and dual-energy source technology, a fuzzy control strategy is designed. And an overall model of composite-energy control system is constructed for DC motors. The simulation results show that with the system energy control strategy, the power distribution and energy management can be reasonably performed, and the driving mileage, the maximum speed, the climbing performance and the acceleration performance on a single charge are improved to different degrees compared to the single-power pure electric vehicle. The charge and discharge current and temperature are significantly reduced, which prolongs the service life of the battery.

Acknowledgments
This work was supported by the Key Research Program of Shaanxi Province, China (Grant No. 2018GY-181).

References

[1] ZHOU M.L, FENG J.L, ZHANG Y. Composite energy storage system and its energy control strategy for electric vehicles [J]. Electric Machines and Control,2019,23(5)51-58.
[2] LI J, YANG D.H, SHU H.B, et al. Matching and Performance Simulation of Dynamic Systems of Pure Electric Vehicles[J].Journal of Huaqiao University (Natural Science),2016,31(9):129-136.
[3] CHEN L, CHENG Y. F, PAN C.F. Study on Driving Range of Electric Vehicle with Energy Recovery Function[J].Journal of Chongqing University of Technology(Natural Science),2016,31(9):129-136.
[4] Wang R, Li X.M， Wei W etc. Hybrid power system of pure electric vehicle based on ADVISOR [J].Journal of Shangdong University of Technology (natural Science Edition),2014,28(1):73-78.
[5] Xing F, Yang S, Xiong F. Research on energy management for hybrid power system electric vehicle [J].Journal of Inner Mongolia Normal University (natural Science Edition),2018,47(6):486-489.
[6] Wang X.Y, Fu T. Position senseless control of BLDC motor based on global fast terminal sliding mode observer [J]. Transaction of CHINA electrotechnical society, 2017,32(9):16-23.
[7] NIKOLAOS A S, YIANNIS S B. Multiple models fuzzy control: a redistributed fixed models based approach [J]. Intelligent Automation & Soft Computing,2014,9(23): 202-206.
[8] LU S, LIU Z.Q, GE J, et al. Simulation study on regenerative braking strategy of pure electric vehicle based on fuzzy control[J]. Journal of Changsha University of Science and Technology (Natural Science),2017,14(3):90-96.
[9] LIU Z.Q, WANG H.L, DU R.H. Fuzzy control method of regenerative braking in electric vehicles considering parameter sensitivity[J]. Journal of Central South University (Science and Technology),2016,47(1):3700-3706.

IOP Conf. Series: Materials Science and Engineering **799** (2020) 012037 doi:10.1088/1757-899X/799/1/012037

Research and Design of Hybrid Scheduling Algorithm Based on LIN Bus

Zhanwei Liang[1], Hongyu Shi[1], Hao Yan[1], Dongmei Ma[1]andJie Li[1]

[1]Chongqing University of Posts and Telecommunications Automation College, Chongqing 400065

E-mail: 2820429366@qq.com

Abstract: In the LIN(Local Interconnect Network) bus communication transmission, there is a problem of data transmission delay in the process of sending messages by the master-slave node ECU(Electronic Control Unit).Aiming at this kind of delay problem, this paper analyzes the characteristics of static scheduling and dynamic scheduling algorithm of LIN bus, optimizes the scheduling strategy of LIN bus message, and proposes a hybrid scheduling algorithm based on time slice scheduling. At the same time, the simulation model of LIN bus master-slave node message transmission is established, and the simulation experiment verification platform of LIN bus hardware-in-the-loop is built. It is verified that the algorithm reduces the delay by about 30%.

1. INTRODUCTION

Nowadays, the traditional data transmission method based on LIN (Local Interconnect Network) bus is the static scheduling method based on the static scheduling table to realize the message transmission. This kind of transmission method is simple and stable, but it has the problem of high transmission delay when the master node has event message to transmit. The LBFP (left-handed optimal matching algorithm) proposed by Jiang Yinfeng[1] mainly solves the problem of response time delay of message transmission between master and slave nodes ECUs, but does not solve the problem of transmission delay between messages in the transmission process. The LBFP-GA (genetic algorithm) [2] proposed by Yin Jianli mainly solves the problem of response time delay in the transmission between messages and the problem of delay time in the receiving of messages by nodes, and also does not solve the problem of transmission delay between messages in the transmission process.

Firstly, through the research and analysis of the traditional static scheduling algorithm data transmission method, the scheduling matrix of the static scheduling table is optimized, and a hybrid scheduling algorithm is designed. The main idea is to static in the scheduling matrix in the traditional static scheduling method. The time slice is divided into a basic time slice and a periodic scheduling time slice. The time slice interleaving method is used, so that when the active node needs to transmit the event type message, the latest time slice can be found, thereby reducing the delay.

2. Analysis of traditional static scheduling algorithm

2.1. Defect analysis of traditional scheduling algorithm

As shown in Figure 1, the data transmission characteristics of the LIN bus are in the form of master-slave. The host computer sends the frame header or frame header plus data in the form of information exchange with the lower computer. After the lower computer receives the message sent by the upper computer, it defines The database analyzes the ID (Identity Document) or ID plus data, performs corresponding operations according to the parsed result, and returns feedback data. The host computer receives the message sent by the lower computer and parses its ID and data. The upper computer functions to define the message transmission sequence and time of the LIN bus.

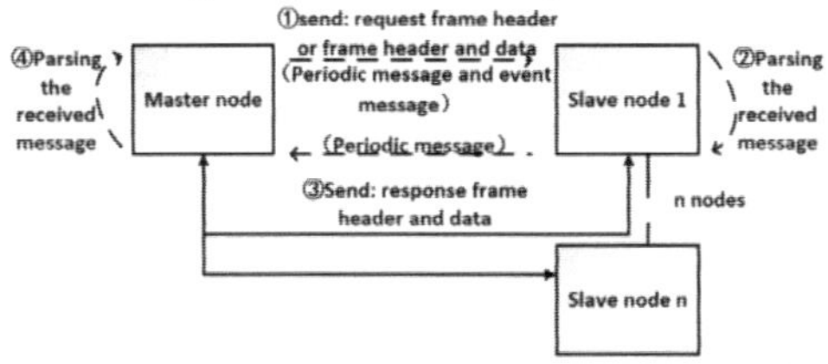

Fig 1. Schematic diagram of data transmission between the master and slave nodes of the LIN bus

The time that a message is transmitted on the LIN bus is determined by two factors[3],delay time and response time. The delay factors of the message mainly include the following parts: the message sending waiting time T_q,the sending average delay time T_d,the message transmission factor is μ_1,and the verification factors of the message delay time are positive or negative. Load rates ρ_a and ρ_b.

2.2. *Delay analysis of traditional scheduling algorithm*

The message sent from the host computer to the lower computer is a forward transmission message, and the message sent from the lower computer to the upper computer is a reverse transmission message. In the LIN network[4],the forward message sent by the upper computer is $M_1, M_2, \dots, M_n, M_{n+1}, \dots M_m$,and consider that all forward packets of $M_1 \sim M_n$ are event-type messages, $M_{n+1} \sim M_m$ all forward messages For the periodic message, the reverse message sent by the lower computer is $S_1, S_2, S_3, \dots, S_n$ and considers that all reverse messages are periodic messages.The transmission factor of the bus message is$\mu_1, \mu_2, \mu_3, \dots, \mu_n$, the transmission factor represents the average waiting delay coefficient of each frame of the message in transmission. The expressions of the positive and negative load rates ρ_a and ρ_b in the LIN network are:

$$\rho_a = 1.4 \times \left\{ \sum_{i=1}^{n} [34 + 10N_i]\,\mu_i + \sum_{j=n+1}^{m} [36 + 10N_i]/\,M_j \right\} \times 100\% \quad (1)$$

$$\rho_b = 1.4 \times \left\{ \sum_{k=1}^{n} [34 + 10(N_k + 1)]\,/S_k \right\} \times 100\% \quad (2)$$

In the above formula, the data length of N_i and N_j forward transmission messages, and N_k is the data length of the reverse transmission message.

3. Hybrid scheduling algorithm design

3.1. *Hybrid scheduling algorithm design ideas*

Since the data transmission mode based on the LIN bus is based on the static scheduling of the static schedule, when the event message needs to be transmitted by the host computer, it still needs to wait for the corresponding sequence in the static scheduling to transmit, which reduces the delay of the event type message, so this paper proposes a hybrid scheduling algorithm to improve this situation. The scheduling diagram of the hybrid scheduling algorithm is as follows:

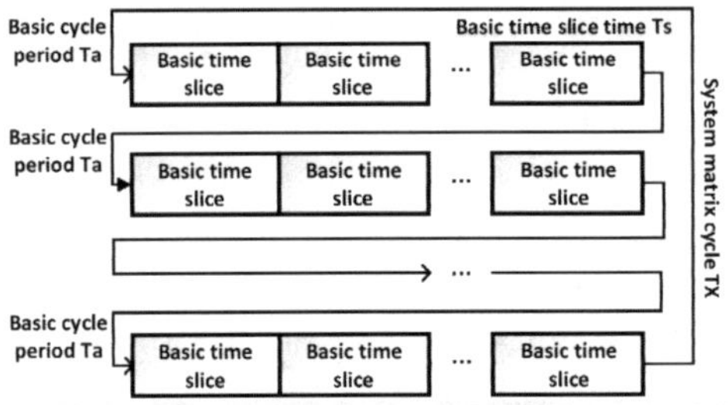

Fig 2. LIN bus hybrid scheduling theory table

Figure 2 is a hybrid schedule. The hybrid scheduling period of the hybrid schedule is T_x. The basic cycle of each packet is T_a,and the time of each base time slice is T_s The object of this method is the common cycle in the project. Type, does not contain all the packets have the same period and the period of the message is a very special case such as prime numbers[5].

The two time slices are respectively a periodic scheduling time slice and a basic time slice. The periodic scheduling time slice sends a forward type transmission message and a periodic type message in the reverse transmission message, and the basic time slice is sent in the forward transmission message. The event-type message, the system scheduling period T_x of the hybrid scheduling is the least common multiple LCM(Lowest Common Multiple) between all periodic-type message transmission periods, and the basic cyclic period T_a of the hybrid scheduling is the greatest common divisor GCD(Greatest Common Divisor) between all periodic-type message transmission periods,the expressions of T_x and T_a are:

$$T_x = \text{LCM}(M_1, M_2, \dots, M_n, \ S_1, S_2, \dots, S_n) \quad (3)$$

$$T_a = GCD(M_1, M_2, \dots, M_n, \ S_1, S_2, \dots, S_n) \quad (4)$$

In the hybrid scheduling process, all time slices in the schedule are basic time slices before the start of scheduling. When the periodic message is sent, the time slice occupied by the periodic message becomes periodic scheduling[6].

After the hybrid schedule starts to be scheduled, the periodic packets are filled in the packets from the upper left corner of the table. The arrangement is arranged according to the packet sending period from small to large. The packets with small periods are listed in the previous basic time slice. And so on, know that all periodic messages are discharged into the hybrid schedule. This type of arrangement will make the periodic message occupy a complete line in the static schedule when $T_a = T_s$, when $T_a = nT_s$ When the two values are in the n-fold relationship,the periodic message occupies a fixed number of base time slices in a row in the static schedule. In the static schedule, the number of time slices of each row is fixed and the same.

3.2. Delay analysis of hybrid scheduling algorithm

In the above formula, f(s) represents the total number of event-type messages and periodic messages sent by the host computer,T_x represents the cycle period of the static schedule, T_s represents the length of the base time slice in the static schedule,and C_i is the message[7].

In the hybrid scheduling process, the periodic type message sent by the upper computer and the periodic type message sent by the lower computer send the message according to the time slice specified in the hybrid scheduling. The following is the analysis of the message waiting time delay of the periodic message and the event type message. According to the queuing theory of bus communication, this paper can calculate the average message transmission waiting time T_q of the hybrid schedule as follows:

$$T_q = \rho_a T_s \times \frac{n}{m} + \rho_b T_s \quad (5)$$

It can be seen from the above equation that the average message transmission waiting time of the hybrid schedule is almost determined by the periodic message transmission time.

The average delay time T_d of the periodic message sent by the hybrid schedule is as follows:

$$T_d = \rho_a T_s \times \frac{n}{m} + \rho_b T_s + \sum_{i=1}^{n} C_i \quad (6)$$

The average delay time T_l of the event type message sent by the hybrid schedule is as follows:

$$T_l = nT_s + \sum_{i=1}^{n} C_i \quad (7)$$

4. Hybrid scheduling algorithm delay optimization verification

The communication mode of LIN bus is the main form. In this paper, the communication between the upper computer and the lower computer is used for analog verification. The upper computer is PC and the lower computer is an ECU controller. This paper takes an ECU controller as an example and the upper computer. The LIN bus communication environment is built with CANcaseXL, and the hybrid scheduling algorithm designed in this paper is used to verify the algorithm feasibility.

In this system, the baud rate is 19200 bps, the base time slice T_s takes 20 ms, the hybrid scheduling table T_x takes 500 ms, and the basic message transmission basic cycle T_a takes 100 ms.

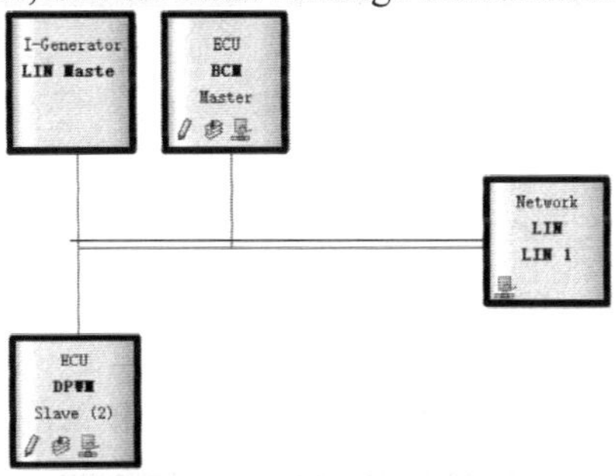

Fig 3 test engineering topology

Figure 3 shows the test sample topology. The master node is BCM, the slave node is DPWM, the BCM optional periodic message has 8 frames, the optional event type message has 5 frames, and the DPWM periodic message has 7 frames. The specific frame contents are listed below.

Table 1 shows the information of the 8-frame period packet sent by the host computer in the system, including the packet name, ID, packet length, and period. The BCM_DDAPWL_01 and BCM_DDAPWL_02 are periodic monitoring request frame headers. Then, the monitoring frame header is sent,BCM_01 is the Global message of the ECU, and is used for the ECU remote controller control, and BCM_Check_02 to BCM_Check_06 are the periodic type verification request frame headers, and if the verification is performed, the verification frame header is transmitted.

Table 1 Host machine cycle type message

Message name	ID	Message length	cycle /ms
BCM_DDAPWL_01	0x27	8	20
BCM_DDAPWL_02	0x28	8	20
BCM_01	0x31	8	20
BCM_Check_02	0x22	8	20
BCM_Check_03	0x23	8	20
BCM_Check_04	0x24	8	20
BCM_Check_05	0x25	8	20
BCM_Check_06	0x26	8	20

Table 2 shows the information of the 5-frame event-type packet sent by the host computer in the system, including the packet name, ID, packet length, and period. The BCM_02 to BCM_06 are event-type request packets.

Table 2 Host computer event type message

Message name	ID	Message length	cycle /ms
BCM_02	0x32	8	
BCM_03	0x33	8	
BCM_04	0x34	8	
BCM_05	0x35	8	
BCM_06	0x36	8	

Table 3 shows the information of the 7-frame periodic packet sent by the lower-level device in the system, including the packet name, ID, packet length, and period. DDAPWL_01 and DDAPWL_02 are monitoring response packets. After the frame header, the monitoring response message is sent, and CLI_01 to CLI_05 are the verification reply message. If the host computer checks the frame header, it sends a verification reply message.

Table 3 lower machine cycle type message

Message name	ID	Message length	cycle /ms
CLI_01	0x18	8	20
CLI_02	0x19	8	20
CLI_03	0x1A	8	20
CLI_04	0x1B	8	20
CLI_05	0x1C	8	20
DDAPWL_01	0x10	8	20
DDAPWL_02	0x11	8	20

Figure 4 shows the actual hybrid scheduling table of the system. Each row contains 5 basic time slices. Each row consists of 3 frame periodic messages and 2 basic time slices. The basic time slice is interspersed with periodic message settings. The basic time slice T_S The value is 20 ms,the hybrid scheduling period T_X of the hybrid schedule is 500 ms, and the basic packet transmission basic period T_a takes 100 ms.

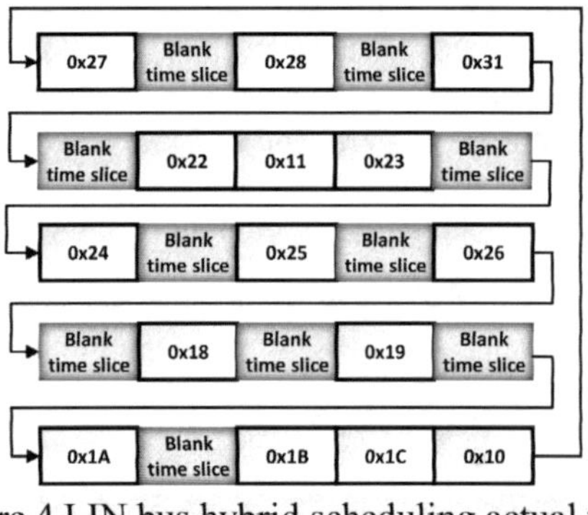

Figure 4 LIN bus hybrid scheduling actual table

Figure 5 is the effect of the delay time of the event type message under the load rate of different periodic types of messages. The figure shows the delay time curve of the event type message during the change of the periodic message load rate from 0% to 95%.The graph shows the changes in the delay profile of event-type messages in three different cases.

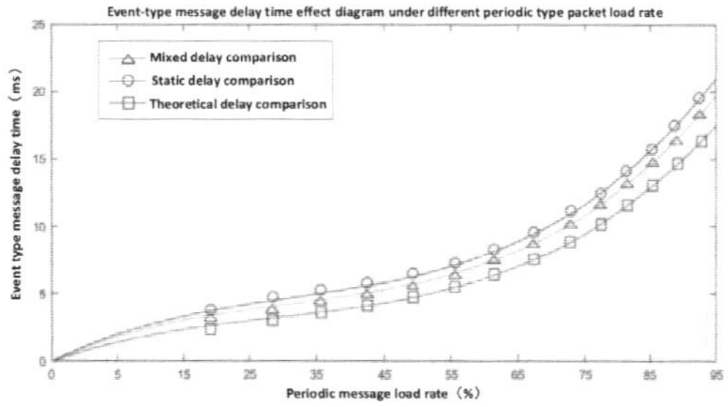

Figure 5 Effect of delay time of event-type packets under different periodic packet load rates

Table 4 shows the average theoretical delay time of periodic packets sent by equations (8), (9), and (10) at different load rates. The average delay time of event-type messages in the above waveform is analyzed. The actual average delay time of the event-type message in the static schedule and the hybrid schedule can be calculated. The calculation can conclude that the hybrid scheduling algorithm in this paper will be compared with the traditional static scheduling algorithm in the event-type packet transmission delay. The delay has increased by about 30%.

Table 4 Average delay time of event-type packets under different periodic packet load rates.

Delay time /ms	Periodic message load rate			
	25%	50%	75%	95%
Theoretical delay	2.5	4.6	7.9	15.1
Static scheduling	3.65	6.05	9.95	18.15
Hybrid scheduling	3.31	5.51	9.21	17.01

5. Conclusion

This paper presents a hybrid scheduling algorithm based on blank time slice and periodic time slice. Through the establishment of hardware in the loop simulation test system, two different scheduling algorithms are compared and tested to verify the timeliness of the hybrid scheduling algorithm. Through the analysis of experimental data, it is concluded that the hybrid scheduling algorithm is 30% more efficient than the static scheduling algorithm in terms of the transmission delay of event-based messages, and in the LIN bus message transmission process, with the increase of the bus load rate The timeliness of data transmission is better.

Acknowledgement

We want to thank Professor Cheng Anyu for his support and tremendous help in this study. This research project was supported by the Chongqing Municipal Education Commission Science and Technology Project (KJ1722377) and the Chongqing Municipal Education Commission's joint construction project.

REFERENCES

[1] Yinfeng. Research on Static Segment Scheduling Algorithm of Vehicle LIN Bus, Master thesis of Hunan University, 2016
[2] Jianli Yin, Research on Dynamic Segment Scheduling and Optimization Configuration of Automotive LIN Network, Master thesis of Hefei University of Technology, 2014
[3] Homchaudhuri B,. Hierarchical control strategies for energy management of connected hybrid electric vehicles in urban roads[J] .Transportation Research Part C Emerging Technologies, 2016,62:70-86

[4] Bu Qiao, Distributed Networked Time Lag Analysis and Control for Distributed Drive Electric Vehicles, Beijing Institute of Technology Master thesis, 2016
[5] Ma C,Ko S Y,Jeong K Y,et al.Design methodology of component design environment for PHEV[J].International Journal of Automotive Technology,2013,14(5):785-795
[6] Lee D H,Kim N W,Jeong J R,et al.Component sizing and engine optimal operation line analysis for a plug-in hybrid electric transit bus[J].International Journal of Automotive Technology,2013,14(3):459-469
[7] Zeng X,Wang J.A Parallel Hybrid Electric Vehicle Energy Management Strategy Using Strochastic Model Predictive Control With Road Grade Preview[J].IEEE Transactions on Crotrol Systems Technology,2015,23(6):2416-2423

IOP Conf. Series: Materials Science and Engineering **799** (2020) 012038 doi:10.1088/1757-899X/799/1/012038

Low-illumination image enhancement algorithm based on multi-feature fusion

Xuewei Zhang[1,2*], Xiaojuan Sun[1], Tianfeng Wang[3]

[1]School of Mechanical and Electrical Engineering, Xi'an Technological University, Xi'an, 710021, China

[2]School of Electronic and Information Engineering, Xi'an Technological University, Xi'an, 710021, China

[3]Xi'an Kaitian electrical reliability laboratory Co., Ltd, Xi'an, 710077, China

[*]Corresponding author's e-mail: xueweizhang163@163.com

Abstract. In order to improve the effect of image acquisition under the condition of low illumination, a low illumination image enhancement algorithm based on multi-feature fusion is adopted in this paper. In this algorithm, the illumination information is extracted by bilateral filtering, and the illumination parameters are processed by adaptive gamma correction and limited contrast histogram equalization algorithm. The final illumination parameters are obtained by the feature weighted fusion of the processed parameters and the original illumination parameters. The RGB image is obtained by combining the illumination and reflection parameters, and the enhancement of the image is realized. The experimental results show that the algorithm used in this paper can not only improve the brightness of the image, but also enhance the characteristic parameters of the image, remove the noise, and make the color information of the image fuller and richer

1. Introduction

Under the condition of low illumination at night, the poor lighting conditions of the surrounding environment and the problems of the image acquisition equipment will affect the quality of the image acquisition. As a result, the gray value and contrast of the whole image are low and the region of interest is difficult to identify. In addition, a part of the image has low gray value due to the lack of light or located in the shadow area, which is fused with the background, making it difficult to extract the information of this part of the original image. Therefore, enhancing the brightness and contrast of low illuminance image has important application value. There have been many methods to solve the problem of low illuminance image enhancement, such as histogram equalization algorithm, Retinex algorithm, MSRCR algorithm, spatial color image enhancement algorithm[1], gradient image enhancement algorithm[2], bionic color image enhancement algorithm[3], homomorphic filtering algorithm based on illumination reflection model[4], HDR algorithm, etc. Histogram equalization algorithm has a particularly good enhancement effect on images with an overall darker or brighter image, it has been widely used in image enhancement. But for low illumination image, it is easy to over enhancement [5-6]. In order to solve this kind of problem, histogram equalization algorithm of keeping brightness mean value and adaptive histogram equalization algorithm with limited contrast are proposed. The two algorithms can restrain over-enhancement phenomenon in the process of enhancement, but can not enhance the image's detail and brightness at the same time [7-8]. Detailed information of dark area of image is lost [9-10]. Wavelet algorithms can enhance image components in different frequency ranges, highlight details in different scales, but the amount of calculation is too large, which is easy to amplify noise [11]. To solve this kind of problem, people proposed the method of combining wavelet with curved wave and the algorithm based on DWT and SVD. Although the noise is removed, the complexity of the algorithm is increased [12]. Retinex algorithm and its derivative algorithm decompose the image into illumination component and reflection component, and then process them. However, this kind of algorithm has a large amount of computation, which is easy to produce halo phenomenon [13]. For this reason, people put forward the Retinex image enhancement algorithm based on gyrosymmetric bilateral filtering algorithm and Markov random field. Although this kind of algorithm avoids halo phenomenon, the noise is not removed, which will produce over enhancement and under enhancement phenomenon. In reference [14], a Retinex algorithm based on bilateral filtering is proposed. This algorithm can remove the "halo" phenomenon and image noise well, and has good visual effect. However, the enhanced image detail information is not clear enough, and the image contrast is also low. Although the enhancement effect of image fusion algorithm is good, it needs to capture multiple images continuously, which is difficult to meet the real-time demand. To solve this problem, literature [15] proposed a weak light image enhancement algorithm of single image fusion, which can enhance the detail information of the image and improve the contrast of the image, but the real-time performance of the algorithm is not good, and the details of the dark area will be lost. Based on this, this paper adopts a low illumination image enhancement algorithm based on multi-feature fusion.

2. Low illumination image enhancement algorithm based on multi-feature fusion

2.1. Algorithm implementation

Based on the light reflection model, this paper combines the contrast enhancement and detail enhancement of low illumination image, and proposes a low illumination image enhancement algorithm based on multi-feature weighted fusion. The algorithm implementation flow is shown in Figure 1.

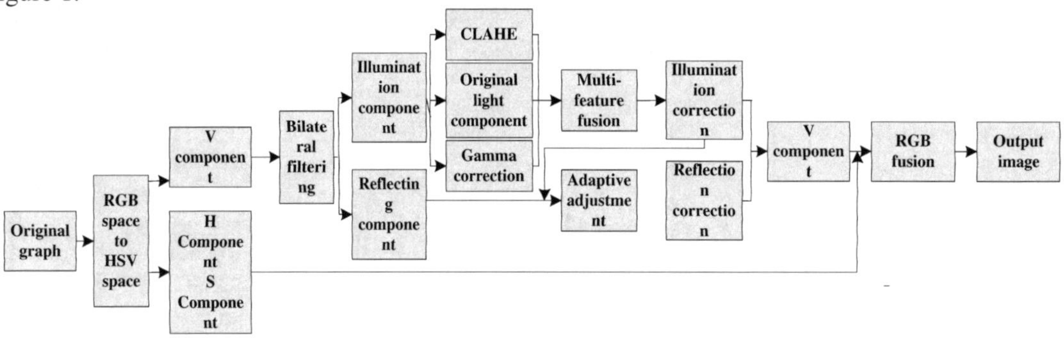

Figure 1. Algorithm flow

2.2. Light component processing

In order to improve the image quality of low illuminance image and reduce the influence of illumination on image, it is very important to extract the illumination information of image accurately. At present, there are many algorithms for extracting illumination information, such as Gauss filtering algorithm, bilateral filtering algorithm and Retinex algorithm of variational frame. Retinex algorithm based on variational frame is used to extract illumination information. In this paper, a bilateral filtering algorithm with edge preserving feature is used to extract the illumination information of the image, which can keep the edge information of the image well and restrain the image noise at the same time. Set A1 is the original light component of the low illumination image. A2 is the result of light component correction by adaptive gamma correction in this paper. The main purpose is to improve the image brightness. Firstly, the parameter g is obtained adaptively by using the mean value of estimated light component A(x, y), and then the parameter of gamma correction G(x, y) is adjusted adaptively. Namely

$$g = 1 - \frac{1}{m \times n}\sum_{x=1}^{m}\sum_{y=1}^{n} A(x, y) \tag{1}$$

$$e(x, y) = \frac{(A(x, y) + g)}{1 + g} \tag{2}$$

$$A_2(x, y) = A(x, y)^{e(x,y)} \tag{3}$$

Among them, m and n are the width and height of the image respectively, and e(x, y) is the adaptive parameter of gamma correction based on the pixel information of the original light component. Then we use the contrast-limited adaptive histogram equalization (CLAHE) to process the light component, and get the A3 component.

$$A_3(x, y) = CLAHE(A(x, y)) \tag{4}$$

According to the above three kinds of light components, this paper adopts a low illumination image fusion algorithm based on pixel level, and each pixel in the fusion image is the weighted average of the corresponding pixels in the three light components. The features of the pixels used can measure the quality of the image and the quality of the detail information. In this paper, the variance, gradient and entropy of the image can be used to reflect the quality of the image. In the experiment, for each pixel (x, y), take a (2d + 1) × (2d + 1) square window with the pixel as the center, count the three image features in the window, and then determine the weight of the pixel in the fusion according to the local features of the region p where the pixel is located. Finally, according to the pixel level fusion method, deduce the three light components of the image In the algorithm, the formula is defined as follows:

$$R_{\mathrm{var}}(x, y) = \frac{1}{f}\sum_{m=x-k}^{x+k}\sum_{n=y-k}^{y+k}[A_p(m,n) - \overline{A}_p]^2 \tag{5}$$

$$R_{grad,p}(x, y) = \frac{1}{f}\sum_{m=x-k}^{x+k}\sum_{n=y-k}^{y+k}(| A_p(m+1,n+1) - A_p(m,n) | + | A_p(m,n+1) - A_p(m+1,n) |) \tag{6}$$

$$R_{s,p}(x, y) = \sum_{m=x-k}^{x+k}\sum_{n=y-k}^{y+k} - p(A_p(m,n)) \log 2 p(A_p(m,n)) \tag{7}$$

$$\overline{A}_p = \frac{1}{f}\sum_{m=x-k}^{x+k}\sum_{n=y-k}^{y+k} A(m,n) \tag{8}$$

$$f = (2d+1) \times (2d+1) \tag{9}$$

Among them: A_p is the p-th light component estimation, $R_{\mathrm{var},p}(x, y)$ is variance, $R_{grad,p}(x, y)$ is gradient and $R_{s,p}(x, y)$ is entropy.

Because the image quality evaluation is related to the above variance, gradient and entropy, the image quality concept is written as:

$$A_p(x, y) = L(A^{\mu}_{\mathrm{var},p}(x, y), A^{\nu}_{grad,p}(x, y), A^{o}_{s,p}(x, y)) \tag{10}$$

The value of μ, ν, o is 0 or 1.

Set the weight coefficients of the above three parameters are:

$$b_p(x, y) = \frac{A_p(x, y)}{\sum_{i=1}^{3} A_p(x, y) + \varepsilon} \tag{11}$$

Among them, $A_p(x, y)$ is the quality degree of light parameter estimation; the weight coefficient satisfies $\sum_{i=1}^{3} b_{i,p}(x, y) = 1$, $0 \le b_p(x, y) \le 1$. ε is a very small positive real number, therefore, after fusion processing, the light components are as follows:

$$A_{zeng}(x, y) = \sum_{p=1}^{3} b_p(x, y) \times A_p(x, y) \tag{12}$$

3. Experiment and analysis

In this paper, two kinds of low illuminance images of different scenes are selected for the experiment. As shown in the figure2-3, a) is the Retinex algorithm, b) is the histogram equalization algorithm, c) is the MSRCR algorithm, d) is the spatial color image enhancement algorithm, e) is the gradient domain image enhancement algorithm, f) is wavelet transform algorithm, g) is the algorithm of this paper. And the 7 methods are used for processing images, then compared and analyzed the experimental results.

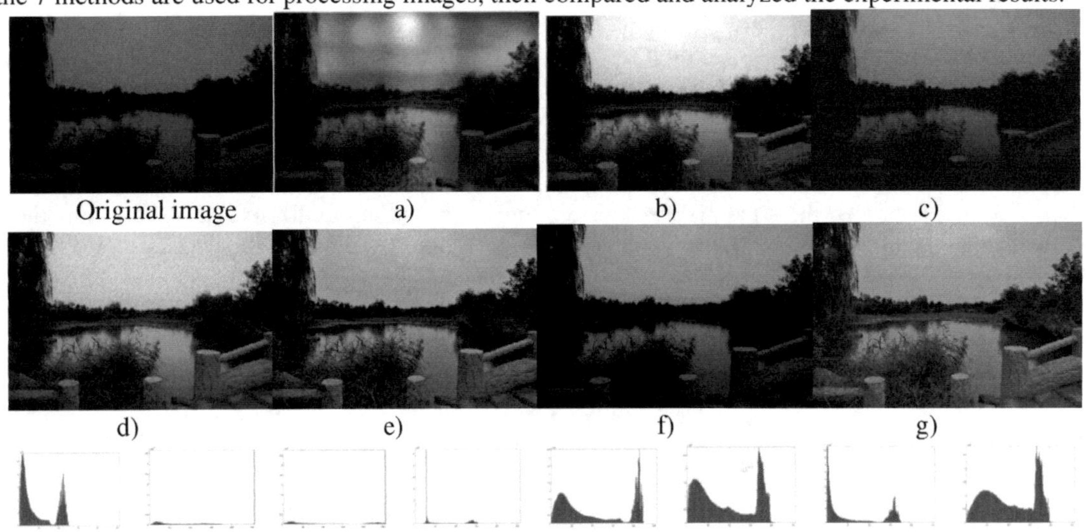

Histograms of original image, a) to g)

Figure 2. Scenery images

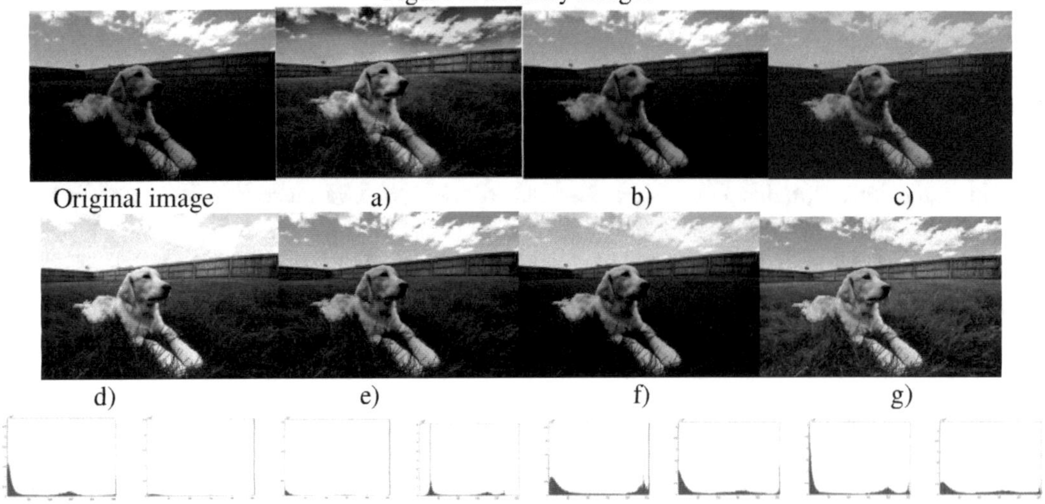

Histograms of original image, a) to g)

Figure 3 Pet images

As can be seen from Figure 2, a) algorithm and b) algorithm can increase the contrast and brightness while increasing the local details of the image, but the image as a whole will be over enhanced, resulting in the visual effect is not natural. After the image is enhanced by c) algorithm and f) algorithm, there is colour distortion. d) algorithm has halo. The results of this paper's algorithm are more natural, no halo, and the image visual effect is more in line with the human visual characteristics. As can be seen from Figure 3, the enhancement effect of c) algorithm and f) algorithm is not obvious enough. e) algorithm in the low illumination area and the contour around the light source are not high, and the enhancement algorithm improved by the algorithm of this paper improves the image quality more obviously than other algorithms. This paper's algorithm has full colour and histogram distribution is well-distribute.

In order to better evaluate the quality of panoramic image processed by different algorithms, the enhancement image is evaluated objectively from two aspects: clarity and the amount of information contained in the image. This paper selects the improved Brenner image sharpness evaluation algorithm to assess definition, which is very sensitive to image sharpness and proposed by literature[16]. The closer the index is to 1, the higher the image sharpness is, and vice versa. In terms of the information content of the evaluation image, this paper selects the information entropy as the evaluation index. The higher the information entropy is, the more information the image contains, and vice versa. The information entropy of the enhanced image corresponding to each algorithm and the image clarity proposed in literature [16] are shown in Table 1 and Table 2 respectively.

Table 1. Entropy Of relevant algorithms

Algorithm name	Fig.2	Fig.3
Original image	5.23	6.04
Retinex algorithm	6.67	7.11
Histogram equalization algorithm	6.43	7.18
MSRCR algorithm	6.01	6.57
Spatial color image enhancement algorithm	5.78	7.23
Gradient domain image enhancement algorithm	5.89	6.85
Wavelet transform algorithm	5.58	6.78
Algorithm in this paper	6.77	7.22

Table 2. Image definitions of relevant algorithms

Algorithm name	Fig.2	Fig.3
Original image	0.53	0.57
Retinex algorithm	0.88	0.78
Histogram equalization algorithm	0.71	0.89
MSRCR algorithm	0.71	0.71
Spatial color image enhancement algorithm	0.78	0.91
Gradient domain image enhancement algorithm	0.79	0.76
Wavelet transform algorithm	0.74	0.75
This paper's algorithm	0.91	0.93

It can be seen from Table 1 and Table 2, compared with the original image, each image enhancement algorithm can improve the amount of information contained in the image. The clarity and entropy of the improved algorithm are higher than those of the traditional algorithm, which shows that the paper's algorithm can effectively improve the image quality.

4. Conclusion

In order to solve the problem of low illuminance image enhancement, this paper proposes a low illuminance image enhancement algorithm based on multi-feature fusion. The experimental results verify that the algorithm in this paper can not only obtain good enhancement effect, but also overcome the shortcomings of colour distortion, halo effect and low contrast in the process of enhancement, so that the overall brightness of the enhanced image is significantly improved, and the details are further highlighted and increased. The colour of the strong image is natural, which effectively improves the visual quality of low illuminance panoramic image. In the process of low illuminance image enhancement, noise often has a great impact on the image. Therefore, it is of great significance to research on the enhancement of low illuminance image by adding the denoising process to the low illuminance image enhancement algorithm to realize the combination of image denoising, contrast enhancement and detail enhancement.

Acknowledgments
This work has been supported by Shaanxi Provincial Education Department (No.18JK0388).

References

[1] Cheng I, Basu A. Contrast enhancement from multiple panoramic images[C]. Int Conf onComputer Vision. Rio de Janeiro: IEEE, 2007: 1-7.
[2] Ding Chang, Dong Li-li. Review of “his-togram” equalization technique for image enhancementJ］. Computer Engineering and Applications, 2017, 53(23): 12-17.
[3] PAIHAR AS, VERMAOP, KIIANNA C. Fuzzy-Contextual contrast enhancement［J］. IEEE Transactions onImage Processing, 2017, 26 (4): 1810-1819.
[4] FANG Shuai, ZHAO Yu-kun. Nighttime haze removal based on illumination estimation[J]. Acta Electronica Sini-ca, 2016, 44(11): 2569-2575.
[5] Lee C, Lee C, Lee Y Y, et al. Power-constrained contrast enhancement for emissive displays based on histogram equalization[J]. IEEE Trans on Image Processing, 2012, 21(1): 80-93.
[6] Arici T, Dikbas S, Altunbasak Y. A histogram modification framework and its application for image contrast enhancement[J]. IEEE Trans on Image Processing, 2009, 18(9): 1921-1935.
[7] Zuiderveld K. Contrast limited adaptive histogram equalization[C]. Graphics Gems. San Diego: Academic Press Professional, 1994: 474-485.
[8] Kim Y T. Contrast enhancement using brightness preserving bihistogram equalization[J]. IEEE Trans on Consumer Electronics, 1997, 43(1): 1-8.
[9] Huang L S, Yu Q Y, Li Y X. Image enhancement method based on bilateral gamma correction and multi-scale image fusion[P]. China: CN104574337A, 2015.
[10] Huang L, Cao G, Yu L. Efficient contrast enhancement with truncated adaptive gamma correction[C]. Int Congress on Image and Signal Processing, Biomedical Engineering and Informatics. Datong: IEEE, 2017:189-194.
[11] Antonini M, Barlaud M, Mathieu P, et al. Image coding using wavelet transform[J]. IEEE Trans on Image Processing, 1992, 1(2): 205-220.
[12] Bhutada G G, Anand R S, Saxena S C. Edge preserved image enhancement using adaptive fusion of images denoised by wavelet and curvelet transform[J]. Digital Signal Processing, 2011, 21(1): 118-130.
[13] Land E H, Mccann J J. Lightness and retinex theory[J]. J of the Optical Society of America, 1971, 61(1): 1-11.
[14] Hu W W, Wang R G, Fang S, et al. Retinex image enhancement algorithm based on bilateral filtering[J]. Graphics Journal, 2010, 31(2): 104-109.
[15] Fu X, Zeng D, Huang Y, et al. A fusion-based enhancing method for weakly illuminated images[J]. Signal Processing, 2016, 129(C): 82-96.
[16] Wang J, Chen H B. An improved Brenner algorithm for image definition criterion[J]. Acta Photonica Sinica, 2012, 41(7): 855-858.

IOP Conf. Series: Materials Science and Engineering **799** (2020) 012039 doi:10.1088/1757-899X/799/1/012039

Research on alarm control system based on voice lock

Wu Rui[1], Yang Zhao[2*], Zhang He[3]

[1]College of Civil and Environment Engineering, Anhui Xinhua University, Hefei, Anhui Province, 230088, China

[2]College of Electronic Communication Engineering, Anhui Xinhua University, Hefei, Anhui Province, 230088, China

[3]College of Civil and Environment Engineering, Anhui Xinhua University, Hefei, Anhui Province, 230088, China

[*]Corresponding author's e-mail: wuuriu@163.com

Abstract. In order to realize the intelligent control and remote control of the safe and improve the efficiency of unlocking, a voiceprint unlocking system is proposed. The research uses voiceprint to control the switch of electronic lock. The core technology of voiceprint lock is speech recognition. Speech recognition technology is to convert speech signal to machine language, so that the machine can recognize the voiceprint. Because of the uniqueness of the voiceprint, compared with other unlocking methods, it has the advantages of high precision, high life, high efficiency, etc. Besides, because the voiceprint does not use the mechanical key to unlock, the efficiency of unlocking is greatly improved. The application of voiceprint recognition technology to the voiceprint lock is a high stability unlocking method, which can greatly improve the safety of the safe. Through mobile app and direct voiceprint input, the remote control of the safe is realized. The recognition rate is over 98%, which greatly improves the unlocking efficiency of the safe.

1. Introduction

The safe box appeared in the 18th century and has a history of more than 300 years. The structure of the original safe box is very simple and the shape is very single. After nearly 300 years of development, its shape and unlocking method have changed a lot, but its most fundamental anti theft function has not changed. People have higher and higher requirements for information security, so the long history of safe will continue[1]. At present, mechanical lock and electronic lock are mainly used in safe at home and abroad. When mechanical lock is opened, it is troublesome to turn around. Once the key is lost, it will cause great trouble for users. Electronic lock has short service life and easy to lose password. At present, there are few research on the application of voiceprint technology in voiceprint lock at home and abroad[2]. Compared with mechanical lock and electronic lock, voiceprint unlock has the advantages of high stability, high efficiency, high precision, etc. it will not have the disadvantages of password leakage, key loss, etc. This system uses the voiceprint to unlock, because the stability and uniqueness of the voiceprint greatly improve the safety of the safe. The image of school enterprise alliance of Internet of things in China regards speech recognition ratio as "machine recognition system"[3].

IOP Conf. Series: Materials Science and Engineering **799** (2020) 012039 doi:10.1088/1757-899X/799/1/012039

2. overall design

The main working principle of the system is to convert the user specific voiceprint or the specific voiceprint signal sent by the mobile app into a specific current, and use the electromagnetic induction to open the lock.

2.1. voiceprint recognition system:

A voiceprint recognition consists of two processes: voiceprint training and voiceprint recognition. The diagram of voiceprint recognition is as follows:

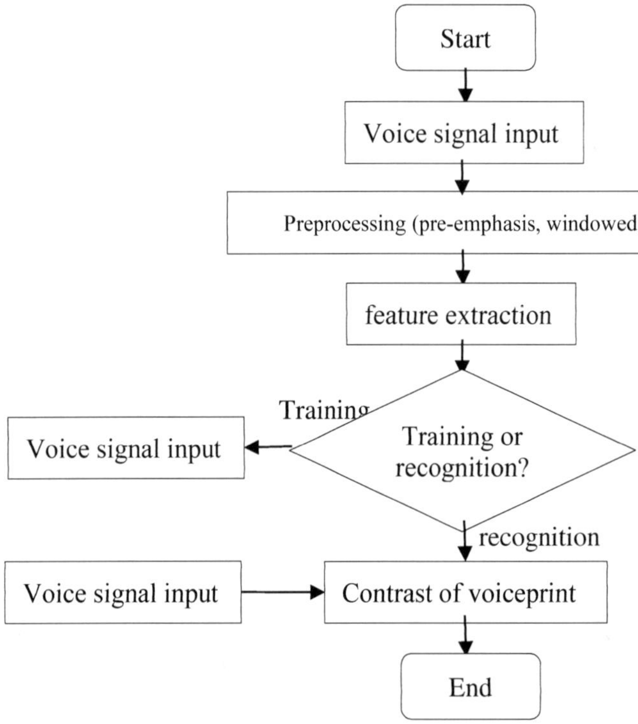

Figure 1. Flow chart of voiceprint recognition system

B denoising:

The input voiceprint signal is denoised, the voiceprint is quantized and sampled, processed into digital voice signal, then the low-frequency interference is removed (using pre weighting technology), and the noisy voice signal is processed to reduce the noise to get clean and clear voice signal.

C endpoint detection

The short-time zero crossing rate and short-time energy are applied to the endpoint detection of the system's voice signal. The data of each frame of the voice signal is 20ms, the sampling frequency is 8kHz, and the total number of sampling points is 160. The short-time energy and short-time zero crossing rate are recorded every 20ms. It can use the detected short-time energy and short-time zero crossing rate to remove white noise and voiceless frame, and keep the voiced signal. The voiced signal is very useful to get the characteristic parameters such as gene LPCC [4].

D voiceprint recognition -- feature extraction

The speech signal is transformed into DCT by adding window, calculating spectrum and Mel. The extraction process is as follows:

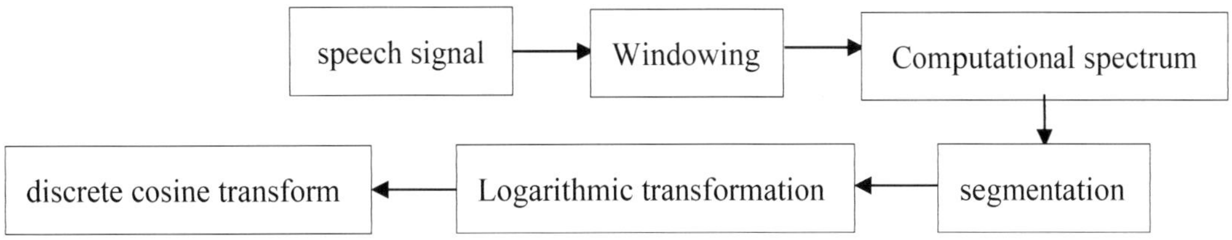

Figure 2. Process chart of feature extraction

E voiceprint recognition -- building acoustic model

VQ vector quantization, which classifies all vectors (set to m), is called vector quantization, assuming that the vector

The quantization error of the nearest vector center is. Then the average quantization error of T vectors is.

By using LBJ clustering algorithm, we can find the vector center set which can make d to be the minimum. A set of codes is composed of this set. This set of codes is the embodiment of the best distribution of T vectors in n spaces, that is, acoustic model.

F voiceprint recognition -- pattern matching

On the basis of obtaining the speaker's personality parameters, we match the parameter model and template waiting for recognition with the template or model base after training, get the similarity distance measurement between feature patterns and select the appropriate distance measurement as the threshold value, then judge the best result of recognition. The VQ clustering method LBJ was used.

2.2. schematic diagram of voiceprint unlocking

The specific voice signal from the mobile app or the voice print from the user will be converted into a certain current through the single chip microcomputer. When the input is correct, the current will generate magnetic force through the winding coil, and the lock will be opened by the principle of mutual attraction of different magnetic poles.

In this study, proteus simulation system is used [5]. The wireless signal or voiceprint signal of mobile phone is preprocessed by windowing, pre weighting, quantization and so on, and then enters ADS1216 analog-to-digital converter for conversion. The converted signal enters SPCE061A single-chip microcomputer and enters the display system after being processed by the single-chip microcomputer [6]. When the master sends a voice print or sends a specific signal through the mobile phone app, the LED displays on and the LED is green. At the same time, the current enters the unlocking system and finally opens the lock. When the signal input other than the master, the LED shows off and the LED is red. There is no current passing through the unlocking system and the lock is not opened at last. The schematic diagram is as follows:

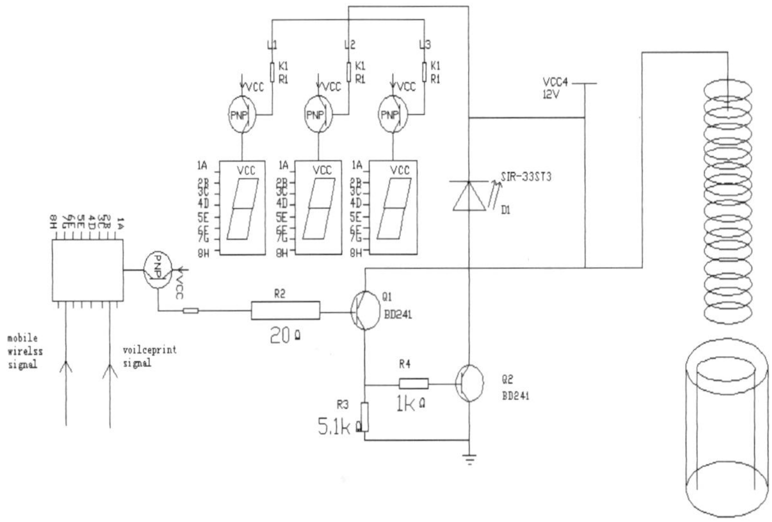

Figure 3. Schematic diagram of voiceprint unlocking simulation circuit

The input voiceprint enters the storage area for temporary storage, and then enters the model library for model matching. When the matching result fails, the recognition process will be interrupted back to the voiceprint input stage. When the matching result is successful, it will enter the electromagnetic trigger stage, and finally the lock will be opened. The flow chart of voiceprint unlocking is as follows:

2.3. Hardware design

The hardware circuit of this design consists of ADS1216 analog-to-digital conversion chip, mobile wireless receiver, SPCE061A single chip, LED display, sir-33st3 led, data acquisition, data processing, identification circuit and some driving circuits. SPCE061A was launched with 61 boards, and the resources of SPCE061A were comprehensively utilized. A safe box with speech recognition function is designed by using single chip microcomputer instead of traditional mechanical unlocking [7].

The ice interface of the online simulation circuit is embedded in the SPCE061A single chip microcomputer, which can process the complex signals quickly [8]. The wireless receiver of the mobile phone can receive the wireless signals sent by the mobile phone. The wireless signals enter the SPCE061A single chip microcomputer for temporary storage. The wireless signals or voiceprint signals of the mobile phone enter the ADS1216 analog-to-digital converter for conversion after pre-processing such as windowing, pre weighting and quantification. The converted signal enters SPCE061A single chip microcomputer, and finally converts into a stable 3.3V voltage signal.

The display part is led display screen, which adopts two-color screen and three position display. The display time is three seconds, so that the visual retention time is kept at three seconds, which will make people's eyes feel more comfortable [9]. When the signal is not sent out by the user, the recognition circuit cannot process the voiceprint signal into a 3.30v voltage signal. The display screen shows off and the ir-33st3 light-emitting diode emits red light. The current will not enter the circuit after the diode. If the input is the user's voiceprint, the display screen will show on and the ir-33st3 light-emitting diode emits green. The safety lock is made of magnet, which is closed when there is no current passing through. When there is current passing through the drive circuit, the current passing through the multi turn coil can produce a larger magnetic force, which makes the lock open under the mutual attraction of magnets. When the current disappears, the magnets attract each other, making the lock closed. The hardware structure is as follows:

Figure 4. Hardware deconstruction diagram

2.4. System analysis

The experimental results show that when the input user's voice print, the test voltage is 3.30v, the LED is green, and the electromagnetic lock is turned on. When other people's voice print is input, the test voltage is not 3.30v, the LED is red, and the electromagnetic lock is not opened. The same result can be obtained by using the signal from the mobile app to unlock the lock. The master receives the voiceprint signal from the mobile app through the wireless receiver, which is converted into 3.30v electrical signal by the analog-to-digital converter and the single-chip microcomputer. The LED is green, and the electromagnetic lock is opened at the same time. The voice print signal of other mobile phones received by the wireless receiver is not converted into 3.30v electrical signal, the LED is red, and the electromagnetic lock is not opened. It can be seen from the results that only the user's voiceprint can be converted into 3.30v electrical signal whether it is direct or indirect voiceprint input. Other people's voiceprint signal will not be converted into the required electrical signal, and then the lock will not be opened. The experimental analysis is as follows:

Table 1. Voiceprint test

test	noise(DB)	distance	output voltage	led display	warning lamp	if open
Master voiceprint	40	1m	3.3V	ON	green	open
	80	3m	3.3V	ON	green	open
A voiceprint	40	1m	3.7V	OFF	red	no
	80	3m	3.7V	OFF	red	no
B voiceprint	40	1m	2.7V	OFF	red	no
	80	3m	2.7V	OFF	red	no
C voiceprint	40	1m	3.6V	OFF	red	no
	80	3m	3.6V	OFF	red	no

3. summary

Voiceprint recognition can be applied to many fields. With the help of the uniqueness of human voiceprint, voiceprint lock can ensure the safety of property. It is a leading new type of security product at home and abroad, which encrypts the safe system and protects the safe. With advanced voiceprint recognition technology, the voiceprint lock can immediately identify the user's identity by the user's voiceprint. There is no digital password in the voiceprint lock, so the user can not remember a large number of passwords, let alone worry about the loss of passwords and the loss of property. Therefore, high efficiency, high precision and long life represents a major trend in the safety lock.

IOP Conf. Series: Materials Science and Engineering **799** (2020) 012039 doi:10.1088/1757-899X/799/1/012039

Acknowledgments
This work was supported by Natural Science Research Projects of Colleges and Universities in Anhui Province No. KJ2017A616, and School-level Quality Engineering Projects in 2018 No.2018sysxx01.

References
[1] Cai gengping, Huang shunzhen, Xu Zhihong, Lan Bo, fan Guohua, Liang fan. Voiceprint recognition system [J]. Journal of Shenzhen University; 2002.02 electromagnetic induction 18-23.
[2] university physics, edited by Zhang Sheqi, China Agriculture Press, 123-125.
[3] Yang Yang, Chen Yongming. Voiceprint recognition technology and its application [J]. Electroacoustic technology; 2007.02.121-135.
[4] Li Yuehua, Gu Zhixin, Liu Shuqing, Wang Shuyang. Voiceprint certification technology [J]. Water conservancy technology and economy; 2005, 06, 65-73.
[5] Bai Shunxian, Ma Suisse. Research on decision mechanism in speech endpoint detection [J]. Journal of Dalian Institute of nationalities; 2010, 03, 128-129.
[6] Ma Zhixin, Wang Hong, Li Xin. Speech recognition technology review [J]. Journal of Changji University; 2006, 03, 136-138.
[7] Zhang Mohan, Zhang Bei, Wang Tianliang. Wireless signal transmission device and transmission method [Z]. Cn101694740a: 2010. 98-101.
[8] Lin Qunyao. Wireless signal transmission device] Z]. Cn301028800: 2007.36-41.
[9] Zeng Shengwu, Wu Chunfa. A wireless signal transmission and reception system] Z]. Cn1662062: 2005.136-139.

IOP Conf. Series: Materials Science and Engineering **799** (2020) 012040 doi:10.1088/1757-899X/799/1/012040

Symmetry detection algorithm to detect forwarded spoofing interference signals of BOC modulation receivers

Zhiying Wang, Yucen Liu and Menglan Wang*

Satellite Navigation Center, Beijing, 100094, China

*Corresponding author's e-mail: 862780529@qq.com

Abstract. In this paper, the symmetry detection algorithm based on the symmetry of binary offset carrier (BOC) modulation signals is proposed to solve the problem of the global navigation satellite system (GNSS) spoofing interference in BOC modulation and acquisition phase which set on the base of the characters of correlation peaks, and achieve the detection of the forwarded spoofing interference signals. The theoretical value and the simulation value of the receiver operating characteristic (ROC) curve of the symmetry detection algorithm can be obtained by the numerical calculation and the experimental simulation. It can be seen from the simulation that with the increase of the accumulation times after the correlation, the signal to noise ratio (SNR) of the spoofing signal compared to the real signal, and the SNR of the real signal, the detection probability of the spoofing signal increases as well. When the power of the spoofing signal is 3 dB greater than that of the real signal, the detection probability can be more than 99% under the condition that the false alarm probability is 0.1%. Given that it only needs to add the module to the software receiver to detect spoofing signals, the algorithm proposed in this paper is of practical significance.

1. Introduction

With the development of the global navigation satellite system (GNSS) and the continuous expansion of its application field, its security has been widely concerned by people from all walks of life[1]. Spoofing as a malicious attack, sends forged or delayed navigation signals to the target receiver to mislead it into wrong position and time information, causing serious consequences[2-3]. Nowadays, the binary offset carrier (BOC) modulation technology is used in many navigation satellite systems, and the existing spoofing detection algorithms are aimed at the binary phase shift keying (BPSK) modulation signals[4]. In this paper, a symmetry detection algorithm of the forwarded spoofing signals is proposed by using the symmetry of correlation peak secondary peak with respect to main peak of the BOC modulated signal under the condition of the small delay.

2. Algorithm principle

BOC modulation symmetry detection algorithm of forwarded spoofing interference works in phase of pseudocode acquisition of navigation receivers. In the condition of small delay, it uses the symmetry of correlation peak secondary peak with respect to main peak of the BOC modulated signal, establishing model by calculating the accumulation of square difference value of correlation peak secondary peaks to achieve the detection of forwarded spoofing interference.

As shown in figure 1, the condition of small delay stands for the procedure of alignment between secondary peaks of the spoofing signal's and real signal's correlation peaks when the relative delay between correlation peaks of the spoofing signal and the real signal is less than 2 chips.

IOP Conf. Series: Materials Science and Engineering **799** (2020) 012040 doi:10.1088/1757-899X/799/1/012040

Based on the knowledge of statistical signals, the binary assumptions can be constructed.

H_0: There is no spoofing interference.

H_1: There is spoofing interference.

Through this binary detection, it can ensure that receiver can move into tracking phase normally after capturing the real signal.

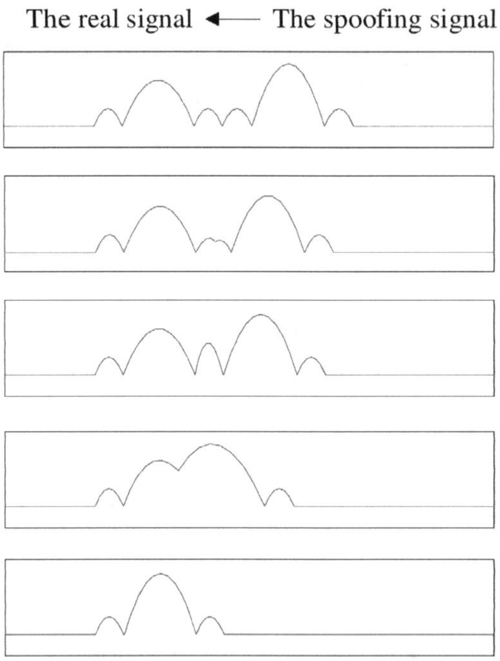

Figure 1. The schematic diagram of the small delay condition.

The premise of adopting the algorithm is that the real signal's main peak of correlation is not deformed.

In the process of signal transmission, the baseline complex signal can be expressed as equation (1), the real signal can be expressed as equation (2), and the spoofing signal can be expressed as equation (3).

$$s_{IF}(t) = s(t) + j(t) + w(t) \tag{1}$$

$$s(t) = Ad(t-\tau)\chi(t-\tau)c(t-\tau)e^{j[2\pi\Delta f(t-\tau)+\theta_s]} \tag{2}$$

$$j(t) = kAd(t-\tau-\tau_0)\chi(t-\tau-\tau_0)c(t-\tau-\tau_0)e^{j\left[2\pi\Delta f'(t-\tau-\tau_0)+\theta_j\right]} \tag{3}$$

The reference signal of receivers can be expressed as

$$h(t) = \chi(t)c(t)e^{j(2\pi ft+\theta)} \tag{4}$$

In the process of signal acquisition, the receiver despreads the carrier doppler and searches the field of code phase, and the output of the correlator can be expressed as

$$x_n = \frac{1}{T_c}\int_{nT_c}^{(n+1)T_c} s_{IF}(t)h(t)^* \,\mathrm{d}t = R_n + w_n, t \in [nT_c, (n+1)T_c], n = 1, \ldots, N \tag{5}$$

In equation (5), R_n is the correlation peak value of the signal, w_n is the Gaussian white noise independent from the signal.

Under the binary assumptions, the correlator output is normally distributed.

Under the assumption of H_0, the received signal is real. The value of the left secondary peak is $R_1^{(1)}$ and the right one is $R_2^{(1)}$. $R_1^{(1)} = R_2^{(1)}$. Then the assumption of H_0 can be expressed as

$$\begin{aligned} x_1(n)\big|H_0 &= R_1^{(1)} + w_1(n) \sim N(R_1^{(1)},\sigma_n^2) \\ x_2(n)\big|H_0 &= R_2^{(1)} + w_2(n) \sim N(R_2^{(1)},\sigma_n^2) \\ x(n)\big|H_0 &= x_2(n) - x_1(n) \\ &= w_2(n) - w_1(n) \sim N(0,2\sigma_n^2) \end{aligned} \tag{6}$$

Under the assumption of H_1, the secondary peak is deformed because of spoofing. The value of the left secondary peak value is $R_1^{(2)}$ and the right one is $R_2^{(2)}$. $R_1^{(2)} \neq R_2^{(2)}$. Then the assumption of H_1 can be expressed as

$$\begin{aligned} x_1(n)\big|H_1 &= R_1^{(2)} + w_1(n) \sim N(R_1^{(2)},\sigma_n^2) \\ x_2(n)\big|H_1 &= R_2^{(2)} + w_2(n) \sim N(R_2^{(2)},\sigma_n^2) \\ x(n)\big|H_1 &= x_2(n) - x_1(n) \\ &= R_2^{(2)} - R_1^{(2)} + w_2(n) - w_1(n) \sim N(\Delta R^{(2)},2\sigma_n^2) \end{aligned} \tag{7}$$

In equation (7), $\Delta R^{(2)} = R_2^{(2)} - R_1^{(2)}$.

Then the above two assumptions can be expressed as

$$\begin{aligned} H_0 : x_n &= \frac{1}{T_c}\int_{nT_c}^{(n+1)T_c} w(t)h(t)^*\,\mathrm{d}t \approx w_n \\ H_1 : x_n &= \frac{1}{T_c}\int_{nT_c}^{(n+1)T_c} A(t)\,\mathrm{d}t + \frac{1}{T_c}\int_{nT_c}^{(n+1)T_c} w(t)h(t)^*\,\mathrm{d}t \approx R_n + w_n \end{aligned} \tag{8}$$

In equation (8), the normalization condition $\int_{nT_c}^{(n+1)T_c} |h(t)|^2\, dt=1$. $A(t)$ represents the channel gain of the synthesized signal.

Based on the above analysis, the receiver can construct such binary assumptions by accumulating the value of the square difference between secondary peaks of the signal to determine whether there is a spoofing interference signal.

So the test statistic of the binary hypothesis can be expressed as

$$T = T(x) = \mathrm{X}^H\mathrm{X} = \sum_{n=1}^{N} x_n^2 \tag{9}$$

In equation (9), the test statistic T represents the accumulation of square difference value between secondary peaks of correlation of the synthesized signal. Based on the hypothesis of module, the test statistic is central chi-square distribution whose degree of freedom is N in the condition of H_0. In the condition of H_1, the test statistic T is non -central chi-square distribution whose degree of freedom is N.

Under the assumption of H_0, the probability density function (PDF) of the test statistic T can be expressed as

$$f_{T|H_0}(T) = \begin{cases} \dfrac{1}{2^{\frac{N}{2}}\Gamma(\frac{N}{2})} T^{\frac{N}{2}-1} \exp(-\dfrac{T}{2\sigma_n}), & T>0 \\ 0, & T<0 \end{cases} \tag{10}$$

In equation (10), the gamma function $\Gamma(u)$ can be expressed as

$$\Gamma(u)=\int_0^{\infty} t^{u-1}\exp(-t)\,\mathrm{d}t \tag{11}$$

Under the assumption of H_1, the PDF of the test statistic T can be expressed as

$$f_{T|H_1}(T)=\begin{cases}\frac{1}{2}\left(\frac{T}{\lambda\sigma_n}\right)^{\frac{N-2}{4}}\exp\left(-\frac{1}{2}\left(\frac{T}{\sigma_n}+\lambda\right)\right)\mathrm{I}_{\frac{N}{2}-1}\left(\sqrt{\frac{\lambda T}{\sigma_n}}\right), & T>0\\ 0, & T<0\end{cases} \tag{12}$$

In equation (12), the non-central parameter $\lambda=\frac{1}{\sigma_n}\sum_{n=1}^{N}\left|\Delta s^{(2)}\right|^2$. The three-order modified Bessel function $I_r(u)$ can be expressed as

$$\mathrm{I}_r(u)=\frac{(\frac{1}{2}u)^r}{\sqrt{x}\Gamma(x+\frac{1}{2})}\int_0^{\pi}\exp(u\cos\theta)\sin^{2r}\theta\,\mathrm{d}\theta \tag{13}$$

We can obtain the probability density map of the test statistic T under the two assumptions, as shown in figure 2.

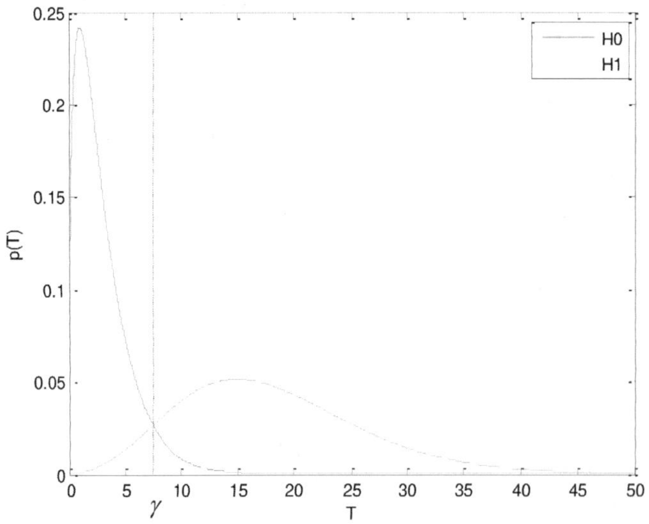

Figure 2. The PDF of the test statistic T under the two assumptions.

Proper selection of the detection threshold γ reasonably is essential for good performance of the detector. On the one hand, if the threshold is too low, it may lead to false alarms, which means that the detector treats noise as a real signal. On the other hand, if the threshold is too high, it may lead to missing detection, which means that the detector cannot detect the existing spoofing signals.

So the false alarm probability P_{FA} should be selected firstly, and then the detection threshold γ can be obtained.

According to Neyman-Pearson lemma[5], the false alarm probability P_{FA} can be expressed as

$$P_{FA}=Pr\{T>\gamma|H_0\}=\int_{\gamma}^{\infty} f_{T|H_0}(T)\,\mathrm{d}T \tag{14}$$

IOP Conf. Series: Materials Science and Engineering **799** (2020) 012040 doi:10.1088/1757-899X/799/1/012040

In equation (14), γ represents the judgement threshold of detection statistics, which can be expressed as

$$\gamma = F_{T|H_0}^{-1}(1-P_{FA}) \tag{15}$$

In equation (15), $F_{T|H_0}(\cdot)$ represents the cumulative distribution function of the detection statistics T under the assumption of H_0, which can be expressed as

$$F_{T|H_0}(T) = \int_0^T \frac{1}{2^{\frac{N}{2}}\Gamma(\frac{N}{2})} T^{\frac{N}{2}-1} \exp(-\frac{T}{2\sigma_n}) \mathrm{d}T \tag{16}$$

As the detection threshold are selected, the detection probability P_D of a real spoofing signal being detected can be expressed as

$$P_D = Pr\{T > \gamma | H_1\} = \int_\gamma^\infty f_{T|H_1}(T)\mathrm{d}T \tag{17}$$

3. Detection method

The detection process is shown in figure 3. If there is a spoofing signal during the acquisition phase of the receiver while the relative delay of the correlation peak of the spoofing signal and the actual signal is less than 2 chips, we can make such a hypothesis statistics by using the deference of secondary peaks of two sides, when these two correlation peaks start contacting and deforming. Comparing detection statistics with threshold fixed, if it is more than threshold receiver will claim there is a spoofing signal, otherwise receiver will switch to the tracking phase according to the normal process.

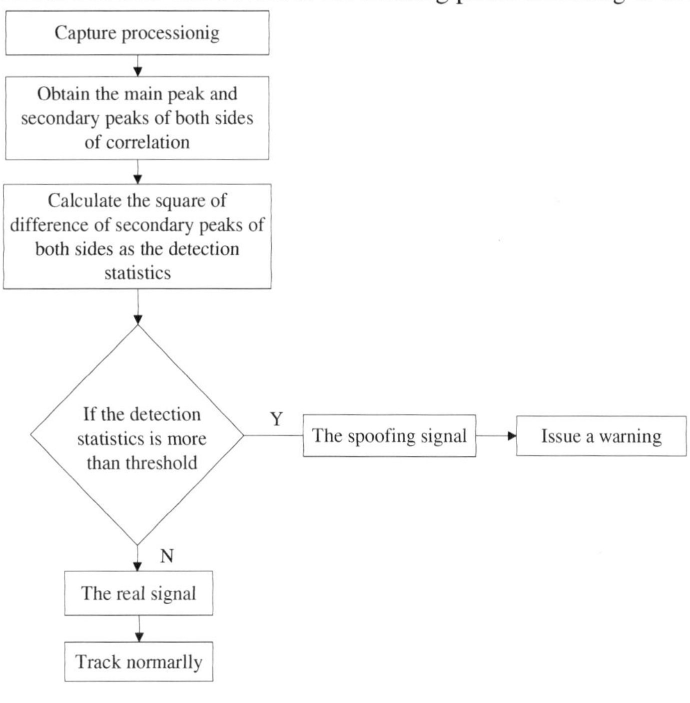

Figure 3. The procedure of symmetry detection algorithm.

The steps of symmetry detection algorithm are as follows:

- Receivers can obtain the complexed baseband signal $x(n) = I(n) + jQ(n)$ by multiplying the received signal with the in-phase and quadrature signal produced by local oscillator to get the fast Fourier transform $X(k)$ of $x(n)$, $n = k = 0,1,2,\cdots,N$.
- Receivers calculate the fast Fourier transform $H(k)$ of $h(n)$ which is the product of local recurrence code and square subcarrier, and we can get the conjugate value $H^*(k)$.
- Receivers can obtain $P(k)$ by multiplying $X(k)$ and $H^*(k)$, and do the reverse Fourier transform to $P(k)$ to get $r(n)$, then modulo $r(n)$ to get $|r(n)|$.
- Based on the last step, receivers can obtain main peak of correlation R and secondary peaks $R^{(1)}$. And then receivers can get the detection statistics T by calculating the square of difference between two secondary peaks.
- Comparing detection statistics T with threshold γ, if T is bigger than γ, which means there is a spoofing signal, receivers will issue a warning, otherwise receivers will step into capture phase which means the received signal is real.

The steps of symmetry detection algorithm are as follows:

4. Simulation verification

Based on such an analysis, expressions of false alarm, threshold, deception detection probability and probability density function and cumulative distribution function of detection statistics V in two hypothesis conditions, now we can obtain the theoretical and simulation value of the receiver operating characteristic (ROC) curve of the symmetry detection algorithm through numeral calculations and stimulation.

According to the literature, the power of the received satellite navigation signals is about the same when signals reach the ground, while the carrier-to-noise ratio is in the range of 35 ~ 55 dB·Hz roughly [6]. Therefore, for a capture process using a coherent integration with an integration step size of 1 ms, the signal-to-noise ratio of the correlator output result is in the range of 5 ~ 25 dB.

Assumed that the BOC modulation receiver was interfered by the forwarded spoofing signal, we would use MATLAB software to verify the effectiveness of algorithm proposed by two simulation experiments. During the simulation, the carrier to noise ratio (CNR) of the received real signal is 35 dB·Hz or 37 dB·Hz, the coherent integration time is 1ms and the number of accumulated N (the degree of freedom) after correlation is 1, 3 or 5.

If the CNR of the real signal is 35 dB·Hz, the signal to noise ratio (SNR) of output is 5 dB after the correlation of 1 ms. If the CNR of the real signal is 37 dB·Hz, the SNR of output is 7 dB after 1ms' correlation. We will do Monte Carlo simulation for 5000 times, and the parameters is shown in table 1.

Table 1. The parameter of symmetry detection algorithm stimulation.

Parameter	The real signal/dB	The spoofing signal/dB
Experiment A	5	7
		8
Experiment B	7	9
		10

In Experiment A, the SNR of the real signal is 5 dB, while the SNR of the spoofing signal is 7 dB and 8 dB respectively.

Under the condition that the SNR of the spoofing signal is 7 dB in Experiment A, the ROC curve of symmetry detection is shown in figure 4. In figure 4, the horizontal axis is the false alarm probability of the spoofing signal, the vertical axis is the detection probability of the spoofing signal. The solid line is theoretical value of numerical calculation, circle is the stimulation value of Monte Carlo simulation. If the SNR of the spoofing signal is 2 dB more than that of the real signal, in the condition of same false alarm probability, with increasing of accumulation times N after correlation, the

IOP Conf. Series: Materials Science and Engineering **799** (2020) 012040 doi:10.1088/1757-899X/799/1/012040

probability of detecting the spoofing signal may get higher. When the degree of freedom is 1 and the false alarm is 2%, the detection probability can only reach over 46%. When the degree of freedom is 3 and the false alarm is 2%, the detection probability will be over 85%. When the degree of freedom is 5 and the false alarm is 2%, the detection probability can will be more than 96%.

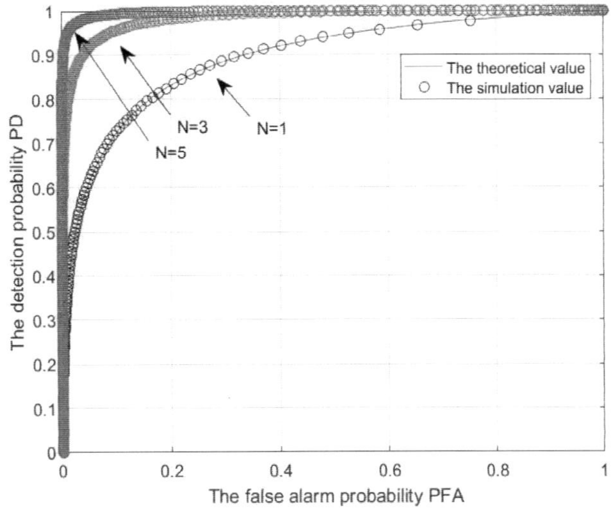

Figure 4. The ROC curve of symmetry detection (the SNR of the spoofing signal is 7 dB).

Under the condition that the SNR of the spoofing signal is 8 dB in Experiment A, the ROC curve of symmetry detection is shown in figure 5. In figure 5, it can be seen that when the degree of freedom is 1 and the false alarm is 2%, the detection probability may be more than 57%. When the degree of freedom is 3 and the false alarm is 2%, the detection probability may be more than 93%. When the degree of freedom is 5 and the false alarm is 2%, the detection probability may be more than 99%.

According to figure 4 and figure 5, it can be seen that the simulation results agree with the numerical calculations. When the SNR of the real signal is the same, the detection probability of the spoofing signal increases as the SNR of the spoofing signal increases.

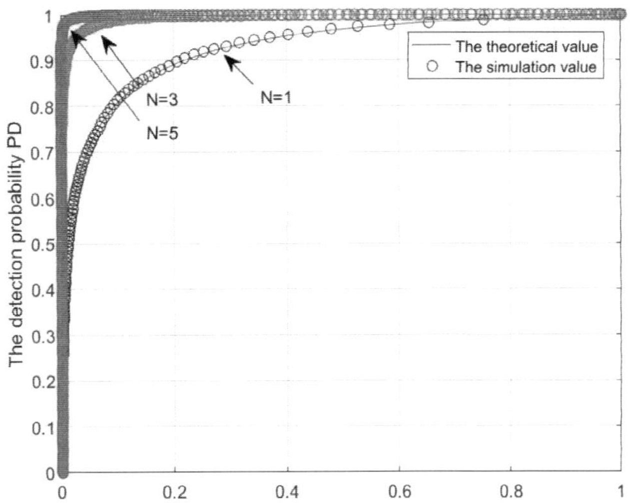

Figure 5. The ROC curve of symmetry detection (the SNR of the spoofing signal is 8 dB).

In Experiment B, the SNR of the real signal is 7 dB, while the SNR of the spoofing signal is 9 dB and 10 dB respectively.

Under the condition that the SNR of the spoofing signal is 9 dB in Experiment B, the ROC curve of symmetry detection is shown in figure 6. When the false alarm probability is the same, the detection probability of the spoofing signal increases as the accumulation times after the correlation increases. When the degree of freedom is 5 and the false alarm is 0.2%, the detection probability is more than 98%.

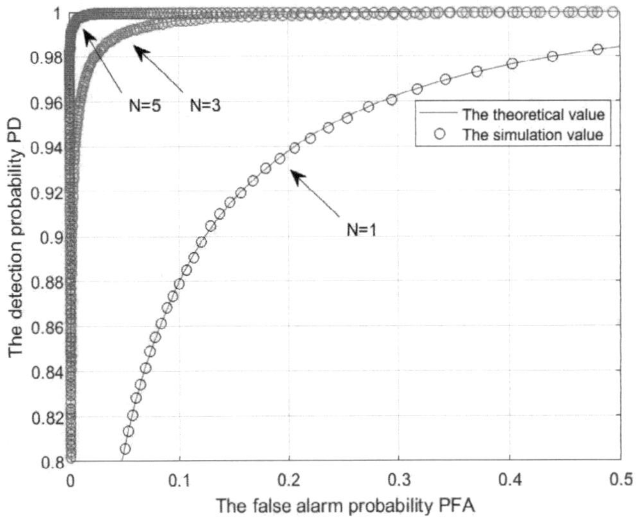

Figure 6. The ROC curve of symmetry detection (the SNR of the spoofing signal is 9 dB).

Under the condition that the SNR of the spoofing signal is 10 dB in Experiment B, the ROC curve of symmetry detection is shown in figure 7. It can be seen that the SNR of the spoofing signal is 3 dB more than that of the real signal, when the degree of freedom is 5 and the false alarm is 2%, the detection probability is more than 99%.

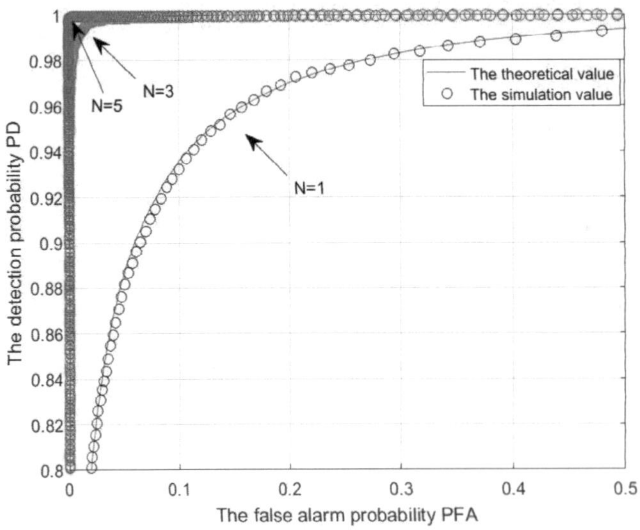

Figure 7. The ROC curve of symmetry detection (the SNR of the spoofing signal is 10 dB).

IOP Conf. Series: Materials Science and Engineering **799** (2020) 012040 doi:10.1088/1757-899X/799/1/012040

According to the figure 4 ~ 7, it can be seen that the detection probability of the spoofing signal increases as the SNR of the real signal increases.

In summary, with the increase of accumulation times after correlation, the SNR of the spoofing signal relative to the real signal, and the improvement of the SNR of the real signal, the detection probability gets higher and the theoretical values of numerical calculation agree with the Monte Carlo simulation. When power of the spoofing signal is 3 dB more than the power of the real signal, the detection probability can achieve 99% under the condition of the false alarm probability is 0.1%.

5. Conclusion

In this paper, the symmetry detection algorithm based on the symmetry of BOC modulation signals is proposed to detect forwarded spoofing interference signals of BOC modulation receivers. The algorithm uses the symmetry of secondary peak of correlation with respect to main peak of correlation of the BOC modulated signal to achieve the detection of forwarded spoofing interference signals. The performance of the algorithm can be seen from the simulation. With the increase of the accumulation times after the correlation, the SNR of the spoofing signal compared to the real signal, and the SNR of the real signal, the detection probability of the spoofing signal increases as well. When the power of the spoofing signal is 3 dB more than the power of the real signal, the detection probability can achieve 99% under the condition that the false alarm probability is 0.1%. Given that there is only need to add the module to the software receivers to detect spoofing signals, the algorithm proposed in this paper is of practical significance.

References

[1] Wullems, C.J. (2012) A spoofing detection method for civilian L1 GPS and the E1-B Galileo safety of life service. IEEE Transactions on Aerospace and Electronic Systems, 48: 2849–2864.
[2] Hatch, R., Sharpe, R., Yang, Y. (2004) An innovative algorithm for carrier-phase navigation. Proceedings of ION GNSS, Long Beach, CA, USA.
[3] Kaplan, E.D. (2006) Understanding GPS principles and applications. Artech House, Boston, USA.
[4] Wei, X. (2011) Research on GPS signal interference technology modulated by BOC. Electronic design engineering, 19:139–144.
[5] Luo, P., Zhang, W. (2014) Basics of Statistical Signal Processing: Estimation and Detection Theory. Publishing House of Electronics Industry, Beijing.
[6] Xie, G. (2009) Principles of GPS and Receiver Design. Publishing House of Electronics Industry, Beijing.

IOP Conf. Series: Materials Science and Engineering **799** (2020) 012041 doi:10.1088/1757-899X/799/1/012041

Research on Digital Pulse Compression Technology of Overlapping Data Splicing

Panpan He

Kunming Shipborne Equipment Research and Test Center, Kunming, Yunnan Province, 650051, China

824577638@qq.com

Abstract. Digital pulse compression technology is an effective technology to solve the contradiction between the maximum operating distance and the range resolution of sonar. However, in practical application, when the amount of data to be processed is large and the segmented processing is needed, the data segmentation may be unreasonable, resulting in the weak peak value of the signal after pulse compression, thus affecting the detection results. This paper introduces a method of overlapping data splicing. By improving the correlation between signal and matched filter, the problem of unreasonable data segmentation is effectively solved.

1. Introduction

With the strategic transformation of various navies and the development of modern science and technology, new underwater weapons are developing in the aspects of long range, fast speed and good concealment performance, which also puts forward higher and higher requirements for sonar system to detect underwater targets. The main technical index to measure the performance of sonar system is the maximum range and range resolution. The maximum operating distance refers to the maximum distance that sonar can effectively find targets and measure their data under certain conditions. The range resolution refers to the resolution of sonar system to two adjacent targets in space[1]. In order to improve the range resolution, the signal is required to have a large bandwidth. In order to increase the maximum operating distance, the signal is required to have a large energy. However, the peak power of the signal transmitter cannot be infinitely increased. In practice, the average power of the transmitted signal is often increased by increasing the pulse width of the signal to achieve the purpose of increasing the maximum operating distance, which requires the signal to have a large Time width, bandwidth product. It can be seen from the theory of signal and system that the time bandwidth product of common signal is a constant, so it is impossible for signal to have large time bandwidth and bandwidth at the same time[2]. In order to solve this contradiction, a new technology pulse compression technology must be adopted.

Pulse compression technology[3-5] can effectively solve the contradiction between the maximum operating distance and range resolution of sonar. By using pulse compression technology, the signal with large time width and bandwidth can be transmitted at the transmitting end to improve the maximum operating distance of the signal. At the receiving end, the wide pulse signal can be compressed into a narrow pulse to improve the range resolution. When sonar detects the target, it needs to process the data received by sonar in real time, which requires data segmentation and splicing. Simple data segmentation processing may result in the same pulse signal being divided into two parts, thus affecting the pulse compression result. In order to solve this problem, this paper takes LFM signal[6-8] as the research object and adopts overlapping data splicing method The feasibility of this method is verified by MATLAB simulation.

2. Principle of dagital pulse compression

Digital pulse compression refers to the use of digital signal processing technology to complete the correlation matching, usually using the matching formula of time-domain pulse compression and frequency-domain pulse compression. Among them, the time-domain pulse compression directly convolutes the echo signal received by sonar. When the convolution speed is less than the AD sampling frequency, the data can not be processed in real time. Therefore, it is necessary to reduce the amount of calculation and adopt the frequency-domain digital pulse compression algorithm with fast operation speed.

Frequency domain digital pulse compression technology is to complete the pulse compression of digital signal through frequency domain matched filtering technology. Firstly, the frequency spectrum of digital input signal is obtained by Fourier transform (FFT), then it is multiplied by the complex number of frequency response of matched filter, and then the multiplied signal is transformed by inverse Fourier transform (IFFT), that is to say, the signal after pulse compression is obtained. The whole process can be expressed as follows:

- $$y(n) = IFFT\left[s(w)\cdot h(w)\right] = IFFT\left[FFT\left(s(n)\right)\cdot FFT\left(h(n)\right)\right] \quad (1)$$

The realization block diagram of frequency domain digital pulse compression is shown in Fig. 1.

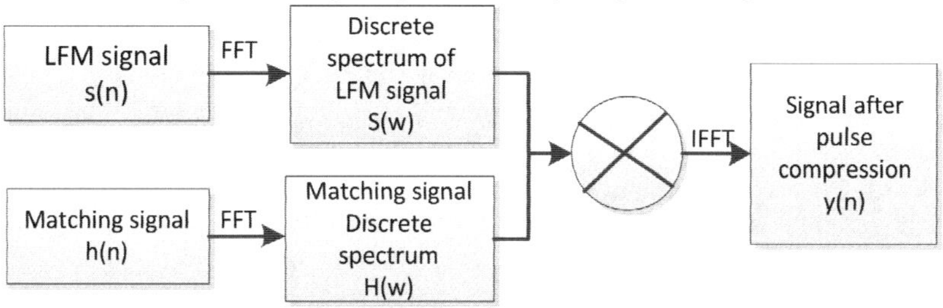

Figure 1. Principle block diagram of digital pulse compression in frequency domain.

3. Pulse compression simulation

Fig. 2 is the simulation result of frequency-domain pulse compression of single cycle ideal believe FM signal, Fig. 2 is the simulation figure of ideal LFM digital signal, Fig. 2 is the simulation figure of frequency-domain pulse compression of LFM signal, Fig. 3 is the simulation result of frequency-domain pulse compression of LFM signal after single cycle superimposed white noise, Fig 3. The figure above is the simulation figure of LFM digital signal with superimposed white noise, and the figure below is the simulation figure of frequency-domain pulse compression of LFM signal with superimposed white noise.

IOP Conf. Series: Materials Science and Engineering **799** (2020) 012041 doi:10.1088/1757-899X/799/1/012041

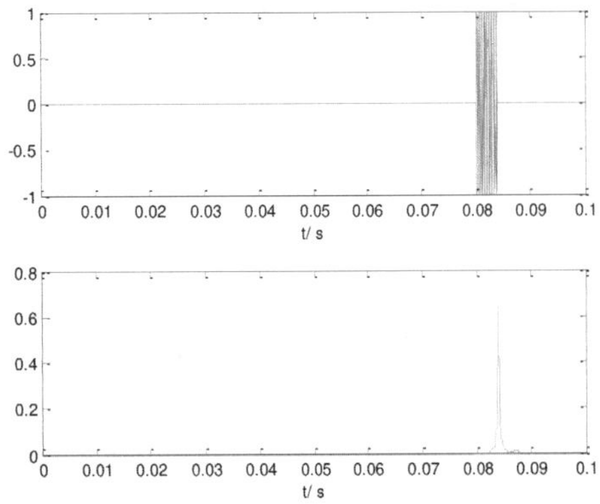

Figure 2. Pulse compression result of ideal digital sampling signal

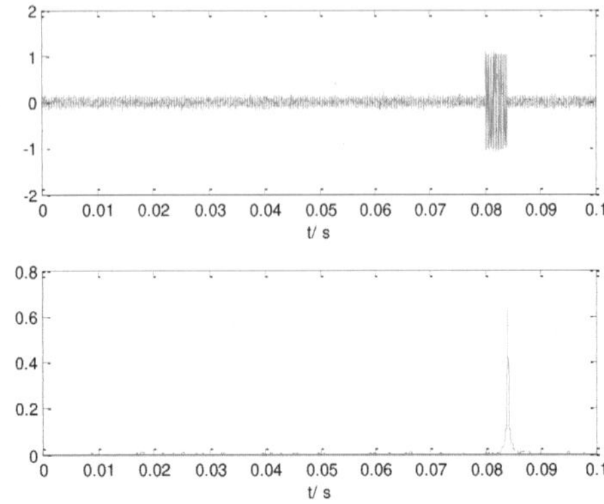

Figure 3. Pulse compression result of superimposed white noise digital sampling signal

It can be seen from Fig. 2 and Fig. 3 that the frequency domain pulse compression technology can effectively compress the wideband LFM signal into a narrowband signal, and the peak position of the narrowband signal is just at the end of the wideband LFM signal.

4. Data splicing method

In practical application, the signal received by sonar generally needs real-time processing, which inevitably requires the use of data segmentation and splicing technology. The conventional data segmentation and splicing method is shown in Figure 4. The same pulse signal may be divided into two segments. When processing the data for pulse compression, the two segments of data are partially related to the matched filter, which will affect the peak position after pulse compression, thus affecting the determination of the peak position after pulse compression.

In order to solve this problem, this paper proposes a method of overlapping data segmentation and splicing, as shown in Figure 5, which is the overlapping part of the two sections of data. Using the method of overlapping data segmentation and splicing can ensure that each complete pulse is in the same data section, fully related to the matched filter, and the peak position energy value is large after pulse compression.

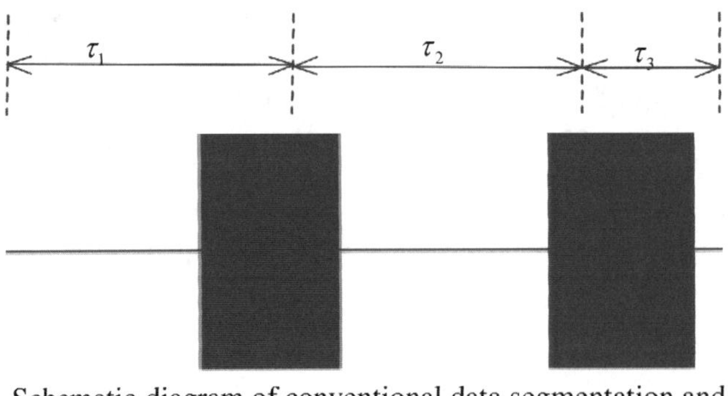

Figure 4. Schematic diagram of conventional data segmentation and splicing.

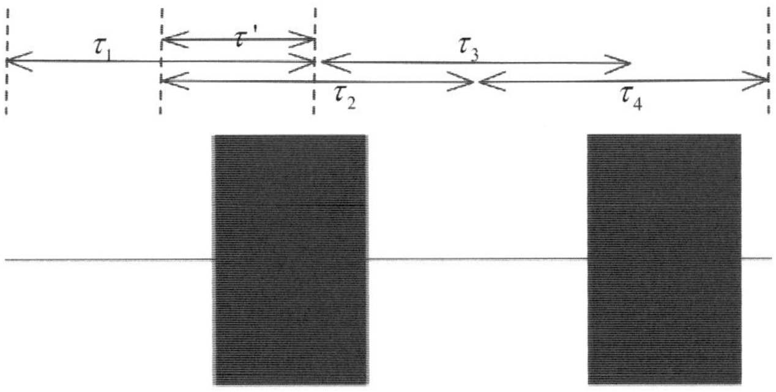

Figure 5. Schematic diagram of overlapping data segmentation and splicing.

5. Data splocing simulation research

Fig. 6 is the simulation result of pulse compression processing for two groups of digital sampling signals using conventional data division and splicing method, and Fig. 7 is the simulation result of pulse compression processing for two groups of digital sampling signals using overlapping data division and splicing method.

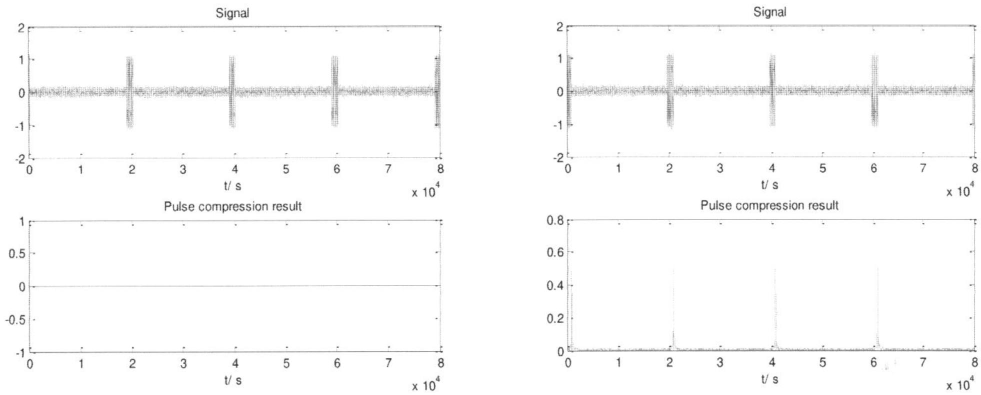

Figure 6. Pulse compression results of conventional data splicing.

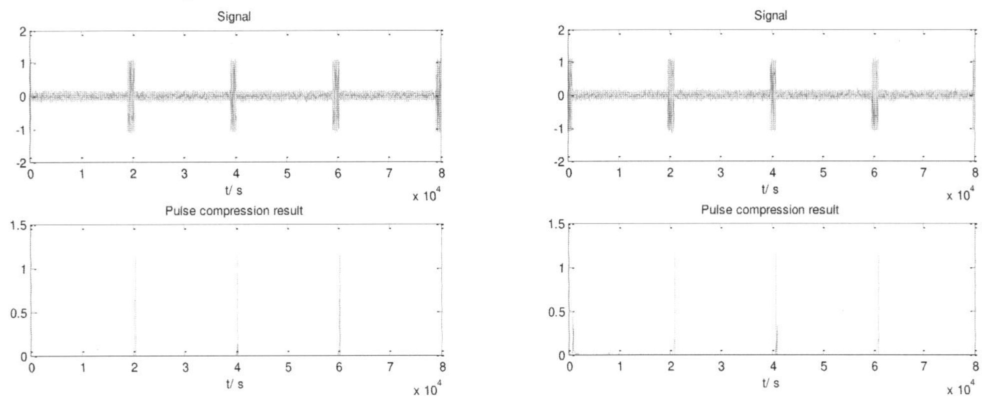

Figure 7. Pulse compression results of overlapping data splicing.

It can be seen from Fig. 6 that when the same pulse signal is divided into two parts, the correlation with the matched filter decreases. When the pulse compression technology is used to compress the narrow-band signal, the energy at the peak of the narrow-band signal decreases or does not exist, so the pulse position cannot be effectively identified.

It can be seen from Figure 7 that when the pulse is in a data processing section, the data is highly correlated with the matched filter, and the peak energy after pulse compression is also large. The pulse position can be effectively identified by judging the peak energy at the adjacent position.

To sum up, the overlapping data splicing method can effectively solve the irrationality of the conventional data segmentation splicing method, improve the correlation between the segmented data and the matched filter, increase the energy value at the peak position after pulse compression, and improve the ability to identify the pulse signal.

6. Conclusion

Pulse compression technology not only reduces the power requirements of sonar transmitter, but also effectively solves the contradiction between the maximum operating distance and range resolution of sonar, which is widely used in sonar signal processing. Using overlapping data segmentation and splicing technology for digital pulse compression processing can improve the matching degree of signal and filter, and improve the resolution ability. In this paper, matlab simulation is used to verify the effectiveness of overlapping data splicing method in the need of segmented pulse compression processing.

Acknowledgments

Thank all the teachers who have guided me, my colleagues who have helped me and my family. With your help, instruction and encouragement, I can work better. I want to express my deep thanks to them here!

References

[1] Tian, T. (2009) Sonar Technology. Harbin Engineering University Publishing, Harbin.

[2] Zheng, L. (2011) Analysis of digital pulse compression technology of LFM signal. Modern Electronics Technique,34: 39–42.

[3] Chen, G. (2011) Simulation of LFM signal pulse compression technology. DA ZHONG KE JI,10: 51-53.

[4] Cao, J. (2008) Research and implementation of broadband real-time pulse compression technology. Huazhong University of science and technology, Hubei.

[5] Lu, X. (2001) A method of matched filtering for LFM continuous wave. Journal of electronic s and information, 24: 1216–1219.

[6] Zhu, W. (2013) Detection and parameter estimation of LFMCW signal. Journal of electronics and information,35: 1562–1568.

[7] Qian, Y. (2006) Parameter estimation algorithm of LFMCW signal. Modern radar, 28: 40-43.

IOP Conf. Series: Materials Science and Engineering **799** (2020) 012042 doi:10.1088/1757-899X/799/1/012042

KSR: Knowledge-based Sequential News Recommendation System

Jiameng Bai[1,2], Xin Huang[1,2], Min Li[1*] and Jia Peng[1]

[1]State Key Laboratory of Information Security, Institute of Information Engineering, CAS, Beijing, 100093, China

[2]School of Cyber Security, University of Chinese Academy of Sciences, Beijing, 100049, China

[*]Corresponding author's e-mail: minli@iie.ac.cn

Abstract. With the popularity of the Internet, some online news reading habits have gradually replaced traditional media devices. People can see a variety of news on mobile phones or web pages. The overloaded information makes it impossible for users to quickly get the news they want to see, so the existence of a news recommendation system is necessary. However, unlike other fields, the recommendation of news is time-sensitive, and the problem of sparse data makes traditional collaborative filtering algorithms invalid. There are many entities in the news, and the knowledge graph is a collection of a large number of entities. In this paper, we propose a KSR model (Knowledge-based Sequential Recommendation) which uses the knowledge graph as side information to enrich the feature representation of the news to calculate the user's probability to click the forecasted news. KSR uses the knowledge graph to represent the entities in the news and uses a recurrent neural network to capture the sequential relationships in the news data to enrich the feature representation of the news. Because the different news that user has clicked have different relations with the news to be predicted, the attention model is introduced to calculate the weights. Finally, we conduct experiments on the existing dataset and results prove the efficacy of KSR over several baselines. Different contrast experiments also prove the effectiveness of each module of the model.

1. Introduction

With the booming popularity of the Internet and the changes in the way people read, traditional media such as newspapers, radios, and other devices are gradually out of sight. At the same time, some online news reading systems have emerged, such as Google News and Bing News. People can see news from the world on their mobile phones or the web. Massive news gives users a lot of opportunities to choose the information they want. However, the overload of information makes it impossible for users to obtain the information they want quickly. The appearance of the recommendation system dramatically slows down the information overload. It can personalize the recommendation according to the user's historical interest [1] to help user find information quickly.

Different from recommendation systems in other fields, such as movies, music, and e-commerce recommendations. Recommendation systems in the area of news often face the following challenges. First, the timeliness of news recommendation is very high. [2] shows that in business news dataset news will soon become obsolete, and the average life of news is only 4.1h. In the recommendation field, collaborative filtering has achieved great success, and the collaborative filtering method is based on the interaction record of user history, which is heavily dependent on the user's item scoring matrix. It will face the problem of sparse and cold start. The problem in news domain makes the traditional collaborative filtering recommendation [3] invalid. Second, when users read the news, interest tends to shift significantly with time flows. Third, the news field has a unique feature that the language used in the title is often highly concise and contains many entity words.

In order to solve these problems, researchers are committed to introducing more side information to make recommendations valid, such as users-item attributes [4] and social networks [5]. Among this side information, knowledge graph appears as a relatively new knowledge information in the sight of researchers. Knowledge graph is a graph network that contains rich semantic information. The nodes in the knowledge graph represent the entity, and the edges represent the relationships between the entities. In recent years, researchers have proposed many knowledge graphs, such as DBPedia, Google Knowledge Graph, and Microsoft Satori. These knowledge graphs have been widely applied to the question-and-answer system [6] and the study of the word vector [7].

Inspired by the idea of applying knowledge graphs to other tasks, as shown in Figure 1, we can also apply the knowledge graph to the recommendation system. Applying the knowledge graph to the recommendation system has the following benefits. First, the knowledge graph contains a large number of semantic relationships between items which can find hidden deep relationships between items and can improve the accuracy of recommendations. Second, the knowledge graph is a heterogeneous graph, which contains many different types of relationships, which can enrich the user's interest portraits and provide more diverse recommendation results. Third, the knowledge graph connects the user history and the objective item, which provides interpretability for recommendations. So we introduce the knowledge graph into the recommendation system and propose the KSR model (Knowledge-based Sequential Recommendation). The KSR model aligns the entity in the news with its entities in the knowledge graph, extracts the subgraphs containing the entities in the knowledge graph, and uses the knowledge graph to represent the learning method for encoding. Since the title's words of the news are sequential, the vector which training from the knowledge graph is introduced into the recurrent neural network for learning the sequential relationships. The user's feature is represented by his historical clicked news. Since the relationship between the news to be predicted and the news that the user has seen has different weights, we use the attention model to calculate those weights.

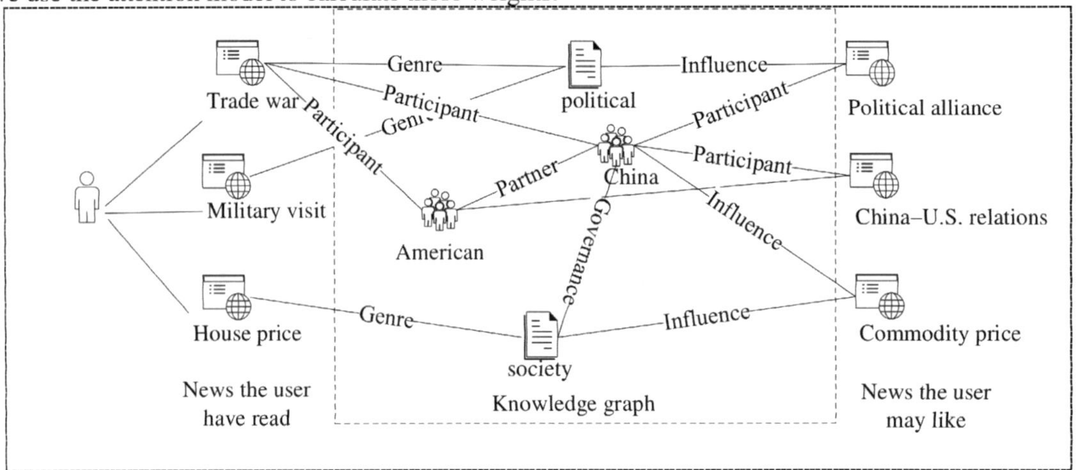

Figure 1. This picture briefly describes how the knowledge graph works.

In summary, this article contributes as follows:

(1) We introduce the knowledge graph into the recommendation system. That is, the feature of the news data is represented as its own word vector, and the feature representation of the entity appearing in the news in the knowledge graph which can enrich the feature representation of the news. There are different degrees of correlations between the news clicked by the user history and the news to be predicted, so we introduced the attention mechanism to calculate that weight.

(2) The words in the news's title contain sequential relationships. The existing knowledge graph-based recommendation model does not capture these relationships. our KSR model uses a recurrent neural network to learn the sequential relationship. Compared with some existing models, it achieves better results. Moreover, in the subsequent comparative experiments, it is also proved that the module for the sequential relationship has a specific improvement for the recommendation effect.

2. Preliminaries

2.1. LSTM cell

The LSTM(Long and Short Time Memory)[8] network is a kind of recurrent neural network that can capture the sequential relationship well. It is a chain network with repeated recurrent neural network cells.

LSTM can well capture sequential relationships, and on the other hand, it can solve long-term dependence problems in recurrent neural networks, such as gradient disappearance or gradient explosion. It solves these problems by introducing a gate mechanism. For example, "Forget Gate", "Input Gate", "Update Gate" and "Output Gate". The Forget Gate controls how much information at the previous moment is retained to the current state. The Input Gate determines how much of the current input is retained. The update gate is used to update the current state. Finally, the Output Gate determines the output of the current layer.

2.2. KGE(Knowledge Graph Embedding Solutions)

A knowledge graph contains a large number of entity-relationship triples (h,r,t), where h, r, t represent the head node, the relationship node, and the tail node, respectively. Knowledge graph representation learning refers to the embedding of entities and relationships in the knowledge graph. The method is used to represent entities and relationships in a low-dimensional dense vector space. In recent years, translation-based graph representation learning methods have received widespread attention due to their outstanding performance. In this paper, we introduce the following methods:

TransE[9] expects $h + r \approx t$ when (h,r,t) is present. TransH[10] handles a one-to-many relationship by mapping the representation of the entity to a relational space. TransR[11] proposed to establish their own relational space for different relationships. TransD[12] overcomes the shortcomings of TransR in using the same transformation matrix for mapping. It uses different vector space to map the node and relation.

3. Model Define

In this section, we will introduce the KSR model in detail. The framework of the model is shown in Figure 2. The model takes the news that the user has clicked and the news that is to be predicted as input. For each piece of new we use the KGE module to process its information level representation, and get the embedding of each piece of news. We enrich the news embedding vector by capturing sequential relationship through LSTM, which is described in detail in 3.1. The user's embedding is represented by the news that he has clicked. In order to obtain the final embedding of the news, we use the attention mechanism to calculate the weight of user history vector and the news vector to be predicted which will be described in detail in 3.2. Finally, we put the user's embedding and the embedding of the news to be predicted into the neural network to get the probability that the user clicks on the news.

IOP Conf. Series: Materials Science and Engineering **799** (2020) 012042 doi:10.1088/1757-899X/799/1/012042

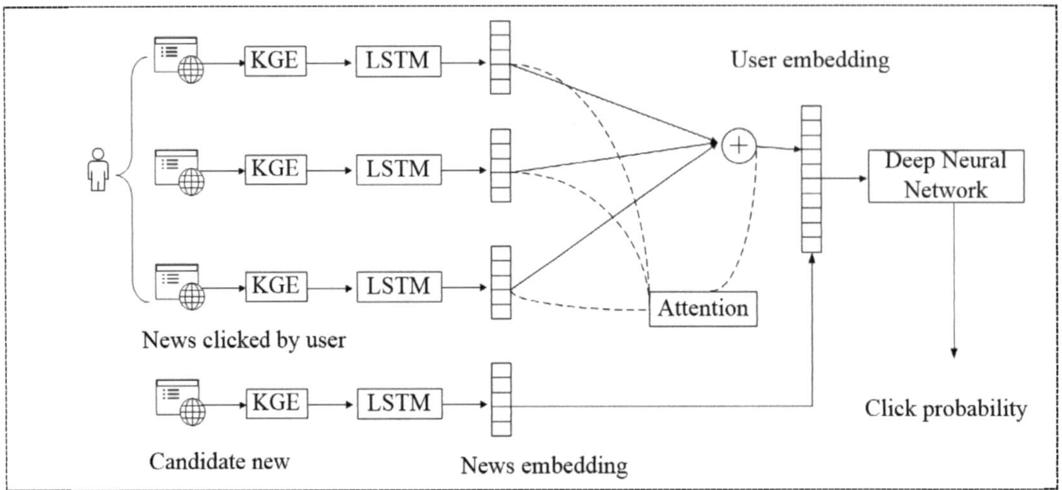

Figure 2. This is the framework of the KSR model.

3.1. News embedding refining

The KGE and LSTM modules work primarily to get the embedding of the news, which is used to enrich the user's characteristics. First, we use the traditional word2vec for the title in the dataset. We use the Gensim library to calculate the word embedding w_i, as shown in Equation 1. Then we get the entities in the title, and extract all the relationship links from the original knowledge graph to construct a subgraph. For the extracted subgraphs, we use the knowledge graph embedding methods TransE, TransH, TransR, and TransD to get the entity embedding. The entity embedding is expressed as e_i, as shown in Equation 2. We found in the experiment that when the entities in the news titles are constructed into small subgraphs, although the structural information of some entities can be obtained, these subgraphs are often sparse. So we use the context entities to enrich the sematic information. The context entity is represented as the average of its neighbor entities, denoted here by c_i, as expressed by Equation 3. We complete the first step of feature refinement. Next, we will use the recurrent neural network to capture the sequential relationship between each part of the vector.

$$w_i = Gensim\ (word_i) \tag{1}$$

$$e_i = Trans(G_i) \tag{2}$$

$$c_i = \frac{1}{|context(e_i)|} \sum_{e \in context(e_i)} e \tag{3}$$

We can do this by using the recurrent neural network, as shown in Section 2.1. In subsequent section, we also used the GRU unit which is variants of LSTM for comparative experiments. We pass the word embedding in the title mentioned above and the entity embedding generated by Trans method into two different LSTM networks to obtain a embedding containing the sequential relationship, such as formula (4)(5), where $[w_1 w_2 \dots w_n]$ is the words contained in the news. The embedding of the entity part is the same. Finally, the three parts of the embedding are connected as an input to the subsequent modules, as shown in Equation 6.

$$[w_1 w_2 \dots w_n] = LSTM(w_1 w_2 \dots w_n) \tag{4}$$

$$[e_1 e_2 \dots e_n] = LSTM(e_1 e_2 \dots e_n) \tag{5}$$

$$W = [w_1 w_2 \dots w_n e_1 e_2 \dots e_n c_1 c_2 \dots c_n] \tag{6}$$

3.2. Attention module

Through the processing in Section 3.1, we obtain the news embedding that the user history clicked. We use the news that the user history clicked to indicate the user's preference. We can average these vectors, but because of the variety of user interests, news that have been clicked by user history may have different effects on the predicted news. In order to indicate the user's preference for different news, we use the attention model [13] to calculate the weight of news clicked by users and news to be predicted. The attention model diagram is shown in Figure 3. We use the news to be predicted, and the news clicked by the user history as input to the deep neural network for training, and finally use the softmax function to learn the weight of each historical news.

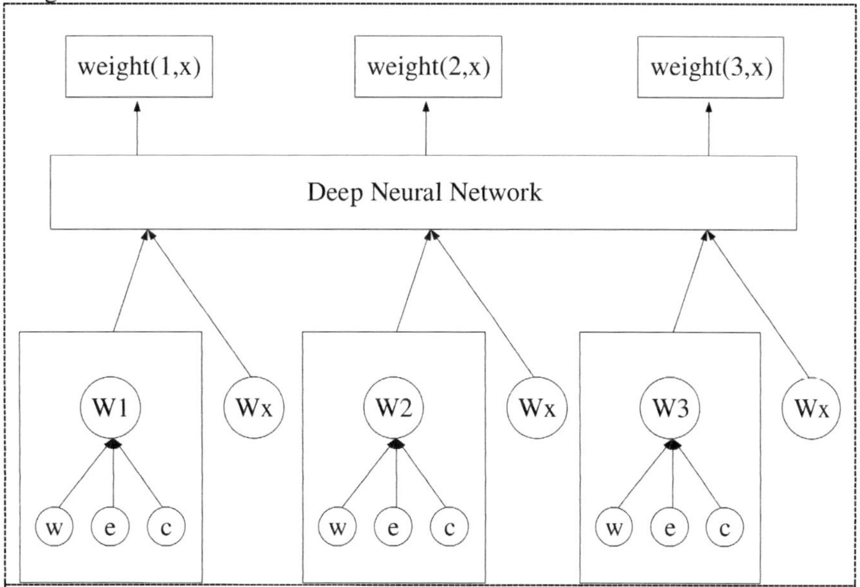

Figure 3. The attention model.

4. Experiment

In this section, we will introduce the experimental part, including the introduction of the dataset and the results of the experiment and the comparative experiment.

4.1. Dataset description

Our data set comes from Microsoft Bing News (for publicly released news data). After desensitization, each log mainly includes the user ID, news title and click status (1 is clicked, 0 is unclicked), entity ID, entity name. The knowledge graph is a graph of all the entities present in the dataset in the Microsoft Satori knowledge graph and their neighbours within several hops.

4.2. Baselines

DMF [14] is a depth matrix factorization model that uses a multi-layer neural network to process user embedding and item embedding for recommendation. We only use the user's implicit feedback for input.

FM [15] is a representative model based on use's history. We use the doc2vec model to encode the news titles for calculation.

DKN [16] takes entity embedding and word embedding as input and encodes it into the KCNN network for encoding.

4.3. Experimental setup

Our experimental environment is Win7 64-bit, Python 3.6 and TensorFlow 1.14. KGE module is Fast-TransX. It is an effective way to implement TransE and its variants, compiling with g++. Our optimization algorithm is Adam. The loss function is the cross entropy loss function plus the loss of the regularization term. The optimization goal is shown in Equation 7. Where y is the label of the sample, and $\bar{y}$ is the probability that the model is predicted to be positive.

$$Loss = \sum -[y * \log(\bar{y}) + (1-y) * \log(1-\bar{y})] + \theta * Loss_{regularization} \quad (7)$$

4.4. Experimental results

We analyze the experimental results with AUC and MSE. AUC is an evaluation value to measure the pros and cons of the model, indicating the probability that the positive example of the prediction is in front of the negative example. Performance of the model is better when this value is higher. MSE is the mean square error, which is the expected value of the square of the difference between the predicted value and the real value, which we expect the lower, the better. As shown in Table 1, the KSR model obtains the highest AUC value and the lowest MSE value and it achieves best results than the baselines.

Table 1. Comparison of different models

	AUC	MSE
KSR	**58.38**	**0.2406**
FM	44.96	0.2428
DKN	57.65	0.2419
DMF	53.90	0.2591

We will analyze the model and conduct comparative experiments from the following four dimensions. 1. In Section 3.1, we mentioned that we not only use the vector of the entity in the news but also introduce the context entity. We will experimentally verify whether this works. 2. There is a well-known variant of LSTM--GRU, we will compare which effect is better. 3. We introduced thc attention model into the model. We will verify this in our experiment. 4. In Section 2.2, we introduced four KGE methods, and we will figure out that which one has best effect. We performed experiments on these four cases on the KSR model, and the experimental results are shown in Table 2. The first row in the table is the standard KSR model, using LSTM to capture sequential relationships and using TransD model to get the knowledge graph embedding. The second and third rows in the table prove the necessity of introducing the context vector and attention module. The fourth row in the table is the result of sequential capture by using GRU method, which proves that LSTM achieves better results than GRU. The fifth, sixth, and seventh rows in the table are the results of using different Trans methods, which proves the superiority of using TransD.

Table 2. Comparison of variants.

	Epoch @5	*Epoch @10*	Epoch @15	*Epoch @20*	Epoch @25
KSR	**0.571**	**0.577**	**0.572**	**0.569**	**0.570**
KSR(-context)	0.568	0.559	0.520	0.525	0.543
KSR(-Attention)	0.568	0.515	0.545	0.528	0.522
KSR(GRU)	0.567	0.550	0.512	0.560	0.544
KSR(TransH)	0.564	0.559	0.556	0.557	0.465
KSR(TransE)	0.564	0.563	0.547	0.547	0.523
KSR(TransR)	0.571	0.569	0.571	0.560	0.524

5. Conclusions

This paper proposes a KSR model for the problem of data sparsity in the field of news recommend-dation. KSR uses a knowledge graph as side information to introduce the recommendation process, enriches the feature representation of news vector, and proposes to introduce the sequential module into the model to capture the sequential relationship between words and entities. Since the relationship between the news clicked by the user history and the new to be predicted has different weights, based on this consideration, we introduce the attention mechanism into the model. Finally, the experiments are carried out on the actual news dataset. it achieves best results than the baselines. In the subsequent experimental analysis, the rationality and effectiveness of the module in this model are also deeply analyzed.

Acknowledgments
I want to thank my mentor, my predecessors and my classmates for their support and help so that I can achieve my present achievements. This work is partially supported by National Key Research and Development Program of China (No.2019QY1704).

References
[1] Son, J. W., Kim, A., Park, S. B. (2013). A location-based news article recommendation with explicit localized semantic analysis. In: SIGIR. Dublin. pp. 293-302.
[2] Zheng, G., Zhang, F., Zheng, Z., Xiang, Y., Yuan, N. J., Xie, X., Li, Z. (2018). DRN: A deep reinforcement learning framework for news recommendation. In: WWW. Lyon. pp. 167-176.
[3] Wang, C., Blei, D. M. (2011). Collaborative topic modeling for recommending scientific articles. In: SIGKDD. San Diego. pp. 448-456.
[4] Mikolov, T., Sutskever, I., Chen, K., Corrado, G. S., Dean, J. (2013). Distributed representations of words and phrases and their compositionality. In: Neural Information Processing Systems. Montréal. pp. 3111-3119.
[5] Covington, P., Adams, J., Sargin, E. (2016). Deep neural networks for youtube recommendations. In: ACM conference on recommender systems. Boston. pp. 191-198.
[6] Cheng, H. T., Koc, L., Harmsen, J., Shaked, T., Chandra, T., Aradhye, H., ... Anil, R. (2016). Wide deep learning for recommender systems. In: Deep Learning for Recommender Systems. Boston. pp. 7-10.
[7] Socher, R., Perelygin, A., Wu, J., Chuang, J., Manning, C. D., Ng, A., Potts, C. (2013). Recursive deep models for semantic compositionality over a sentiment treebank. In: EMNLP. Seattle. pp. 1631-1642.
[8] Zhang, P., Ouyang, W., Zhang, P., Xue, J., Zheng, N. (2019). SR-LSTM: State Refinement for LSTM towards Pedestrian Trajectory Prediction. In: CVPR. Long Beach, CA. pp. 12085-12094.
[9] Bordes, A., Usunier, N., Garcia-Duran, A., Weston, J., Yakhnenko, O. (2013). Translating embeddings for modeling multi-relational data. In: NeurIPS. Montréal. pp. 2787-2795.
[10] Wang, Z., Zhang, J., Feng, J., Chen, Z. (2014). Knowledge graph embedding by translating on hyperplanes. In: AAAI. Québec City. pp. 1112-1119.
[11] Lin, Y., Liu, Z., Sun, M., Liu, Y., Zhu, X. (2015). Learning entity and relation embeddings for knowledge graph completion. In: AAAI. Austin. pp: 2181-2187.
[12] Ji, G., He, S., Xu, L., Liu, K., Zhao, J. (2015). Knowledge graph embedding via dynamic graphping matrix. In: ACL. Beijing. pp. 687-696.
[13] Wang, X., Yu, L., Ren, K., Tao, G., Zhang, W., Yu, Y., Wang, J. (2017). Dynamic attention deep model for article recommendation by learning human editors' demonstration. In: SIGKDD. Halifax. pp. 2051-2059.
[14] Xue, H. J., Dai, X., Zhang, J., Huang, S., Chen, J. (2017, August). Deep Matrix Factorization Models for Recommender Systems. In: IJCAI. Melbourne. pp. 3203-3209.
[15] Rendle, S. (2010, December). Factorization machines. In: ICDM. Sydney. pp. 995-1000.
[16] Wang, H., Zhang, F., Xie, X., Guo, M. (2018, April). DKN: Deep knowledge-aware network for news recommendation. In: WWW. Lyon. pp. 1835-1844.

IOP Conf. Series: Materials Science and Engineering **799** (2020) 012043 doi:10.1088/1757-899X/799/1/012043

The Cognitive Communication based Anti-jamming Method for UAV Swarm Communication

Jian Wu, Rang Liu* and Sen Zhao

China Academy of Electronics and Information Technology, Beijing, 100041, China

*Corresponding author's e-mail: liurangcetc@163.com

Abstract. The communication environment of intelligent UAV swarm is complex and easy to be interfered by various electromagnetic signals. An anti-jamming method of UAV swarm communication based on cognitive communication is proposed to realize the communication networking function of UAV swarm in certain interference environment. The main approaches include spectrum sensing in electromagnetic environment, anti-jamming based intelligent decision making, and series typical anti-jamming techniques such as adaptive interference and signal separation, dynamic spectrum access, high reliability waveform enabled.

1. Introduction

The intelligent wireless communication network system based on environment awareness is an important application field of UAV swarm communication anti-jamming technology. It can sense the external environment and use artificial intelligence technology to learn from the environment. By changing some parameters (such as transmission power, carrier frequency and modulation technology) in real time, its internal state can adapt to the statistical changes of the received wireless signal, so as to achieve the purpose of highly reliable communication and effective use of spectrum resources at any time and any place. The environment sensing ability of the communication system enables it to capture or perceive information from the wireless environment, so that it can identify the unused spectrum resources in specific time and space, and select the most suitable spectrum and working parameters. Cognitive communication anti-jamming technology requires that the communication system can adjust the working parameters of the transmitter without modifying the hardware part and without interrupting the work. It can be carried out at the beginning of transmission or in the process of transmission. It is the specific performance of cognitive radio communication system to adapt to the dynamic wireless environment.

Reference [1] makes a visual analysis of battlefield electromagnetic situation from multiple perspectives. The method of electromagnetic environment visualization in Artillery Battlefield is given in reference [2]. In reference [3], the communication parameter adjustment method combining genetic algorithm and simulated annealing is proposed and compared. In reference [4], a swarm based dynamic multichannel network strategy is introduced and a spectrum cognitive interference routing algorithm is proposed. In reference [5], a dynamic network coding technique in weak communication environment is proposed. In reference [6], an artificial intelligence communication anti-jamming system based on radial basis function neural network algorithm is designed. In reference [7], the intelligent anti-jamming communication technologies such as interference recognition and waveform reconstruction are analyzed. In reference [8], two intelligent anti-jamming decision engines are simulated and the results are analyzed.

2. Spectrum sensing of UAV swarm communication to electromagnetic environment

2.1. Distributed electromagnetic environment cooperative detection

Based on the mechanism, process and algorithm of distributed task collaborative allocation, the dynamic allocation of detection tasks is carried out in a unified way, so as to achieve the purpose of fast acquisition of all frequency, spatial and domain electromagnetic environment signals.

Based on the distributed network collaboration, make full use of the electromagnetic environment detection resources on the platform of the collaborative network to allocate the detection area, frequency band and target, so as to achieve the full-time, wide area and wide spectrum cooperative detection effect. The following figure shows the cooperative detection of distributed electromagnetic environment.

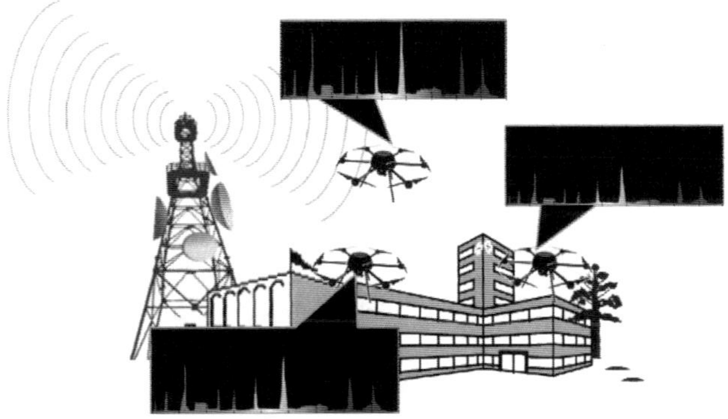

Figure 1. Distributed electromagnetic environment cooperative detection.

2.2. Field location and identification of multi UAV cooperative radiation sources

Based on the communication network cooperation of multiple UAVs, multi-sensor intelligence data fusion can be realized. Through situation sharing and data fusion, the confidence of target attribute discrimination, threat level assessment and activity situation awareness can be improved. The direction finding technology of amplitude comparison method, interferometer and MUSIC are studied.

First, a feature database of target radiation sources is established. Then, the signal recognition algorithm is designed, and the template in the template database is used to quickly compare, match and identify the extracted target features. Based on the network cooperation, the enemy radiation sources are cooperatively positioned to improve the positioning accuracy and shorten the positioning convergence time. It also has the ability to quickly identify and coordinate with the target, and it can achieve real-time situation information acquisition and sharing.

2.3. The knowledge base of electromagnetic radiation source and the construction of spectrum map form the electromagnetic situation of battlefield

Through the acquisition, analysis, prediction and classification of the electromagnetic environment elements of UAV swarm, the electromagnetic radiation source knowledge base is formed. Through sensing the electromagnetic field data around the target from time domain, frequency domain, spatial domain, energy domain and other aspects, and also through processing and analysis, using the spatial discrete inversion electromagnetic field distribution technology, the dynamic electromagnetic field is displayed and visualized, which forms the battlefield electromagnetic situation and multi-dimensional spectrum map.

3. Cognitive decision based on genetic algorithm

The UAV swarm intelligently can sense the electromagnetic environment, obtain the battlefield electromagnetic situation and spectrum map, and finally conduct intelligent decision-making for communication anti-jamming.

As shown in the following figure, this paper conducts research on anti-jamming intelligent decision-making technology based on cognitive results. Real-time monitoring of the spectrum environment of the battlefield, achieving interference avoidance or elimination through active and passive methods, and improving the comprehensive anti-jamming capability of swarm communication. Active evasion mainly adjusts the system transmission parameters in real time according to the electromagnetic environment, realizes frequency domain avoidance by changing the frequency hopping pattern, or improves the anti-jamming tolerance by changing the modulation and coding mode, the symbol width, and the length of the spreading code. Passive cancellation mainly aims at different interference types, using adaptive notch or fractional Fourier transform to eliminate the influence of interference signals on the system.

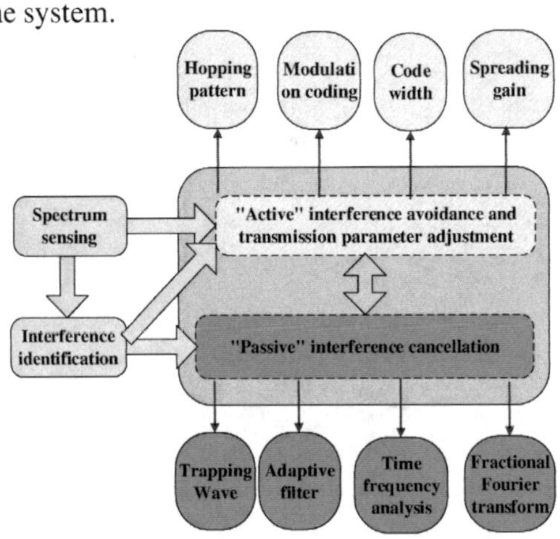

Figure 2. Intelligent anti-jamming decision.

On the basis of accurate estimation of spectrum situation, this paper uses genetic algorithm to solve the problem of cognitive parameter configuration and improve the anti-jamming ability and spectrum utilization. An improved algorithm combining simulated annealing algorithm with genetic algorithm is proposed, which can greatly improve the search efficiency of the algorithm without changing the genetic algorithm. By controlling whether to accept the new individuals after crossover and mutation of genetic algorithm, the shortcomings of genetic algorithm are improved. The combination of simulated annealing algorithm and genetic algorithm improves the effectiveness of cognitive decision engine. For the implementation of genetic algorithm, the scheme first uses parameter coding to connect n communication parameters (modulation mode, center frequency, symbol rate, bandwidth, transmission time percentage, etc.) in frequency hopping anti-jamming system into chromosomes, as shown in the figure below. Then the objective function is obtained as:

a_1	a_2	...	a_{n-1}	a_n

Figure 3. Chromosome of genetic algorithm.

The objective function is:

$$f' = f^{10} = (\sum_{i=1}^{m} w_i f_i)^{10} \quad (1)$$

As a measure mechanism of chromosome fitness, genetic algorithm is characterized by survival of the fittest and survival of the fittest. Individuals with high fitness are left to the next generation, and those with low fitness are eliminated, so cross exchange genes to generate new individuals; variation can enhance the global search ability of genetic algorithm by increasing individual diversity. In order to make up for the deficiency of the efficiency of genetic algorithm, the simulated annealing algorithm is introduced to decide whether to accept the new individuals after the crossover operation. The objective function values of pre-crossing chromosome are expressed as f_{1old} and f_{2old} , and the

IOP Conf. Series: Materials Science and Engineering **799** (2020) 012043 doi:10.1088/1757-899X/799/1/012043

objective function values of post-crossing chromosome are expressed as f_{1new} and f_{2new}. P is the probability of whether a new chromosome will replace a crossed chromosome, as follows:

$$p=\begin{cases}1 & \Delta f<0\\ \exp(-\Delta f/T_g) & \Delta f\geq 0\end{cases} \tag{2}$$

$$\Delta f=(f_{1old}+f_{2old}+f_{1new}+f_{2new})/4-f_{1new}(f_{2new}) \tag{3}$$

T_g is the temperature, and the temperature attenuation function is as follows:

$$T_g=T_0\alpha^{g-1} \tag{4}$$

where T_0 is the initial temperature, and G is the evolution algebra, $0<\alpha<1$.

According to the above formula, with the increase of evolution algebra g, T_g decreases gradually, so the probability of accepting new individuals with lower value than the expected objective function decreases. In the early stage of algorithm evolution, most new individuals can be accepted to expand the value space. While in the middle and late stage of algorithm evolution, the possibility or probability of accepting new individuals with higher value than the expected objective function is greater.

4. Anti-jamming mode of cognitive communication

After obtaining the spectrum environment information, the swarm communication system takes active defense or evasion measures according to the interference characteristics to improve the communication quality and reliability. The measures include adaptive interference signal and separation, gradual dynamic spectrum access, parameter automatic control, high reliable waveform self enabling.

4.1. Adaptive interference and signal separation

When the communication system is affected by narrow-band interference, resulting in some frequency points being affected, the cognitive communication system can extract the characteristic information of the interference signal, such as modulation mode, power size, etc. At the receiving end, the interference signal is separated by adaptive notch or matched filter. And the impact of the interference is reduced to improve the communication quality.

When using adaptive interference and signal separation, it is assumed that n of m frequency points are interfered. The probability of n frequency points demodulation error caused by interference is p_i respectively, and the frequency points can be demodulated correctly after using interference and signal separation, then the anti-jamming gain of dry signal separation is:

$$G_f=10\log\left(\frac{m}{m-\sum_{i=1}^{n}p_i}\right) \tag{5}$$

4.2. Gradual dynamic spectrum access

The communication system is interfered by some frequency bands, which results in some frequency points seriously affected. If the number of affected frequency points does not exceed 1 / 3 of all the frequency points, the cognitive communication system can optimize the frequency hopping pattern. The affected frequency points in the frequency hopping pattern are replaced by other frequency points in the frequency set to avoid the interference actively. If the frequency point of the interference changes dynamically, the frequency hopping pattern will change accordingly, so as to realize the dynamic spectrum access and improve the communication quality.

In the process of frequency hopping pattern optimization, only some disturbed frequency points in the pattern are replaced, and no significant changes are made to the frequency hopping pattern. Because of the uniform probability of frequency points in the pattern, more than 2 / 3 of the changed pattern is the same as the original pattern, which is a gradual process. The advantage of keeping the pattern gradual change is that the nodes in the communication system that do not optimize the pattern in time can still receive and receive messages in the network normally, but the communication quality is not high, which is the same as the effect when the whole network does not change the pattern in the case of partial band interference.

When using gradual dynamic spectrum access, assume that n of 51 frequency points are seriously interfered, where $n \leq 17$. The interference causes the demodulation error of N frequency points, then the anti-jamming gain of gradual dynamic spectrum access is:

$$G_a = 10\log\left(\frac{51}{51-n}\right) \tag{6}$$

4.3. High reliability waveform autonomous operation

High reliability waveform includes long code spread spectrum and high redundancy channel coding. When all or most of the frequency points of the communication system are seriously affected, the frequency hopping gain disappears. At this time, the hopping speed can be reduced or even the fixed frequency can be used, the pulse interval time can be reduced, and the pulse duration can be increased. When the baud rate of the channel is unchanged, a longer spread spectrum code can be used for frequency expansion communication, and the spread spectrum gain can be increased to resist strong interference. For example, the communication system currently uses 32-bit code to expand 5-bit original information. If the length of spread spectrum code is increased and the original information of 2^k-bit spread k-bit is used ($k > 5$), the increment of spread spectrum gain is:

$$G_k{}' = 10\log\left(2^k/k\right) - 10\log(32/5) \approx 10\log\left(2^k/k\right) - 8 \tag{7}$$

At present, the communication system uses RS (31,15) channel coding, and the coding efficiency is 1/2. Using a higher redundancy coding method, reducing the coding efficiency, it can obtain a greater error correction ability, thus improving the anti-jamming ability. Assuming that the new coding rate is r, $r < 1/2$, the increment of spread spectrum gain is:

$$G_r{}' = 10\log(1/r) - 10\log(2) \approx 10\log(1/r) - 3 \tag{8}$$

Therefore, the total anti-jamming gain of high reliability waveform is:

$$G_w = G_k{}' + G_r{}' \approx 10\log\left(2^k/k\right) + 10\log(1/r) - 11 \tag{9}$$

The high reliable waveform greatly improves the anti-jamming ability, at the cost of the decrease of the transmission rate, which is as follows

$$R = 0.2kr2^{7-k} \tag{10}$$

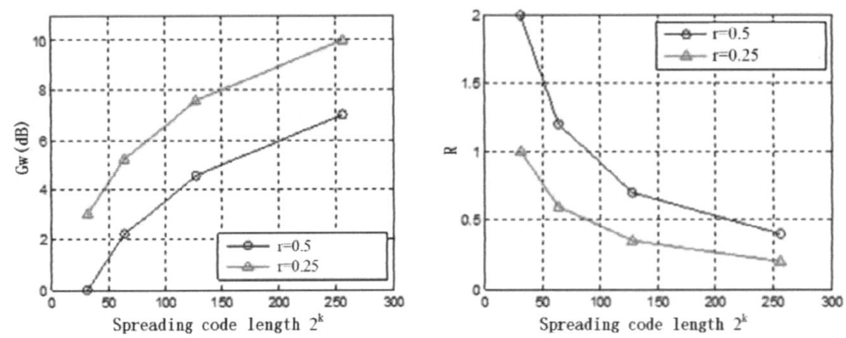

Figure 4. Estimated anti-jamming gain and decrease ratio of signal transmission rate.

Cognitive communication system nodes can use high reliable waveform transmission independently without informing the receiver. Therefore, communication system nodes need the dual-mode receiving ability. Dual mode reception refers to that the communication system node can receive and process high reliable waveform while receiving conventional waveform.

5. Conclusion

This paper discusses the anti-jamming technology of UAV swarm based on cognitive communication, which can detect, locate and identify the communication signals in the electromagnetic environment, and construct the electromagnetic situation map and spectrum map according to the knowledge base of radiation sources. An innovative genetic algorithm based intelligent cognitive decision-making is proposed to choose the appropriate communication anti-jamming technology. Finally, several typical cognitive communication anti-jamming technologies are given and compared.

References

[1] Nie J.L., Tian M.C., Guo D.L. (2013) Visual analysis of multi-dimensional electromagnetic situation using interactive parallel coordinates. Journal of Yanshan University, 553 (6):542-546.

[2] Zhang H. (2010) The investigation into the visualization of the artillery battlefield electromagnetic environment. Electronic Quality,10:76-79.

[3] Zheng S.L,, Zhao Z.J., Shang J.N.(2008) Cognitive Radio Decision Engine Based on Genetic Algorithm and Simulated Annealing. Computer Simulation, 25(1):192-195.

[4] Liu B.X., Jiang T.(2010)Cognition-based anti-interference routing algorithm for wireless sensor networks. Digital Communication,37(1):66-70.

[5] Xu D.F.(2019)Anti-jamming Transmission Technology Based on Network Coding [J]. Command Information System and Technology, 10(2):56-61.

[6] Liu M., Qi. H.Y., Wang J.N., Liu L.Z. (2018) Design of Intelligent Anti-jamming System Based on Neural Network Algorithm. Computer Measurement & Control,26(10):155-159.

[7] Li S.Q., Cheng Y.F., Dong B.H., Tang X.M. (2012) Research on Intelligent Anti-jam Communication Techniques. Radio Communications Technology, 38(1):1-4.

[8] Liu X.L. (2016) The Research on Intelligent Anti-jamming Decision Technology of NC-OFDM System. University of Electronic Science and technology.

IOP Conf. Series: Materials Science and Engineering **799** (2020) 012044 doi:10.1088/1757-899X/799/1/012044

Research on Numerically Solving the Inverse Problem Based on L1 Norm

Guo Yanan[1,2], Cao Xiaoqun[2*], Liu Bainian[2], Peng Kecheng[1,2], Wang Guangjie[3], Gao Mei[1,2]

[1]College of Computer, National University of Defense Technology, Changsha Hunan 410073, China

[2]College of Meteorology and Oceanography, National University of Defense Technology, Changsha Hunan 410073, China

[3]Chinese People's Liberation Army 61741, Beijing 100000, China

[*]Corresponding author's e-mail: caoxiaoqun@nudt.edu.cn

Abstract. In this paper, the numerical solution method based on L1 norm for inverse problem is studied. First, according to the theory of the L1 norm and the characteristics of the problem to be solved, a cost function is constructed. Further, for complex parameter estimation problems and derivative discontinuities, the regularization method is used to construct the cost function and the Huber function is used in the derivative discontinuity. Finally, the problem is solved numerically by the semi-smooth Newton method. The experimental results show that the method based on L1 norm is an effective method under the interference of non-Gaussian noises. For the parameter estimation of complex models, the results based on the L1 norm method are very closer to the real parameters when there is impulse noise. For parameter estimation under non-Gaussian noise, the L1 norm estimation method has significant advantages over the L2 norm method.

1. Introduction

There are a large number of physical model parameter estimation problems in the fields of geophysical sciences such as atmosphere, ocean, and space [1,2]. Although the parameter estimation method based on the L2 norm has been widely used in solving the above problems, there are still major limitations. For example, in the data assimilation of atmospheric observation data, researchers have found that when severe weather changes, observation data often cannot be effectively used, and problems such as poor prediction results often occur. The main reason is that in the data assimilation, it is usually assumed that the observation error obeys the Gaussian distribution, that is, the cost function in the form of L2 norm; and the actual observations (such as satellite remote sensing data) usually contain significant non-Gaussian errors, which leads to one of the important reasons for low data utilization and poor assimilation effect. In recent years, a large number of studies by scientists have shown that the L1 norm has many advantages that the L2 norm does not have. For example, when the observation data is affected by gross errors, the method of minimizing the L1 norm is more stable and reliable, and has better anti-interference performance [3,4]. Because the cost function in the form of L1 norm has non-differentiable computational problems during minimization, its application range is far less extensive than that of L2 norm. However, with the continuous development of mathematical calculation methods and calculation conditions, the parameter estimation method based on the L1 norm has attracted more and more attention from researchers. At present, many solving algorithms have been proposed, such as the alternating direction multiplier method (ADMM), the Homotopy method [5-7], which greatly promoted the application of L1 norm optimization method, and achieved good results in many fields.

2. Parameter estimation method based on L1 norm

2.1. The concept and properties of L1 norm

The L1 norm is also called the Manhattan distance and is defined as the sum of the absolute values of the elements in the vector. At present, the L1 norm has attracted more and more attention from researchers due to its sparseness and other properties. The first is in the fields of machine learning and pattern recognition [8]. When constructing a model, if only a few features contribute to the model and most of the features are not contributing or the contribution is small, researchers often use L1 regularization. L1 norm can achieve sparseness to extract main features, so L1 norm is also called the sparse rule operator. In addition, the L1 norm sparsification feature is also widely used in the fields of compressed sensing, image denoising, and signal processing. Taking oil and gas exploration as an example: geophysicists use geological echo signals to invert geological structures. During the inversion process, the L1 norm sparsity can be used to extract as much information as possible from the limited echo signals. In addition, a large number of studies have found that cost functions based on the L1 norm form can effectively resist the interference of abnormal observation data. For example, in the multiple regression problem, when there is abnormal observation interference, the experimental result of the least square method based on the L1 norm which reflects the good resistance of noisy is obviously better than that the least square method based on the L2 norm.

2.2. Problem definition

In many scientific and engineering problems, it is often necessary to establish a physical model to describe complex physical movements or phenomena, which usually includes two important parts: model structure and model parameters. In most cases, model parameters cannot be directly observed or detected, but need to be estimated based on the observation information of the model state quantities. Such parameter solving problems are often proposed in the form of inverse problems [9,10]:

$$M(v) = y^{\delta} \tag{1}$$

Where v is the target parameter and y^{δ} is the observation data with noise. M is an integral operator or partial differential operator. M maps the parameter v to the function value y. Calculating the value of the parameter v based on the observation data y^{δ} has important applications in the data assimilation of atmosphere, geophysical sciences, and image processing.

2.3. Parameter estimation algorithm based on L1 norm

The above inverse problem is different from the forward problem. It calculates the target parameters in the model through observation. Because of the errors in the observation information and the nonlinearity of the system itself, the inverse problem is often ill-posed. Therefore, methods such as Tikhonov regularization are widely used to solve the problem. The main idea of regularization is to use the loss term and regularization term in the form of L2 norm to construct a cost function, and then obtain the optimal model parameter value through minimization. Considering that the L1 norm has many advantages such as robustness, this article attempts to use the L1 norm to estimate parameters in complex physical models, and constructs a new cost function as follows [11,12]:

$$J(v,\alpha) = \left\| M(v) - y^{\delta} \right\|_{L_1} + \frac{\alpha}{2}\left\| v \right\|_{L_2}^{2} \tag{2}$$

If M is strictly differentiable, the generalized gradient sum rule and chain rule are used to convert the problem into a problem of solving the equation:

$$\nabla_v J = \alpha \|v\|_{L_2} + M'(v)\eta = 0 \tag{3}$$

Where η is determined by Huber function:

$$\eta = \begin{cases} 1 & if\ (M(v)\text{-}y^{\delta}) > \beta \\ -1 & if\ (M(v)\text{-}y^{\delta}) < -\beta \\ (M(v)\text{-}y^{\delta}) * \beta^{-1} & if\ \left|M(v)\text{-}y^{\delta}\right| \le \beta \end{cases} \tag{4}$$

In the above formula, β is the adjustment factor of the Huber function. The value of β has an important influence on the effect of the L1 norm estimation. If the value of β is too large, the effect of the L1 norm estimation is not obvious. If the value of β is too small, the convergence radius of the semi-smooth Newton method will shrink to 0 as β decreases. At the same time, the β value is limited in regularization conditions. When the semi-smooth Newton method is used to solve formula (2), the value of α depends on the magnitude of the noise. But in general, the magnitude of the noise is unknown. Therefore, in the parameter estimation process, the regularization term coefficient α and the adjustment factor β of the Huber function need to be iteratively updated at the same time to obtain the optimal α and β values .

The specific steps of the parameter estimation algorithm based on the L1 norm are explained as follows:

Step 1: Transform the cost function and introduce Huber function for non-differentiable points. A new equation to be solved is obtained by gradient operation, and the initial guess value v_0 and initial regularization coefficient α_0 of the equation are given.

Step 2: According to the semi-smooth Newton method, give the initial value β_0 of the adjustment factor in the Huber function, and then start to solve iteratively until the number of iterations is greater than the maximum number or the gradient norm is less than the set threshold. Then use the automatic parameter selection method to reduce the value of the adjustment factor β in the Huber function, update the objective equation and repeat the above-mentioned solution process. If the number of iterations exceeds the maximum number of iterations, or β is less than the preset minimum, the optimization result is taken from the previous result, and the loop exits to the next step.

Step 3: When the number of updates of α is less than the maximum number of updates, update the regularization coefficient α . Specifically, update α by $\alpha_{k+1} = (\mu - 1)\dfrac{\left\|M(v_{\alpha_k}) - y^{\delta}\right\|_{L_1}}{J'(\alpha)}$, where $J'(\alpha) = \dfrac{1}{2}\left\|v_{\alpha_k}\right\|_{L^2}^2$. Set $\mu > 1$ and μ fixed in the experiment. The value of μ depends on the smoothness of the data. If the difference between the α update result and the previous result is small (the change value is less than the threshold value), then the loop goes out to the next step, otherwise it moves to the second step.

Step 4: Output the solution result and the final regularization coefficient α and the value of the adjustment factor β in the Huber function.

3. Numerical experiment results and analysis

In order to verify the applicability of the parameter estimation algorithm based on the L1 norm in complex systems, this section estimates the unknown coefficients in the variable coefficient Helmholtz equation. Helmholtz equation is a kind of elliptic partial differential equation, which mainly describes a kind of physical phenomenon of harmonic propagation. It has a wide range of applications in many fields, such as electromagnetic wave propagation, acoustic wave scattering, seismic wave propagation, and ocean wave propagation [13,14]. Therefore, studying the coefficient estimation problem of the Helmholtz equation with variable coefficients is very important to deal with many physical problems in the fields of electromagnetics, acoustics, and so on. First give the equation as follows:

$$-\frac{\partial^2 \omega}{\partial x^2}+\varphi(x)\omega=C \tag{5}$$

Set the real coefficients $\varphi(x)$=2-cos$(x^2)+|\sin(x)|$ and C = 1 in the experiment. First, the interval [-3,3] is equally divided into N = 1000 parts, and the partial differential equation is discretized by using the finite element method. The solution interval includes N + 1 nodes and N units.After discretization, the problem of estimating unknown parameters of partial differential equations is transformed into solving equations $M(\varphi(x_i))=y_i$, $i=1,\mathrm{L}\ ,N+1$. Use the following methods to construct observations with impulsive noise:

$$y^\delta=\begin{cases} y+\tau\psi & \textit{with probability } \gamma \\ y & \textit{otherwise} \end{cases} \tag{6}$$

Where y is the real data and ψ is a random variable that obeys the standard normal distribution. The value of τ depends on how much the observation data is affected by noise, where τ is set to $\|y\|_{L_\infty}$. $\gamma\in(0,1)$ is the proportion of abnormal data. The value of γ in the experiment is 0.3. In the following, the cost function is constructed based on the L1 norm and L2 norm respectively, and regularization method is introduced to calculate the coefficient $\varphi(x)$:

$$J(\varphi)=\left\|M(\varphi)-y^\delta\right\|_{L_1}+\frac{\alpha}{2}\|\varphi\|_{L_2}^2 \tag{7}$$

$$J(\varphi)=\left\|M(\varphi)-y^\delta\right\|_{L_2}^2+\frac{\alpha}{2}\|\varphi\|_{L_2}^2 \tag{8}$$

When solving (7), the solution algorithm in Section 2.3 is adopted. The problem is first transformed, and then the initial iteration value v_0 and the initial regularization coefficient α_0 are given. The fixed point iteration method is used to update α , and the semi-smooth Newton method is used for iterative calculation. When solving (8), perform gradient calculation to obtain the optimization target firstly: $2M'(\varphi)\left\|M(\varphi)-y^\delta\right\|_{L_2}+\alpha\|\varphi\|_{L_2}=0$, where α is 0.001, use Newton's method to solve the equation. In each iteration, calculations are performed using the biconjugate gradient stabilized method. The following experimental results were obtained:

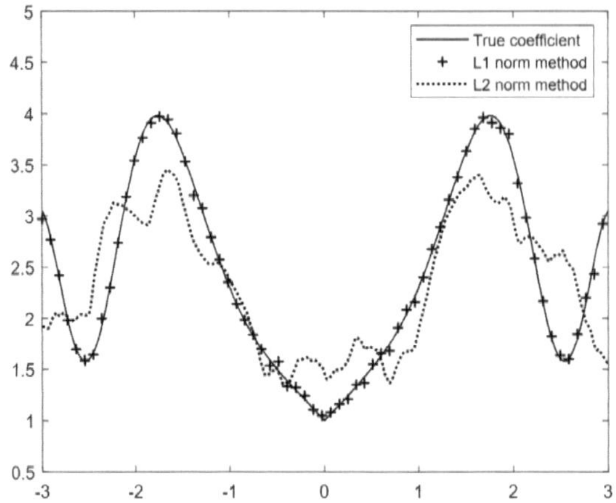

Figure 1 Comparison of estimation results based on two different methods in variable coefficient Helmholtz equation model

The solid black line in Figure 1 is the true coefficient value, the dotted line is the estimated value based on the L2 norm method, and the plus sign is the estimated value based on the L1 norm method. By comparison, it can be seen that the results obtained by the method based on the L1 norm are closer to the true value, while the results obtained by the method based on the L2 norm estimation are greatly affected by the impulse noise, and there are obvious fluctuations. The advantages and disadvantages of the two methods are analyzed by comparing the estimated error sizes. Taking the difference between the estimated result and the true value on the interval [-3, 3] as the standard. By calculation, the estimation error err_2 = 15.1113 based on the L2 norm method, and the estimation error err_1 = 0.1796 based on the L1 norm method. Obviously, the results obtained based on the L1 norm method have smaller errors and better results. This result indicates that in the parameter estimation problem, the method based on the L1 norm is superior to the method based on the L2 norm when the error of the observation data follows a non-Gaussian distribution. It also shows that for data with different error distributions, it is important to choose the appropriate data analysis method.

In the experiment, an automatic update method of regularization coefficients is used. The following specifically analyzes the update method of α and its convergence process:

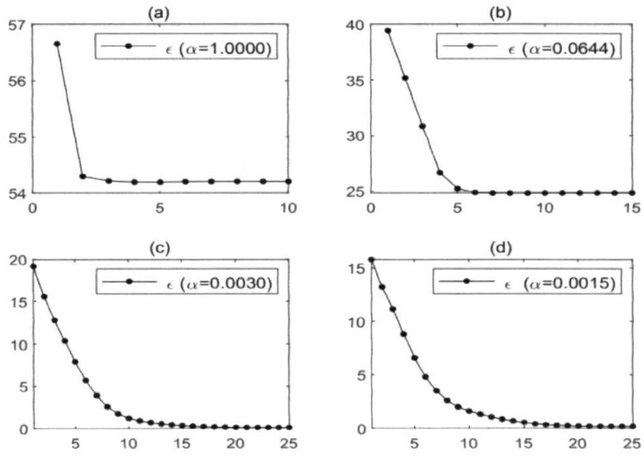

Figure 2 Variation of estimation error with the number of iterations under different regularization coefficients

Figure 2 shows the variation of the estimation error ε with different iteration times for different regularization coefficients, where $\varepsilon = \|v - v_e\|_{L^2}$, v represent the estimation result, and v_e represents the true coefficient. When iterating with fixed regularization coefficient α, the initial value β_0 of the adjustment factor in the Huber function is taken as 1.0. After each iteration, β is reduced to 0.5 times the previous result, so Figure 2 essentially reflects that when the regularization coefficient is determined, the estimation error varies with β changes. It can be seen from Figure 2 that when the regularization coefficient is determined, the estimation error gradually converges to a stable value as β decreases. In Figure 2 (a), α = 1, the estimation error has stably converged to around 52.2 at the third iteration. In Figure 2 (b), α = 0.0644, the estimation error has stably converged to around 25.0 when the iteration is performed for the sixth time. In Figure 2 (c), α = 0.0030, the estimation error converges to around 0.2 at the 20th iteration. In Figure 2 (d), α = 0.0015, the estimation error also converged to around 0.2 at the 20th iteration. In the experiment, because the difference between the fourth update result and the third result of α is less than the given threshold, the loop breaks out. It can be concluded from the above results that the estimation error decreases rapidly with the update, indicating that the automatic update method is very effective for reducing the estimation error.

4. Conclusion

Aiming at the problem that the parameter estimation method based on the L2 norm is not effective in the case of non-Gaussian errors, a parameter estimation method based on the L1 norm is proposed. The minimization algorithm research and numerical experiments are carried out. In this paper, experiments are performed to estimate unknown coefficients of partial differential equations. Aiming at the problem of impulse noise interference, a loss function based on the L1 norm is introduced. Huber function is introduced at the discontinuity of the derivative. Then, the new cost function is solved by using the semi-smooth Newton method. By using the method based on the L1 norm, better experimental results are obtained. The experimental results show that the method based on L1 norm is more robust than the method based on L2 norm under non-Gaussian noise interference.

In addition, the regularization method has an important effect on the parameter estimation effect. In this paper, an L2 regular term is added to the L1 norm loss term to prevent overfitting problems. Taking into account the sparseness of the L1 norm, L1 regular terms will be added to the cost function in the future to study the impact of sparseness on the parameter estimation problem.

Acknowledgement

National key research and development program of China (Grant of No. 2018YFC1506704)

References

[1] Schillings, C., & Stuart, A. M. (2017). Analysis of the ensemble kalman filter for inverse problems. SIAM Journal on Numerical Analysis, 55(3), 1264-1290.

[2] Yang, Y. X., & Zeng, A. M. (2009). Adaptive filtering for deformation parameter estimation in consideration of geometrical measurements and geophysical models. Science in China Series D: Earth Sciences, 52(8), 1216-1222.

[3] Yanyong Xiang, N.R. Thomson, & J.F. Sykes. (1992). Fitting a groundwater contaminant transport model by l1 and l2 parameter estimators. Advances in Water Resources, 15(5), 303–310.

[4] Seyoung Oh, Sunjoo Kwon, & Jae heon Yun.(2006) . A parameter estimation method for model analysis. Journal of Applied Mathematics & Computing, 22(1-2), 373-385.

[5] Zhu, L., Fu, C. W., Jin, Y., Wei, M., Qin, J., & Heng, P. A. (2016). Non-local sparse and low-rank regularization for structure-preserving image smoothing. Computer Graphics Forum, 35(7), 217-226.

[6] Shi, B., Pang, Z. F., & Yang, Y. F. (2012). Image restoration based on the hybrid total-variation-type model. Abstract and Applied Analysis, 2012(13-16), 97-112.

[7] Wang, T., Yang, Z., & Wu, Y. (2010). A maximum entropy function method for signal reconstruction in compressed sensing. Mathematica Applicata, 23(2), 345-352.

[8] Wang, Y., Li, D., Du, Y., & Pan, Z. (2015). Anomaly detection in traffic using L1-norm minimization extreme learning machine. (Vol.149, pp.415-425).

[9] Aster, R. C., Thurber, C. H., & Borchers, B. (2012). Parameter estimation and inverse problems. International Geophysics, 67, 411-421.

[10] Fu, H., Han, B., & Liu, H. (2012). A wavelet multiscale iterative regularization method for the parameter estimation problems of partial differential equations. Neurocomputing, 104, 138-145.

[11] Christian Clason, Bangti Jin.(2012).A Semismooth Newton Method for Nonlinear Parameter Identification Problems with Impulsive Noise. SIAM Journal on Imaging Sciences, 5:2, 505-536

[12] Christian Clason, Tuomo Valkonen.(2017) Primal-Dual Extragradient Methods for Nonlinear Nonsmooth PDE-Constrained Optimization. SIAM Journal on Optimization 27:3, 1314-1339.

[13] Chen, Y., & Rokhlin, V. (1992). On the inverse scattering problem for the helmholtz equation in one dimension. Inverse Problems, 8(3), 365-391.

[14] Barceló, Juan Antonio, Fanelli, L., Ruiz, A., & Vilela, M. (2013). A priori estimates for the helmholtz equation with electromagnetic potentials in exterior domains. Proceedings of the Royal Society of Edinburgh: Section A Mathematics, 143(01), 1-19.

IOP Conf. Series: Materials Science and Engineering **799** (2020) 012045 doi:10.1088/1757-899X/799/1/012045

Review of Crop Disease and Pest Image Recognition Technology

Shaopeng Jia, Hongju Gao*

College of Information and Electrical Engineering, China Agricultural University, Beijing 100083, China

*Corresponding author's e-mail: 2993133691@qq.com

Abstract. Throughout the history of agricultural development, crop diseases and pests have been one of the main obstacles hindering the development of agricultural economy, which not only caused irreversible economic losses to farmers, but also hindered the development of economic and social. The crop pets and diseases identification system based on digital image processing technology has the characteristics of fast, accurate and real-time, which can assist farmers to take effective control measures in time. In this paper, we have done some review on different digital image processing techniques to detect the plant diseases. We compared traditional machine learning and deep learning and discussed the future research trends and directions .

1. Introduction

With the development of the economy and society, the environment has become worse and worse, the occurrence of diseases and the variation of bacteria affect people's lives all the time. The incidence of crop diseases and insect pests is increasing. The higher the rate, the more complex the disease is, the measures to prevent crop diseases are gradually put on the agenda, and it is particularly important to find the disease in time and diagnose the disease and then make corresponding remedial measures. The traditional method of manually detecting pests and diseases depends entirely on the farmer's own experience, or asks experts to come to the door. Such methods are slow, inefficient, costly, subjective, low-accuracy, and time-insensitive. With the rise of the Internet industry, the use of information technology has provided new methods and ideas for improving the identification efficiency of crop pests and diseases. Crop pest and disease image recognition technology refers to the use of image processing technology to identify crop pests and diseases. Image recognition technology is used to classify and identify images to enhance recognition efficiency and speed. The cost is small and the accuracy is high. In order to improve the accuracy and speed of image recognition, experts and scholars at home and abroad have made a lot of research, making an indelible contribution to the agricultural production of farmers and the development of agricultural economy.

2. Traditional image processing technology

Traditional image processing technology has been the mainstream technology until the wave of deep learning has not yet emerged. As shown in Figure 1, it includes graphics collection, image preprocessing, image segmentation, feature extraction, and classifier classification.

2.1. Image collection

Image collection work is generally performed by using an electronic device such as a video camera or a mobile phone to shoot a diseased diseased leaf. Generally, it is chosen to shoot during the day or in good light to ensure that the pictures are clearly visible.

2.2. Image preprocessing

Image preprocessing includes image enhancement and noise reduction. The methods of noise reduction are mainly include median filtering and mean filtering. Image enhancement is performed by image binarization and histogram equalization in order to remove uncorrelated noise and highlight related parts.

2.3. Image segmentation

Image segmentation is the technique and process of dividing an image into specific regions with unique properties and proposing objects of interest. It is a key step from image processing to image analysis. Image segmentation has always been one of the most challenging tasks in digital graphics processing. The existing image segmentation methods are mainly divided into the following categories: threshold-based segmentation methods, region-based segmentation methods, edge-based segmentation methods, and segmentation methods based on specific theories.

In [1], the author used edge detection techniques to segment RGB images. In [2], the author used the histogram threshold segmentation method of optimal threshold to successfully separate the diseased leaf from the background and the effect was good. In [3], the author used statistical mode for image segmentation. In [4], the author used the fuzzy C-means algorithm to segment the color space, and the effect is significant. In [5], the author converted RGB to HIS, where H is the hue, I is the intensity, and S is the saturation value, which is segmented using the K-means clustering technique. In [6], the author used the maximum inter-class variance method to segment the color graphics, segment the intensity of the image (I), and improve the accuracy of the classification.

At present, most of the researches on traditional image segmentation methods are based on simple backgrounds and pictures with good illumination conditions, which are quite different from the complex conditions in the field. In the future, new segmentation algorithms should be studied to segment disease images of complex backgrounds. The use of multiple segmentation strategies to work together will be the development trend of research.

2.4. Feature extraction

Feature extraction refers to the use of a computer to extract image information and determine whether a point of each image belongs to an image feature. Feature extraction includes the description of features and the extraction of features. The result of feature extraction is to divide the points on the image into different subsets, which tend to belong to isolated points, continuous curves, or continuous regions. The main feature extraction methods include morphological feature extraction, color feature extraction and texture feature extraction.

2.4.1. Morphological feature extraction. In [7], in the research of image recognition of stored grain pests, the author extracted multiple morphological features for the binarized image of grain worms, and calculated the optimal features of seven features such as area and perimeter by optimization algorithm. Subspace. Experiments show that with this extraction method, the recognition rate is over 95%. In [8], the author applied the theory of invariant moments to feature extraction, defines the parameters of invariant moments, and extracts the morphological features of wheat leaf disease images. The author applied this method to the identification of wheat diseases, which has a good effect.

2.4.2. Color feature extraction. The following examples show how to format a number of different figure/caption combinations. The color feature is a global feature based on image pixel points. The color feature extraction method is also a commonly used method in image extraction of crop pests and diseases. In [9], the author used the RGB color system to perform feature extraction on diseased leaves for classification. In [10], the author extracted the histogram statistics of R, G, B, H, I, and S, which are represented by two color systems of RGB and HIS, in the process of research on wheat. Experimental results show that this feature extraction method effectively extracted image features and increased recognition accuracy. This proves that the color feature extraction method is also efficient and feasible. The color feature extraction method is intuitive and efficient, but the orientation of the image and the size and other changes are not sensitive and do not represent the target in the image well. Local features have their limitations in agricultural image recognition research.

2.4.3. Texture feature extraction. The following examples show how to format a number of different figure/caption combinations. Texture feature is also a global feature, which means recurring local patterns and their alignment rules. The extraction is the research central issue of the image feature extraction field. In [11], the authors used color symbiosis matrix (CCM) which based on the HIS chromaticity space to extract the texture features of citrus leaves, and used the least square matrix to construct the classifier. The recognition accuracy of the experiment reached 96%. In [12], the author used spatial gray-scale co-occurrence matrix extracts the texture feature of corn disease image, combined with wavelet transform to enhance wavelet decomposition texture map. The experiment achieved good results.

2.4.4. Feature extraction method combining form, color and texture. In [13], the author proposed to extract the color characteristics of cucumber downy mildew disease, extract the characteristics of disease texture using gray level co-occurrence matrix, and extract the morphological features of the lesions. The experimental results proved the validity of the combination of the three. In [14], the authors extracted the color features by calculating the one-to-third-order matrix of the histogram of the HSV color system of the corn disease leaves and extracted the texture features of the image by calculating the gray level co-occurrence matrix. Then authors used these extracted features to construct a classifier for recognition, and the effect is remarkable.

2.5. *Image classification and recognition*

The classification and recognition of images refers to the process of constructing a classifier with extracted image features to achieve target recognition. Classification and recognition is the final step of the whole image processing. The ultimate goal is to effectively identify the disease. The high accuracy of classification and recognition is the goal that experts and scholars at home and abroad have been striving for. Among the classification methods, there are SVM (Support Vector Machine), neural network and fuzzy clustering.

In [15], the author improved the genetic algorithm and used the improved algorithm to classify and identify the diseased leaves of soybeans. The accuracy of the final classification identification reached more than 90%. In [16], the author used the compensated fuzzy neural network to train and identify the extracted color features, and constructed a pest and disease recognition model. The experimental results show that the model has a good recognition effect. In [17], the author proposed an SVM algorithm based on balanced decision tree, which achieved a good classification effect. In [18], the author used radial basis to identify and identify cotton pests, and the recognition accuracy of the experiment reached 88%. In [19], the author extracted the morphology, texture and color feature of rice disease images, and then used stepwise discriminant analysis to classify and identify images, the final accuracy rate reached 97.2%. In [20], the author proposed to use regional growth method to classify and recognize on the basis of hyperspectral image, and the experimental effect was remarkable.

At present, the difficulty in classification and identification of crop pests and diseases is that crop diseases show different symptoms in different growth stages. However, the current classification and recognition algorithms have low recognition efficiency and have great limitations. Most of the diseases are classified when they are obvious. The disadvantage is that the universality is poor, and the recognition and classification effect on complex images is poor.

3. Image recognition technology in deep learning

The craze for deep learning has been enduring for a long time in recent years. The emergence of deep learning has reduced people's workload, so that people don't pay too much attention to how to improve the efficiency of image segmentation and feature extraction. Give all the work to the machine to do. Complex network structure and huge data samples are its biggest features. The emergence of deep learning technology provides a powerful technical guarantee for image recognition.

In [21], the authors trained a convolutional neural network using 54306 ill and healthy plant leaves collected under controlled conditions, and used trained model to identify 14 crops and 26 diseases with an accuracy of 99.35%. In [22],the authors proposed a concept of a deep learning meta-architecture that combines Faster R-CNN, SSD, RFCN, and feature extractors such as vgg net and resnet. The classification and recognition effect by this method is remarkable. In [23], the authors used a deep convolutional network to train image classification recognition models. Experiments show that the classification effect is significant. In [24], the author proposed a model based on deep convolutional neural network to identify rice diseased leaves, achieving an accuracy of 95.48%.,and this accuracy is much higher than the traditional machine learning model. In [25], the authors used Faster R-CNN to perform brown spot, blisters, and brown spot on tea. The preliminary results of the model showed that the average accuracy of the three diseases were 63.58%, 81.08%, and 64.71%.

4. Results and discussion

This paper reviews the literature on image processing technology based on traditional machine learning and deep learning. The purpose of this paper is to show the reader some techniques and progress in the field of image processing as well as to compare the advantages and disadvantages of traditional machine learning and deep learning in the field of image processing. However, most of the images recognized by traditional machine learning methods and deep learning methods are clear pictures taken with high-definition cameras. These pictures are not extensive. Most of the crop growers are farmers, and their means of catching pests and diseases are taken by their own mobile phone cameras, which has a disadvantage is that their mobile phones are basically ordinary mobile phones with low configuration and pixels. The picture taken with such equipment is relatively low-definition. these pictures are difficult to upload to the expert system for identification, and the recognition accuracy is lower. Therefore, future image processing techniques should focus more on images with complex backgrounds and blurred textures, improving the accuracy of recognizing such images.

5. Conclusion

This paper reviews the literature on image processing technology based on traditional machine learning and image processing recognition technology based on deep learning. Then compares the advantages and disadvantages of these two research methods. Although there are many research results in the field of image recognition, but there are still many problems need to be solved. Moreover, the effects of these studies in practical applications are not significant. Future research on image recognition should focus on images of pests and diseases with complex backgrounds randomly found in farmland, improving existing algorithms, and using the advantages of deep learning to identify images.

All in all, using image recognition technology to help farmers or growers identify pests and diseases has far-reaching and bright prospects for safeguarding agricultural production and promoting agricultural economic development.

References

[1] P.Revathi, M.Hemalatha,"Classification of Cotton Leaf Spot Diseases Using Image Processing Edge Detectio Technique",ISBN, 2012, 169-173, IEEE.
[2] Camargo A.Smith J S.An image-processing based algorithm to automatically identify plant disease visual systems[J].Bio-Systems Engineering,2009(102):9-21.
[3] Y. Tian, L. Wang and Q. Zhou, "Grading method of Crop disease based on Image Processing", Computer and computing technologies in agriculture 427-433, 2011.
[4] E. Hitimana & O. Gwun, Automatic estimation of live Coffee leaf infection based on Image processing techniques, Second International Conference on Signal, Image Processing and Pattern Recognition, Sydney, Australia, pp. 255–266, 2014.
[5] Smita Naikwadi, Niket Amoda," Advances In Image Processing For Detection Of Plant Diseases," International Journal of Application or Innovation in Engineering & Management (IJAIEM), Vol2, Issue 11, November 2013.
[6] W. Abudullakasim & J. Unartngam, "Assessment of the Severity of Brown Leaf Spot Disease in Cassava using Image Analysis", The International conference of the Thai Society of Agricultural Engineering [J], 2012.
[7] Zhang Hong-tao,Mao Han-ping,Qiu Dao-yin.Feature extraction for the stored-grain insect detection system based on image recognition technology[J].2009,25(2):126-130.
[8] Diao Zhi-hua,Zheng An-ping,Wu Yuan-yuan.Shape feature extraction of wheat leaf disease based on invariant moment theory[J].IFIP Advances in information and Communicrtion Technology,2012(369):168-173
[9] Jafari A,Mohtasebi S S,Jahromi H E.et al. Color feature extraction by means of discriminant analysis for weed segmentation[J].Computers and Electronics in Agriculture 2004
[10] Zhu Hong, Zheng Limin, Yin Jianling, et al. Small color extraction based on canopy image evaluation of wheat nitrogen nutrition level [J]. Journal of Agricultural Engineering, 2010, 26(2): 16-20.
[11] Pydipati R,Burks T F,et al. Identification of citrus Disease using color texture features and discriminant analisis[J] .Computers and Electronics in Agriculture 2006(52):49-59
[12] Gu Qingkui. Corn leaf disease identification system based on compute image processing [D]. Shenyang: Shenyang Institute of Technology, 2008.
[13] Liang Jiaxi. Experimental research and application of cucumber downy mildew based on image processing [D]. Changchun: Jilin University, 2011.
[14] Patil J K. Feature extraction of diseased leaf images [J].Journal of Signal and Image Processing,2012,3(1):60-63
[15] Qi Guangyun, Ma Xiaodan, Guan Haiou. Image extraction of soybean leaf lesions using improved genetic algorithm [J] Journal of Agricultural Engineering, 2009, 25(5): 142-145.
[16] Wen Zhiyuan, Cao Leping. Image recognition of different diseases and insect pests of navel orange based on compensated fuzzy neural network [J]. Journal of Agricultural Engineering, 2012, 28(11): 152-157.
[17] Yan Zhihua. Research and application of intelligent diagnosis system for field wheat leaf disease [D]. Hefei: University of Science and Technology of China, 2010.
[18] Zhang Jianhua, Yan Ronghua, Yuan Xue, etc. Cotton pest identification based on radial basis support vector machine [J]. Journal of Agricultural Machinery, 2011, 42(8): 178-183.
[19] Guan Zexin, Tang Jian, Yang Baojun, etc. Image-based rice disease identification method research [J]. Chinese Rice Science, 2010, 24(5): 497-502.
[20] Zhang Hongtao, Mao Hanping, Han Green. Near-infrared hyperspectral imaging technique for detecting granary live insects [J]. Journal of Agricultural Engineering, 2012, 28(8): 263-268.
[21] Sharada P. Mohanty, David P. Hughes and Marcel Salathé Using Deep Learning for Image Based Plant Disease Detection[J] Frontiers in plant science, 2016
[22] Alvaro Fuentes D, Sook Yoon, Sang Cheol Kim,et al.A Robust Deep-Learning-Based Detector for Real-Time Tomato Plant Diseases and Pests Recognition. Sensors, 2017,17,2022

[23] Anandhakrishnan MG Joel Hanson, Annette Joy, Jerin Francis. Plant Leaf Disease Detection using Deep Learning and Convolutional Neural Network[J].International Journal of Engineering Science and Computing, 2017.

[24] Yang Lu, Yi Shujuan, Yong Zhang, et al. Identifification of rice diseases using deep convolutional neural networks[J]. Neurocomputing, 2017 - Elsevier.

[25] Sheng-Hung Lee, Shih-Fang Chen. Devel- opment of Image Recognition and Classification Algorithm for Tea Leaf Diseases Using Convolutional Neural Network .2018 ASABE Annual International Meeting,2018.

IOP Conf. Series: Materials Science and Engineering **799** (2020) 012046 doi:10.1088/1757-899X/799/1/012046

Adaptive power line communication for intelligent separate layer production with single-core cables

Liu Changzan[1], Dang Bo[2*], Yang Ling[2], Xu Linkang[2], Dang Ruirong[2]

[1]School of Marine Science and Technology, Northwestern Polytechnical University, Xi'an, 710072, Shaanxi, China

[2]Key Laboratory of Education Ministry for Photoelectric Logging and Detecting of Oil and Gas, Xi'an Shiyou University, Xi'an, 710065, Shaanxi, China

[*]Corresponding author's e-mail: bodang521@126.com

Abstract. Power line communication (PLC) techniques have been proven to be efficient for the intelligent separate layer production (ISLP) system with single-core cables. However, the permanent mode of the ISLP may cause a continuous change of the resistance and capacitance of the single-core cables due to the long-term effect of high temperature, high pressure as well as the vibration of the bad downhole conditions, which will have great influence on the data transmission efficiency. In this paper, we present an adaptive PLC approach for the data transmission of ISLP with single-core cables. Instead of the precise slide rheostat, program-controlled potentiometer is used to achieve an adaptive decoding algorithm for fast initialization of the ISLP transmission. Moreover, a long-term adaptive method is also proposed to adjust the gain and filter coefficients in real time according to the bit error rate of data transmission systems. Finally, experiments were conducted and the results demonstrate the effeciveness of the proposed adaptive PLC data transmission for the improvement of the efficiency and stability of the ISLP system.

1. Introduction

By using the downhole permanent reservoir monitoring system, intelligent separate layered production (ISLP) techniques [1-3] can achieve adaptive adjustment of the oil production as well as water injection schemes for enhanced oil recovery and efficient development of the oilfield. In recent years, the test and control of ISLP systems have attracted lots of attentions and interests. However, in order to ensure the stability of the ISLP system, there still have many problems to be solved in the field of data transmission [4-6]. As one of the most popular data transmission techniques for downhole application, power line communication (PLC) with single-core steel pipe cables [7-9] offer several advantages over the acoustic and mud pulse wireless transmission technologies, such as higher reliability and transmission rate, better trafficability and lower cost. However, since the ISLP systems are placed in the well permanently, the cables are affected by the harsh environmental factors such as high temperature, high pressure and vibration, whose resistance and capacitance will fluctuate correspondingly and may result in unstable data transmission [10]. The traditional high-speed PLC data transmission with single-core cables relies on manual adjustment of the gain and filter coefficients of the ground decoding system to compensate for the change of resistance and capacitance of the single-core cables [11,12]. However, the manual adjustment requires high experienced operators, which is inefficient for the unpredictable change of the bad downhole conditions. Especially in the process of permanent underground monitoring, the manual adjustment method cannot solve the problem of data

IOP Conf. Series: Materials Science and Engineering **799** (2020) 012046 doi:10.1088/1757-899X/799/1/012046

transmission in time and effectively, which will have great influence on the performance of PLC data transmission. In order to solve this problem, this paper presents an adaptive PLC system for the data transmission of ISLP with single-core cables. The program-controlled potentiometer is used instead of the traditional precise sliding rheostat. Using the adaptive decoding algorithm, the gain and filter coefficients of the transmission system are adjusted in real time according to the bit error rate (BER) of the signal transmission, so as to realize the adaptive data transmission and improve the data transmission efficiency and stability of the cabled ISLP system.

2. Data transmission model of cabled ISLP system

The cabled ISLP system adopts the permanent downhole system to monitor the production status of each layer, and the monitored data such as temperature, pressure and the oil nozzle cone are transmitted to the ground processing system in real time through PLC with single-core steel pipe cables. After evaluating the production parameters of all layers, the ground system will send the configuration parameters to downhole system, so that a closed-loop automatic adjustment intelligent system can be formed. The model of the cabled ISLP system with PLC data transmission is shown in the Fig. 1. It should be noted that the uplink data transmission and the downlink command transmission of the cabled ISLP system have low requirements on the transmission rate, where the uplink 20 Kbps transmission rate can meet the actual demand generally, and the downlink command transmission rate only needs to ensure a fixed number of commands to be issued. By contrast, the key of PLC data transmission in the ISLP system is the guarantee of the long-term stable communication.

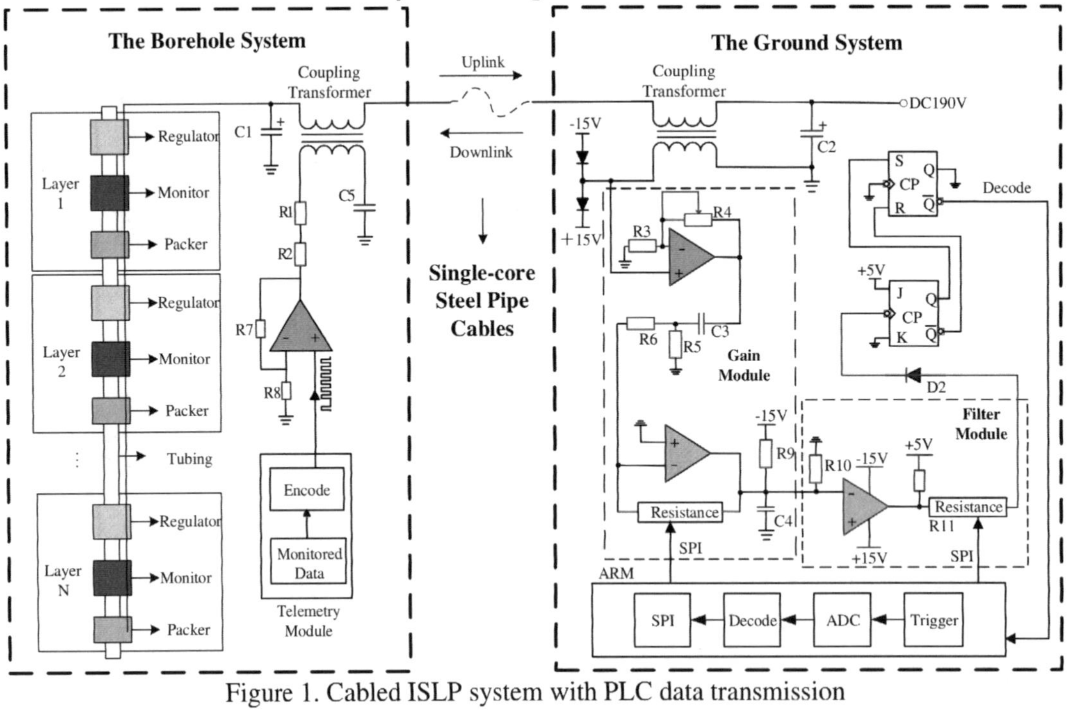

Figure 1. Cabled ISLP system with PLC data transmission

PLC data transmission system with single-core cables in the Fig.1 comprises two parts: power supply and data transmission. The data is transmitted by DC power line carrier, including borehole coupling transformer, single-core cables, ground coupling transformer, gain module, filter module and decoding module. The ground system supplies DC power to the downhole permanent reservoir monitoring system through the single-core cables to ensure the normal operation. The downhole data collected by the permanent reservoir monitoring system are coupled to the single-core cables through the borehole coupling transformer and uploaded to the ground system, and the ground system decouples the signal with the ground coupling transformer. Taking the uplink data transmission as an example, the transmission signal is amplified by the gain module, and the filter module filters out the low-frequency signal and compares it with the reference voltage. The processed signal is put into two JK flip-flops for hardware decoding. At last, the analog-to-digital convertor (ADC) module is used to sample the data to realize the final software decoding. It should be noted that the attenuation of the signal on the actual logging cables is influenced by resistance, capacitance, inductance as well as the insulation of the cables [13].

For the conventional logging system with single-core cable, since the cable of each logging winch is fixed, the gain and filter coefficients in Fig. 1 do not need to be adjusted after the initial setting. However, for the permanent ISLP system which needs to work underground for quite a long time, the parameters of cable, such as resistance and capacitance, will change with the underground temperature, pressure, vibration and other environmental factors. Thereby, the gain and filter coefficients must be adjusted in real time to ensure the stable transmission of the ISLP system.

2.1. *Design of program-controlled decoding system*

Both the gain and filter modules of the PLC decoding system is directly related to the values of the feedback resistances. The traditional manual operations that use two 20K ohm high-precision sliding rheostat for signal regulation is inefficient. As an alternative, the program-controlled potentiometer is utilized instead of the sliding potentiometer, so that the regulations can be controlled by software accurately to ensure the signal quality. Taking the program-controlled potentiometer AD5291 as an example, the ground software writes parameters to RDAC register through SPI communication, so as to adjust the output resistance between W and B ports, whose resistance value R_{WB} can be written as

$$R_{WB} = \frac{D}{256} R_{AB} \tag{1}$$

where R_{AB} is the value of the full range of the program-controlled potentiometer. D is the integer value written into the RDAC register in AD5291, and the resolution of the resistance value is 78.125 Ω. However, there are still some problems in the manual software adjustment, such as the requirement of experienced operators, low efficiency and unable to adjust in time. Therefore, on the basis of the above model, this paper presents an adaptive fast decoding algorithm for the cabled ISLP system.

2.2. *Adaptive fast decoding algorithm for PLC initialization*

In the initialization of communication, the permanent downhole monitoring system circulates a fixed section of data, and the ground processing system decodes the received signal. In this paper, we employ the BER as the criterion to evaluate the appreciation of the gain and filter coefficients, where the BER is defined as the number of error data during a certain monitoring time. The BER $W_{i,j}$ of the current configuration of the two program-controlled resistors can be calculated by

$$W_{i,j} = \frac{C_{i,j}}{T} \tag{2}$$

where i is the programable resistor of gain module, j is the programable resistor of filter module, T is the time duration, and C is the total number of error bits. According to Equation (2), we can analyze the practicability of the current values of the program-controlled resistor. For inappropriate BER, the values of the programmable resistor should be changed correspondingly. We can repeat the above operations to realize the scanning of all possible combinations of two programmable resistors to achieve the best BER performance. However, this would be inconvenient for PLC data transmission. Assuming the total settable number of each program-controlled resistor is E, there will be E^2 possible combinations. If we follow the scheme above, the time required for full traverse is

$$T_b = \mathrm{E}^2 T \tag{3}$$

Taking AD5291 with E = 256 and T = 100ms as an example, it will cost T_b = 1.82 hours to go through all possible combinations, which is substantially unacceptable for practical application. In this paper, we present an adaptive fast decoding algorithm to reduce the time consuming. In the initialization of the PLC data transmission, we use a training set to realize the BER evaluation, where the resistance of two programmed rheostats is scanned by applying the dichotomy to find the appropriate gain and filter coefficients. The scanning is divided into M stages, in which the step length of the m-th stage scanning is La_m and Lb_m for gain and filter resistors with the number of steps of Na_m and Nb_m, respectively. Then we have

$$La_m = Na_{m+1} \times La_{m+1} \tag{4}$$

$$Lb_m = Nb_{m+1} \times Lb_{m+1} \tag{5}$$

The structure of the adaptive fast decoding algorithm is shown in Figure 2, where the m-th stage fast scanning can be divided into two parts as follows:

(1) Scan with step length of La_m and Lb_m, respectively, and then find two groups of values corresponding to the best two BERs, with which the scanning can be terminated as long as the BER satisfies the threshold.

(2) If $m < M$, two (m+1)-th stage scannings should be carried out by taking the two groups of values above as the scanning center, with the range and step of La_m and Lb_m, and La_{m+1} and Lb_{m+1}, respectively. Note that the ranges of the current stage are just the step lengths of the last stage.

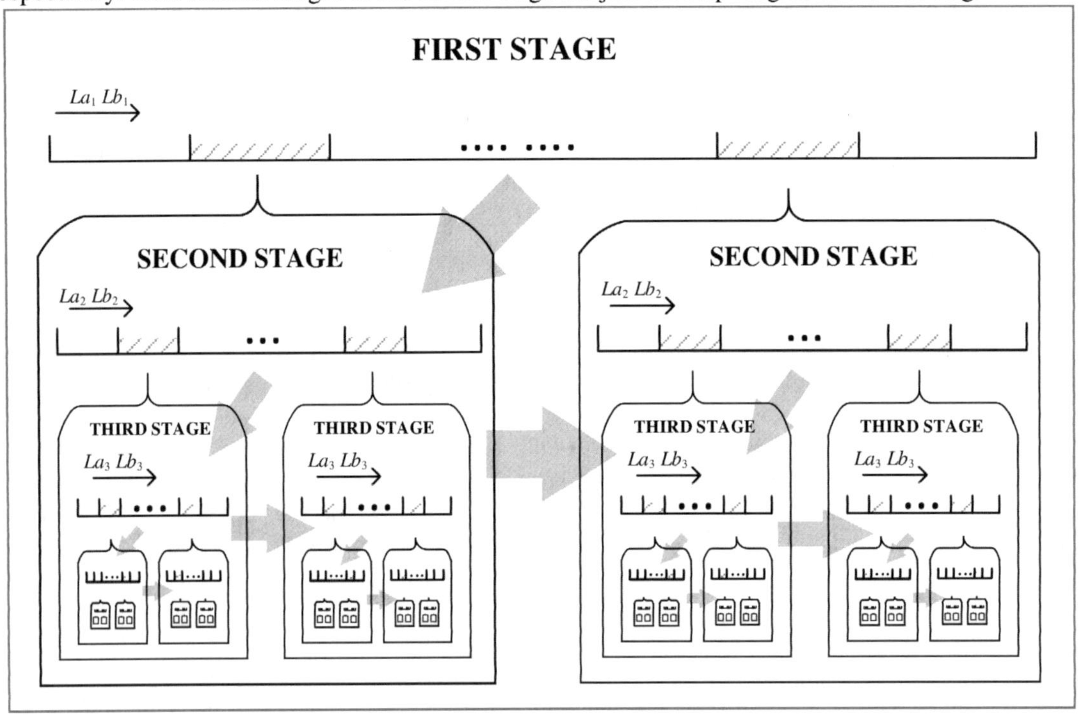

Figure 2. Adaptive fast decoding algorithm structure

Specifically, the procedure of adaptive fast decoding algorithm is shown in Figure 3. Taking the first row with M stages as an example, the first stage scanning is carried out to obtain two groups of values corresponding to the best two BERs within all possible combinations of the gain and filter resistance with the steps of La_1 and Lb_1, respectively. And if the BER of the best two groups reaches the threshold, the scanning will be terminated. Otherwise, the second stage scanning should be conducted with the range of La_1 and Lb_1, respectively. Similarly, the fast scanning of the m-th stage can be divided into two parts (steps). If the BER still cannot satisfy the threshold after all scannings, the coefficients corresponding to the minimum BER obtained should be adopted.

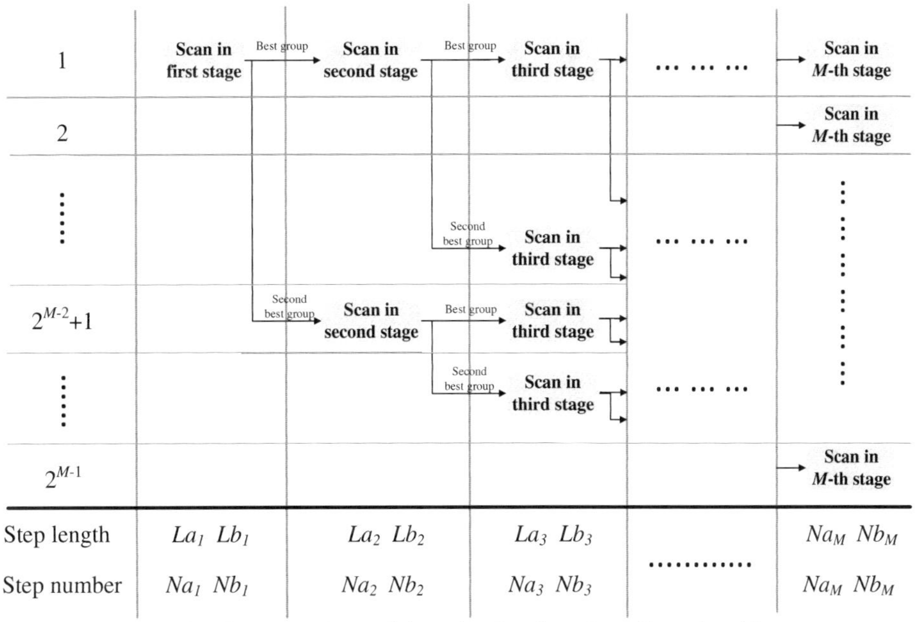

Figure 3. The procedure of the adaptive fast decoding algorithm

The scanning time of the proposed adaptive fast decoding algorithm shall not be larger than:

$$T_b = \sum_{m=1}^{M} 2^{m-1} Na_m Nb_m T \tag{6}$$

Taking the program-controlled resistor AD5291 (E = 256), T = 100 ms as an example, if M = 4 and steps $Na_m = Nb_m$ =4 (m=1,2,3,4), the longest time to complete such a scan will be 24 seconds, which is far less than the time of a full traverse.

3. Long-term adaptive PLC for ISLP systems

Through the adaptive fast decoding algorithm, we can get the appropriate combination of gain and filter coefficients for the initialization of the PLC data transmission. However, the permanent mode of the ISLP may cause a continuous change of the resistance and capacitance of the single-core cables due to the long-term effect of high temperature, high pressure as well as the vibration of the bad downhole conditions, which will have great influence on the data transmission efficiency. And the combination of gain and filter coefficients needs to be adjusted with the continuous change. Unfortunately, the traditional manual adjustment method cannot handle the problem of data transmission in time and effectively. To solve this problem, a long-term adaptive data transmission method is proposed in this paper.

During the long-term monitoring, intermittent transmission of training set will occupy transmission resources. Moreover, since little change of cable parameters in a short time, it is unnecessary to adjust the gain and filter coefficients in a too large range. In this paper, we take the number of frames successfully transmitted in a certain time as the standard of the BER to evaluate the stability of data transmission in the long-term monitoring for the cabled ISLP system. The real-time BER can be expressed as:

$$\overline{W}_{i,j} = 1 - \frac{V_s P_{i,j}}{T_s} \tag{7}$$

where i is the programable resistor of gain module, j is the programable resistor of filter module, T_s is the observation time, V_s is the transmission cycle of each frame, and P is the number of frames successfully transmitted in T_s time.

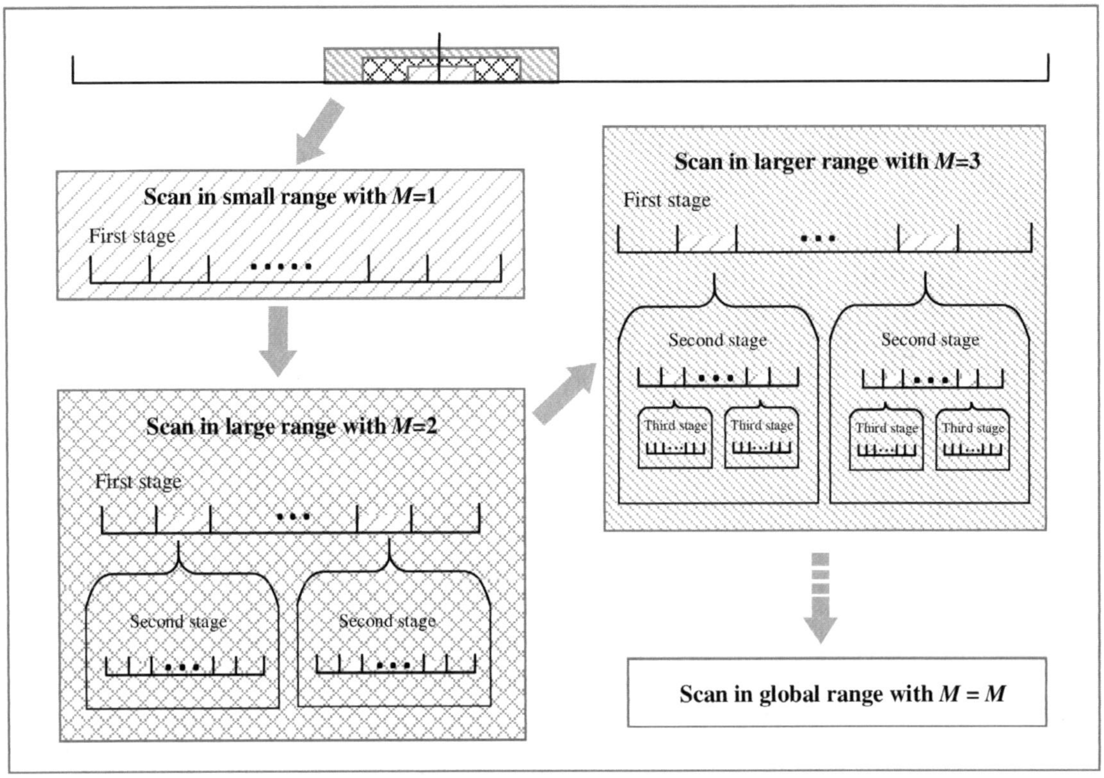

Figure 4. The scheme of the adaptive method

To ensure the stability of the PLC data transmission, the scanning procedure of the long-term adaptive transmission should be made with in a suitable range. The scheme of the adaptive method as shown in Figure 4 can be decomposed to several individual procedures of adaptive fast decoding in Section 2.2. As long as the BER is larger than the threshold during the monitoring, the fast-adaptive decoding scanning with M=1 in Section 2 should be carried out. Furthermore, if the BER satisfies the threshold, the coefficients can be set as the values corresponding to the minimum BER. Otherwise, if the appropriate value is not found during the M=1 scanning, the fast-adaptive decoding scanning with M=2 will be carried out within a larger range. Similarly, we can repeat the above procedures with an increasing M in larger range for scanning until an appropriate BER can be obtained or all the scanning is completed, and then the coefficients values corresponding to the lowest BER will be adopted.

By adaptive adjusting the gain and filter coefficients to maintain the BER in a proper range, the long-term and stable data transmission of the cabled ISLP system can be realized. It is noted that although a small scanning range of each stage may have little impact on the current data transmission, the probability of obtaining appropriate coefficients are also reduced. Thereby, suitable range and step length still need to be optimized according to the current conditions of well and single-core steel pipe cables.

4. Experimental analysis

In order to verify the efficiency and accuracy of the proposed adaptive fast decoding algorithm and the long-term adaptive transmission method, five different parameters of single-core cables for the ISLP system are utilized, and the equivalent resistance and capacitance are shown in Table 1:

Table 1. Parameters of the single-core cables for experiments

Number	**1**	**2**	**3**	**4**	**5**
Resistance (Ω)	96.3	103.5	107.5	113.6	121.3
Capacitance (μF)	1.08	1.11	1.16	1.24	1.27

IOP Conf. Series: Materials Science and Engineering **799** (2020) 012046 doi:10.1088/1757-899X/799/1/012046

The manual adjustment and adaptive fast decoding algorithm are applied to the initialization of the PLC data transmission of the ISLP with the cable parameters of each case in Table 1 for 10 times, respectively. The BER versus adjustment time of the two methods are shown in the Figure 5. It can be seen that the manual adjustment method scatters out for different cable conditions. By contrast, the proposed adaptive fast decoding algorithm performs well, where the experiment results are centralized. It can be seen from the comparison that both the speed and accuracy of the adaptive fast decoding algorithm are better than that of the manual adjustment for efficient ISLP system.

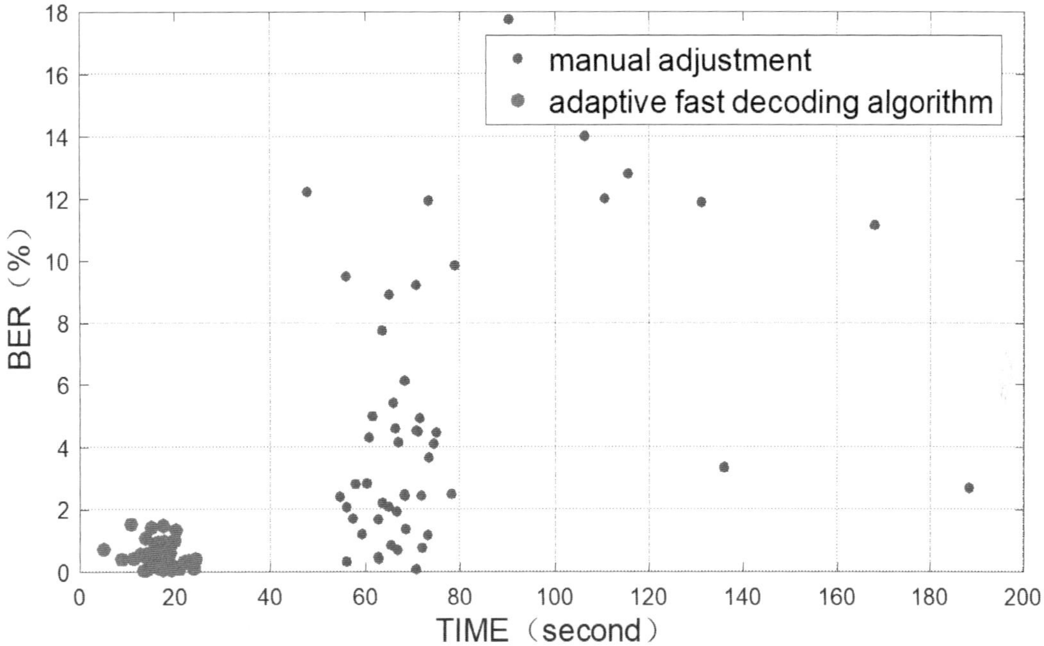

Figure 5. Comparison of the manual adjustment and adaptive algorithm for PLC initialization

In order to verify the performance of the long-term adaptive transmission method, the BER of the PLC data transmission is experimented for a long time, and the BER with fixed coefficients, manual adjustment and long-term adaptive transmission method are compared in the Figure 6. The experimental time is set to be 350 minutes, and the parameters with respect to the resistance and capacitance of the single-core cables are set to be changed from the first case in Table 1 to the fifth case at a uniform speed. It can be seen from Figure 6 that the BER using fixed coefficients increases gradually versus the monitoring time, where the BER performance cannot meet the normal data transmission requirement for a long-term monitoring. Using the manual adjustment, it is obvious that the decrease of the BER performance can be corrected only when the manual operations were conducted (90 minutes approximately in the experiments). Although the performance decrease can be corrected, the manual adjustment method requires a long-time concerning, which would be inefficient and may cause data transmission interruption. The proposed long-term adaptive transmission method in this paper can adjust the coefficients according to the current BER in real time, and the BER can be maintained in a suitable range so that the stable PLC data transmission can be ensured.

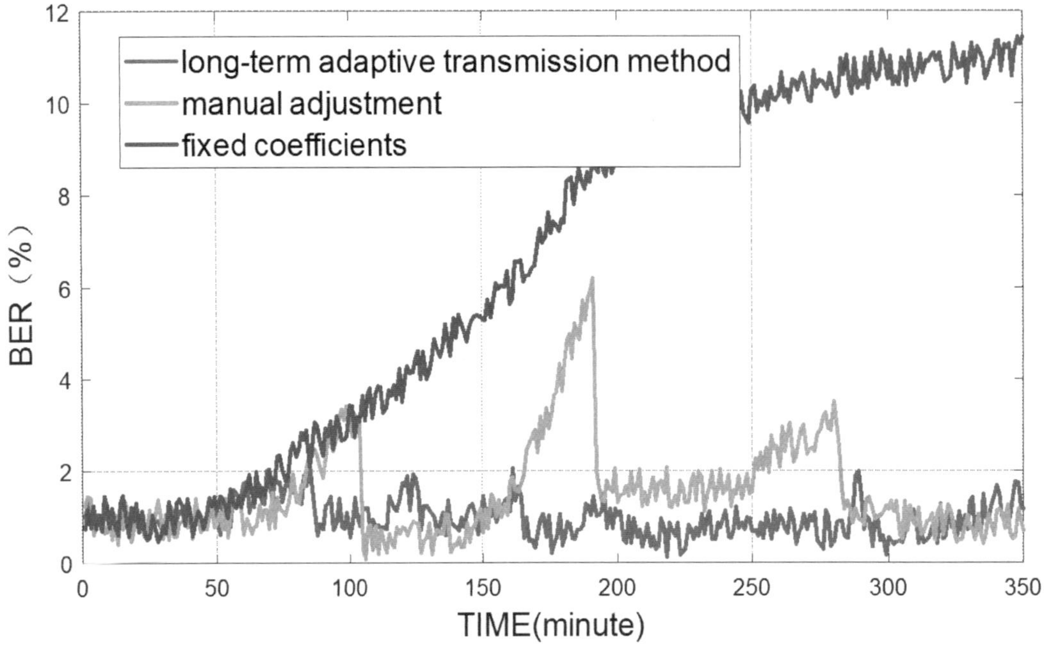

Figure 6. Long-term transmission stability comparison

5. Summary

In this paper, an adaptive PLC approach for the data transmission of ISLP with single-core cables is proposed to improve the stability of the permanent downhole monitoring system. On the basis of the program-controlled potentiometer, an adaptive decoding algorithm for fast initialization of the PLC data transmission is first introduced. Additionally, a long-term adaptive method is also proposed to adjust the gain and filter coefficients in real time according to the BER of data transmission systems. Experiments were conducted and the results demonstrate the effectiveness of the proposed methods for the improvement of the data transmission efficiency and stability of the ISLP system.

Acknowledgements

This work was supported by the National Science and Technology Major Project (Grand No. 2016ZX05028-001), the National Natural Science Foundation of China (Grant No. 51974250, 41874158), the Natural Science Basic Research Plan in Shaanxi Province of China (Grant No. 2018JQ5133), and the Natural Science Foundation of Shaanxi Provincial Department of Education (Grant No. 17JS106).

References

[1] Liu H, Gao Y, Pei X H, et al. (2018) Progress and Prospect of downhole cyclone Oil-water Seperation with Single-well Injection-production Technology, Acta Petrolei Sinica, 39(4):463-471.

[2] Liu Y G, Chen Z, Meng X H, et al. (2019) Cable Implanted Intelligent Injection Technology for Separate Injection Wells in Bohai Oilfield, Petroleum Drilling Techniques, 47(3):133-139.

[3] E. M. Artykaeva, V. S. Genin, V. A. Nesterin. (2011) Prospects for advancement in the energy efficiency of oil-producing sucker rod pumps. Russian Electrical Engineering, 82(10):513-517.

[4] M. A. Castillo, A. C. Troyman, R. H. Gutiérrez, et al. (2018) Modal parameter identification of an electrical submersible pump using impact test and time domain methods. Marine Systems & Ocean Technology, 13(1):1-8.
[5] Quintaes, F, Salazar, A.O, Maitelli, A.L, et al. (2011) Magnetic Sensor Used to Detect Contamination of Insulating Oil in Motors Applied to Electrical Submersible Pump[J]. IEEE Transactions on Magnetics, 47(10):3756-3759.
[6] Zhu J J , Zhu H W , Wang Z H, et al. (2018) Surfactant effect on air/water flow in a multistage electrical submersible pump (ESP). Experimental Thermal and Fluid Science, 98(5):95-111.
[7] Li D L. (2016) Intelligent Layered Polymer Injection Technology with Preset Cable, China Petroleum Machinery, 44(10):93-96.
[8] Dai D, Zhang, X H, Wang J J. (2014) Calculation of AC Resistance for Stranded Single-Core Power Cable Conductors. IEEE Transactions on Magnetics, 50(11):1-4.
[9] Dang B, Yang L, Liu C, et al. (2018) A Uniform Linear Multi-Coil Array-Based Borehole Transient Electromagnetic System for Non-Destructive Evaluations of Downhole Casings. Sensors, 18(8):1-9.
[10] Zhang Q J, Huo Y H. (2017) Design of High Temperature Permanent Down-Hole Data Acquisition System[J], Process Automation Instrumentation, 38(11):40-43.
[11] Qu Y, Chen W, Gu Q S, et al. (2015) Design and Implementation of the Downhole of High-speed Data Transmission System Based on Single-core Logging Cable, Well Logging Technology, 39(2):209-212.
[12] Yang Z H. (2011) Through Casing Resistivity Logging Data Transmission and Decoding Technology, Xi'an Shiyou University.
[13] Yan J F, Zhou Q Y, Liu D J. (2017) Fitting and Circuit Modeling to the Attenuation Characteristics of Logging Cable, Well Logging Technology, 10(05):1-6.

IOP Conf. Series: Materials Science and Engineering **799** (2020) 012047 doi:10.1088/1757-899X/799/1/012047

The effect factors analysis of the cave formation based on the LWD resistivity imaging method

Zhengming Kang* **and Fei Li**

School of Electronic Engineering, Xi'an Shiyou University, Xi'an, Shaanxi, 710065, China

*190720@xsyu.edu.cn

Abstract. The cave formation has strong non-homogeneity and the well logging response mechanism of it is very complex, which is of great significance to be studied. Based on the 3D finite element method, the logging response of the Logging While Drilling (LWD) resistivity imaging tool is simulated. The effects of cave diameter, cave extension and resistivity contrast between cave formation and background formation on ratio of measurement signals are investigated using a single cave model. The results show that the existence of the borehole has a great influence on the measurement signal, so it needs to be corrected or the measurement electrode of the tool should be close to the borehole. When the diameter of the cave is larger than a certain value, the relationship between the cave extension and the measurement signal can be quantitatively described. However, the resistivity contrast between the cave formation and the background formation shows a power index relationship under different cave diameter. The simulation results can provide a theoretical basis for the identification and interpretation of LWD resistivity imaging tool.

1. Introduction

In the field of petroleum exploration, resistivity imaging logging is a technology that directly measures the images of geological structures of the borehole. It can visually reflect the structure near the borehole, such as fracture, cave, etc. With the development of technology, resistivity imaging logging tool is also successfully transitioning from wireline measurement to LWD measurement. Uploading the measurement data to the ground while drilling, real-time geological environment monitoring can be realized, which ensures safety drilling and realizes geosteering [1-3].

Carbonate and volcanic rock have strong non-homogeneity, and logging interpretation is difficult. For these kinds of reservoirs, the tool has better applications effect. Because the cave can not only be used as a storage space, but also as a kind of oil and gas connection channel. Interpretation people can discover the existence of the cave easily by using the resistivity images of the borehole. Based on the advantages of the above-mentioned LWD resistivity imaging technique, the tool is more and more used in cave and fracture formations [4-6]. But there are few research reports on the mechanism of the logging response in the cave formation, which hinders us using the logging data in the cave formation. So it is necessary to study the mechanism using numerical simulation.

In this paper, 3D FEM is used to systematically simulate the influence of cave diameter, cave extension and bedrock resistivity on the measurement signal. The results reveal the response law of LWD resistivity imaging tool to the caves, and provide a theoretical basis for the identification and evaluation of the caves.

2. Simulation results

Many people have done a lot of simulation work about the logging response in the cave formation, but most of them focus on the dual laterolog tool and electromagnetic wave propagation logging tool[7-11]. There are few reports about the logging response mechanism of the LWD resistivity imaging tool, so it is necessary to use numerical simulation method to study its mechanism.

The dual laterolog has a good identification and evaluation effect for the cave with a diameter of more than 1 m, while the LWD resistivity imaging method can show the cave images when the diameter of cave is less than 1 m. Therefore, this paper focuses on caves with a diameter of less than 1 m.

Cave formations are generally asymmetric, so they need to be simulated using 3D models. 3D FEM is used in this paper. Figure 1 shows a cross-section of the formation model of a single cave. The model consists of LWD resistivity imaging tool, borehole, background formation and cave. The button electrode is used to collect current, and it consists of four electrodes, named 1, 2, 3, 4. Each electrode is 90 degrees apart in circumferential direction. Electrode 1 is right towards the cave. The cave is vertical to the borehole.

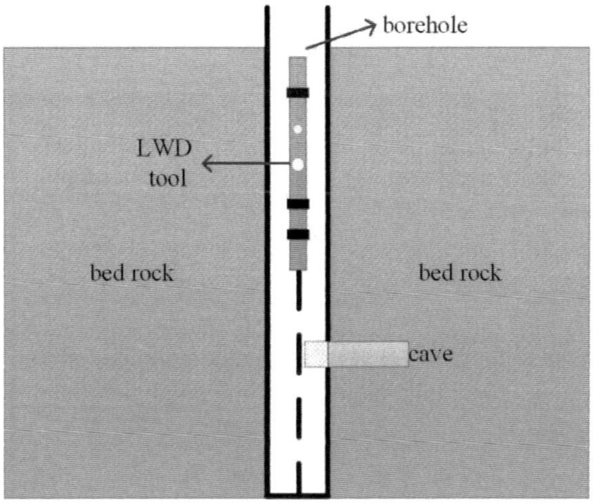

Figure 1. Schematic of single cave model

When analyzing the influence of cave extension, cave resistivity and bedrock resistivity on the measurement signal, the cave is divided into two sizes: medium cave (0.01 m in diameter) and large cave (0.05 m in diameter). Table 1 are simulation parameters of single cave. The influences of cave diameter, cave extension and resistivity contrast on the measurement signal are investigated.

Table 1. Simulation parameters of single cave

Effect factor	Diameter (m)	Extension (m)	Mud resistivity (Ω·m)	Cave resistivity (Ω·m)	Background formation resitvity (Ω·m)
Diameter	0.002-0.1	0.5	1	1	100
Extension	0.01 0.05	0.02-0.5	1	1	100
Resistivity contrast	0.01 0.05	0.5	0.2-20	0.2-20	100

2.1. Cave diameter

The shape of the cave is a cylinder in the 3D FEM model, which is orthogonal to the borehole and drilled through by it. The diameter of the cave is smaller than that of the borehole, so the intersection line is irregular. When the diameter difference between borehole and cave is large, the intersecting line is close to a circle. Geologically, pores with a diameter of more than 0.002 m are called caves. Among them: the diameter of 0.002 m-0.005 m is small hole, the diameter of 0.005 m-0.01 m is medium hole, and the diameter of more than 0.01 m is large hole. The diameter of the cave studied in this paper is between

0.01 m and 0.1 m. The influence of cave with different diameter on the measurement signal is different. In Figure 2, Imax is the current signal measured by button electrode when the button electrode center and cave center are in the same longitudinal position. Imin is the current signal when the button electrode is facing the background formation, which is almost unchanged.

The existence of the borehole is investigated first. It can be seen from Figure 2a that the borehole has a great influence on the button electrode 1 measurement signal when the borehole exists. When the cave diameter is less than 0.01 m (area A), the measurement signal is larger. The measurement signal becomes gradually smaller with the increase of the diameter, and the measurement signal almost comes from the borehole. Affected by the borehole, the button electrode 1 can not identify the cave; when the cave diameter is greater than 0.01 m (area B), the measurement signal increases to a stable level with the increase of the cave diameter. Figure 2b shows the relationship between the measurement signal and the cave diameter without the borehole. As the cave diameter changes from 0.002 m to 0.1 m, the measurement signal is gradually increasing with the increasing of the cave diameter. The relationship between the cave diameter and Imax/Imin is linear. Figure 2a and Figure 2b show that the measurement signal from other directions is not sensitive to the cave, and the value of Imax/Imin equals to 1. Therefore, effectively eliminating the influence of borehole is the key to accurately identify caves. The main method is to put the button electrodes near the well wall or correct the effect of the borehole.

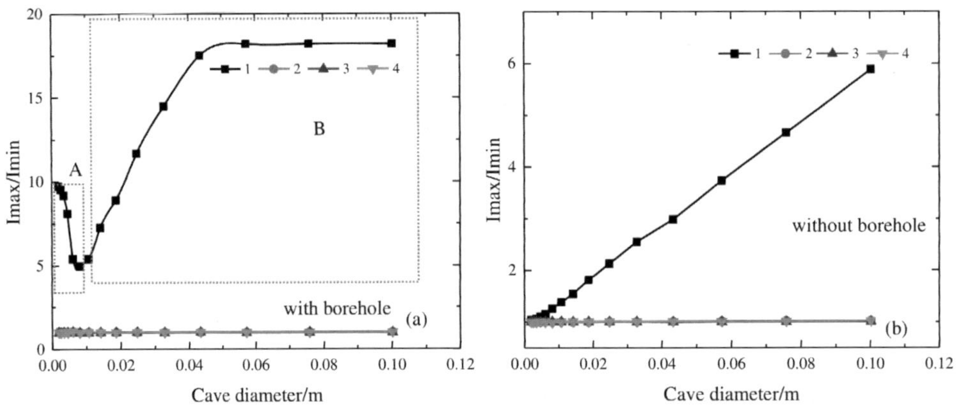

Figure 2. The effect of cave diameters on ratio of measurement signals

2.2. *Cave extension*

Wireline microresistivity imaging tool can directly display the distribution characteristics of caves, but its depth of investigation (DOI) is limited just providing the cave information of the borehole wall. Therefore, it is difficult to detect the radial extension characteristics of caves. The DOI of dual laterolog is deeper than that of the above tool, but it is not sensitive to caves with diameter less than 1 m. Compared with the above two tools, the DOI of LWD resistivity imaging tool is deeper than that of wireline type tool. And it has a better recognition of caves with smaller diameter.

Figure 3a shows that the measured signals change with the increase of the extension in the medium cave. When the extension varies from 0.1 m to 0.5 m, the signals of electrodes 2, 3, 4 are almost unchanged, and the ratio of Imax and Imin is close to 1, indicating that the cave has little impact on the measured signals. When the cave extension is less than 0.15 m, the measured signal of button electrode 1 has a nonlinear relationship with the cave extension, and the Imax/Imin is far greater than 1. Considering the small diameter of the cave, the result is mainly caused by the borehole effect. When the cave extension is greater than 0.15 m, the current signal measured by the button electrode 1 does not change with the cave extension, which shows that the effect of the borehole on the measurement results tends to be stable.

Figure 3b shows the relationship between the extension of the large cave and the ratio of measurement signal. It can be seen that with the increase of cave extension, the measurement signal of button electrode 1 shows a power exponential relationship. Using the formula $I_{max} / I_{min} = 15.24 \cdot E_x^{0.6}$ to fits the relationship between cave extension Ex and Imax/Imin, table 2 compares the relative error between the simulation data and the calculated data by it. It can be seen that the error of most data is within 5% except for individual points, which indicates that the radial extension of the large cave can be obtained by applying the above Formula. According to the theoretical calculation of different cave diameter, we can get the corresponding extension calculation formula of different cave diameters. A large number of simulations of the relationship between the cave extension and the measured signal with different diameters demonstrate the above conclusions, but it should be noted that the cave extension can only be calculated by this model when the diameter of the large cave is between 0.01 m and 0.05 m.

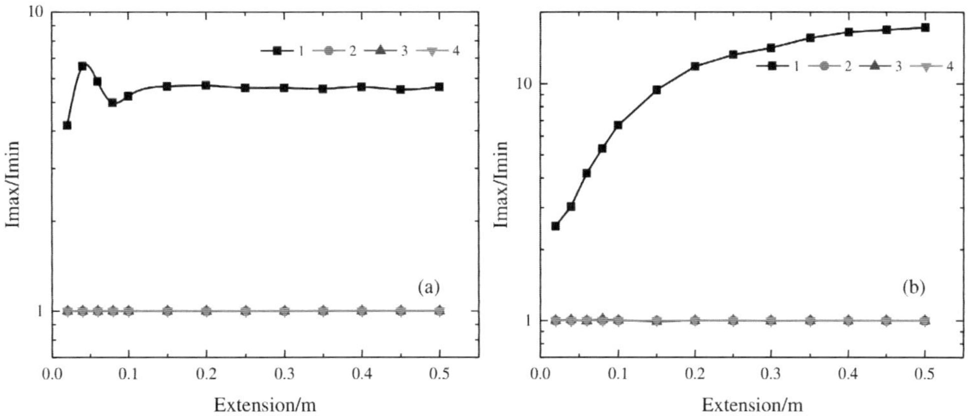

Figure 3. The effect of cave extension on ratio of measurement signals

Table 2. Fitting error analysis of the relationship between large cave extension and ratio of measurement signals

Extension (m)	Imax/Imin (simulation data)	Imax/Imin (fitting data)	Relative error (%)
0.02	1.59	1.46	8.36
0.04	2.23	2.21	0.63
0.06	2.64	2.82	6.95
0.08	3.24	3.35	3.38
0.10	3.71	3.83	3.30
0.15	4.75	4.89	2.90
0.20	5.7	5.81	1.39
0.25	6.59	6.64	0.76
0.30	7.40	7.41	0.07
0.35	8.21	8.12	1.09
0.40	9.00	8.80	2.25
0.45	9.66	9.45	2.18
0.50	10.41	10.06	3.39

2.3. Resistivity contrast of the cave formation and background formation

Tan et al. (2011) simulated the response of different fillers in the cave to dual laterolog. In the simulation, the resistivity of fillers ranged from 1 Ω·m to 500 Ω·m [12]. In this paper, it is assumed that the cave is filled with mud. And the resistivity of the fillers Rm ranges from 0.2 Ω·m to 20 Ω·m, and the resistivity of background formation Rb is fixed at 100 Ω·m. Figure 4 shows the logging response of button electrode 1 to the direction of cave under the single cave model. Figure 4a and Figure 4b show the response characteristics of medium cave and large cave respectively. It can be seen that in the double logarithmic coordinate system, with the increase of Rm/Rb, the current contrast measured by medium cave and large cave shows a linear decreasing trend (actually power index relationship), which is consistent with the response characteristics of fractured formation.

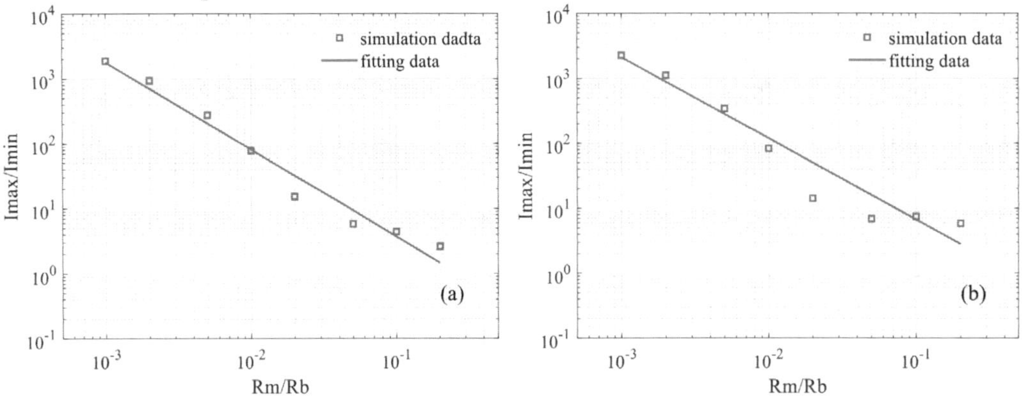

Figure 4. The effect of cave filler on the ratio of measurement signals

3. Conclusion

Using 3D FEM method, the logging responses of LWD resistivity imaging tool in cave formation are simulated. From the results, the following conclusions are drawn.

1. The borehole has obvious effect on the measured current signal.
2. The factors of the cave diameter and cave extension have different relationships with ratio of measurement signal. And the relationship depends on the type of cave (medium cave or large cave).
3. Both the cave diameter and cave extension have a linear relationship and power exponent relationship with ratio of measurement signals respectively when the cave is large cave. Compared with the large cave, the medium cave does not have these certain relationships with ratio of measurement signals.
4. The relationship between the resistivity contrast and the ratio of measurement current is power exponent no matter the size of the cave is medium or large.

Acknowledgments

The authors would like to thank Sinopec Research Institute of Petroleum Engineering for allowing using the data in this study and professor Shizhen Ke for his constructive comments.

References

[1] Allouche, M., Chow, S., Dubourg, I., et al. (2010) High-resolution images and formation evaluation in slim holes from anew logging-while-drilling azimuthal laterolog device[C]SPE EUROPEC/EAGE Annual Conference and Exhibition. Barcelona. SPE-131513-MS.

[2] Ortenzi, L., Dubourg, I., Van, Os. R., et al. (2011) New azimuthal resistivity and high-resolution imager facilitates formation evaluation and well placement of horizontal slim boreholes[C] SPWLA 52nd Annual Logging Symposium. Colorado Springs. SPWLA-2011-LLL.

[3] Kang, Z. M., Ke, S. Z., Li, X., et al. (2018) 3D FEM simulation of responses of LWD multi-mode resistivity imaging sonde. Applied Geophysics, 15(3-4): 401-412.

[4] Prammer, M. G, Morys, M, Knizhnik, S., et al. (2009) A high-resolution LWD resistivity imaging tool–field testing in vertical and highly deviated boreholes. Petrophysics, 50(01).

[5] Wang, B. W., Zhang, S. D., Ji, R., et al. (2017) Application of microscope high resolution LWD resistivity image logging in carbonate reservoirs in Sichuan Basin. Well Logging Technology. 41(3): 358-363.
[6] Yang, S. D., Lei, X., Cai, J., et al. (2010) Integrated solution of LWD resistivity image logging in carbonate reservoir in Beibu Gulf, China. Well Logging Technology. 34(2): 177-182.
[7] Deng, S. G., Mo, X. X., Lu, C. L., et al. (2012) Numerical simulation of the dual laterolog response to fractures and caves in fractured-cavernous formation. Petroleum Exploration and Development, 39(6): 751-757.
[8] Fan, Y. R., Wang, L., Ge, X. M., et al. (2016) Response simulation and corresponding analysis of dual laterolog in cavernous reservoirs. Petroleum Exploration and Development, 43(2): 261-267.
[9] Liu, D. J., Ma, Z. H., Xing, X. N., et al. (2013) Numerical simulation of LWD resistivity response of carbonate formation using self-adaptive hp-FEM. Applied Geophysics, 10(1): 97-108.
[10] Liu, D. J., Ma, Z. H., Zhang, Y. Y., et al. (2013) Numerical simulation of the logging-while-drilling for fracture-vuggy reservoirs using adaptive FEM. Progress in Geophysics, 28(2):0928-0936.
[11] Wang, L., Fan, Y. R., Gao, Y. W., et al. (2017). Numerical simulation and characteristics analysis of dual laterolog in cavernous reservoirs on the basis of 3D-FEM. Chinese Journal of Geophysics, 60(5): 2011-2019.
[12] Tan, M. J., Gao, J., Wang, X. C., et al. (2011) Numerical simulation of the dual laterolog for carbonate cave reservoirs and response characteristics. Applied Geophysics, 8(1): 79-85.

Post-job Data Analysis by Python for a Miniature Recorder in Drilling Motors

Fei Li, Haolan Song

School of Electronic Engineering, Xi'an Shiyou University, Xi'an, Shaanxi, 710000, China

*Corresponding author's e-mail: lif@xsyu.edu.cn

Abstract. A vibration recorder has been deployed to capture the dynamics data of drilling motors for their operation and health evaluation. The recorder is usually inserted inside the rotor catcher of a conventional drilling motor in such a way that no extra length is introduced to the planned bottom hole assembly (BHA). The recorder can log data over the full operation cycle of a drilling motor, including vibration during surface activities and downhole operation. A post-job data analyzer was specifically designed based on Python. The analyzer processes the recorded data and generates statistics, time plots, and the various statistical plots such as histograms. These statistics and plots help the operators and maintenance technicians to gain insights into the motor operation and performance. The analyzer can provide adequate information on a drilling motor to asset owners for maintenance and operation optimization without analysis training or qualification from operators.

1. Introduction

Positive displacement motor (PDM) uses drilling fluid, which referred to as drilling mud, or just mud, to generate eccentric motion in a transmission section, which is then transferred to concentric torque and weight to drive drill bits. Drilling motors can be of different configurations of rotor and stator to provide optimal performance for the desired drilling operation. Typically the increase of the number of lobes and length of the power sections can produce higher output power [1].

The main application of a motor is directional drilling. During downhole operation, a drilling motor is typical of two operational modes, sliding (also called building) or rotating (also called drilling straight). Drilling motors are of high dogleg capacity and relatively cheap cost, and therefore, they have been widely used in the drilling industry. The additional torque provided by a downhole drilling motor allows for a more aggressive drill bit design to be selected and achieve a higher Rate of Penetration (ROP) [2]. However, drilling activities are often challenged with inefficiency and delays from original plans. The O&G (oil and gas) industry becomes more concerned with the understanding of drilling inefficiency and identifying factors that reduce drilling performance caused by adverse drilling dynamics, such as shock and vibration. It is well established that adverse dynamics, such as shock and vibration [3], is a major cause of the Bottom Hole Assembly (BHA) tool damages and undermines drilling performance [4]–[6]. Drillstring shock and vibration (S&V) can be categorized into three types: axial, lateral, and torsional [7], and they are destructive to drilling processes.

Downhole drilling dynamics or vibration have, therefore, become a subject of intensive research [8]–[10]. The study has focused on the downhole measurement of drilling vibration, its consequences, and tool and operation optimization processes for mitigation [11], [12]. Nowadays, drilling vibration can be measured by shock and vibration sensors in MWD tools, logging-while-drilling (LWD) tools, rotary steerable systems, or specific BHA drilling dynamics measurement tools [4], [13]. Dynamics of drilling motors are always indirectly measured from the sensors located either above or below them even though MWD or LWD tools in drilling motor bottom hole assemblies (BHAs) measure dynamics of drill collars, considerably above the motor, often without sufficient resolution to understand adverse drilling dynamics on the motor and to design remedial actions for the operations.

A low-cost and miniature vibration measurement recorder was developed to meet the needs of providing data to motor maintenance and operation companies. This paper presents the development and specifications of the vibration recorder, followed by the post-job data analysis. One case study is presented to show the downhole drilling vibration of drilling motors captured by the recorder.

2. The Vibration Recorder

The vibration recorder is designed approximately 7-in long and 0.75-in in diameter cylindrically, and capable of being easily installed inside of a modified rotor catcher. The vibration recorder consists of a print wire assembly (PWA) of 3-in long by 0.5-in wide, a tubular mounting chassis, and an AA size battery. The PWA consists of a three-axis vibration accelerometer, a three-axis shock accelerometer, a rotation gyroscope, a temperature sensor, and a timestamp IC besides a digital signal processor (DSP). The PWA has been qualified for downhole environments in terms of temperature ratings and shock conditions. The mounting chassis was designed to fit inside a rotor catcher of a drilling motor, and it can be quickly taken out for services. The lithium battery of AA size was qualified for the temperature profile of drilling applications. Figure 1 shows how the vibration recorder is installed in a rotor catcher.

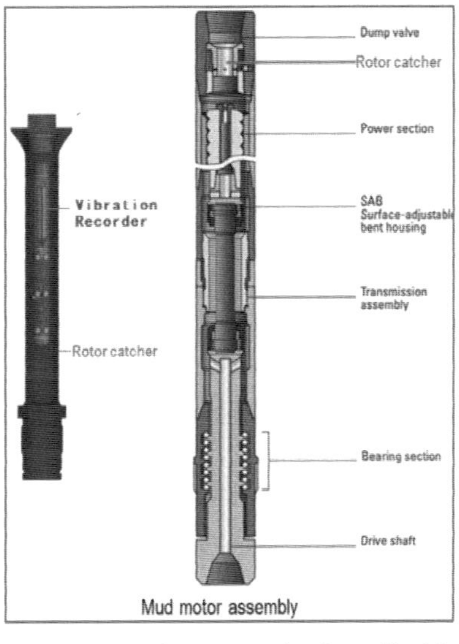

Figure 1. A vibration recorder installed in a motor

In Figure 2, on the left-hand side, a block diagram of digital signal processing, the sensors, and data recording within the vibration recorder is shown. The sensor measurements include vibration (amplitude < 5 gn) and (amplitude >= 5 gn), rotation speed, temperature, and timestamp. These measurements are filtered and sampled, and then the corresponding parameters are calculated; direct measurements and statistical data are stored to the memory on the PWA for the post-job analysis.

The coordinating system of the measurement sensors is shown on the right-hand side in Figure 2, where X is the axial direction, which is along the path of drill strings, and Y and Z are the lateral directions, which is perpendicular to the direction of drill strings.

Table 1 summarizes the sensors, direct parameter measurements, and statistics, and memory recording of the vibration recorder. Because the recorder is installed on a rotor catcher, which is directly connected to the drill bit through the rotor of a drilling motor, the bit status (bit rotation speed and stalling) can also be measured. Direct measurements of the sensors include 3-axis shock and vibration, 3-axis rotation speeds, temperatures, and timestamps.

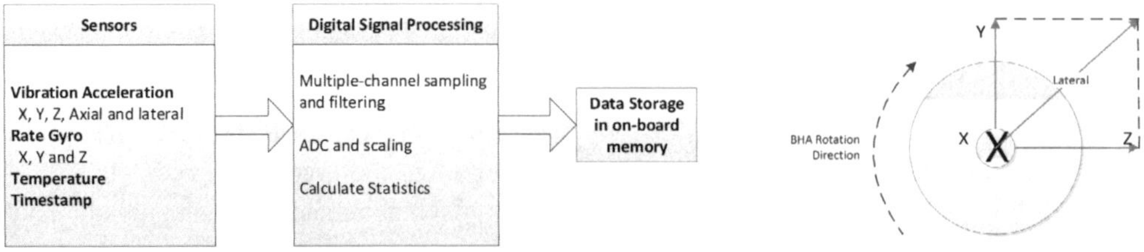

Figure 2. Block diagram of the processing of the sensors and coordinating system

Based on measurements, statistical vibration information is computed at the rate of every 6 minutes for surface data and at the rate of every 10 seconds for downhole data. The statistics information consists of peak axial acceleration (gn), average axial acceleration (gn), and peak lateral acceleration (gn), axial shock counts in the bin from 5–100 gn, axial shock counts in the bin from 100–200 gn, lateral shock counts in the bin from 5–100 gn and shock counts in the bin from 100–200 gn. Based on shock counts, one can compute shock frequency based on the logging frequency of the shock count channel, which can be vital to the understanding of downhole vibration. Based on measurements, statistical rotation speed information consists of minimum rotation speed (rpm), maximum rotation speed (rpm), average rotation speed (rpm), bit stick-slip severity (%), and bit stalling. The stick-slip severity is calculated by equation (1).

$$stick\ slip\ severity = \frac{v_{max} - v_{min}}{2 \times v_{average}} \times 100\% \tag{1}$$

Table 1. The sensors, direct measurements, and statistics and memory recording

Sensors	Direct Measurement	Statistics and Memory Recording
Data and Time	Real-time Clock and Calendar	Not Available (N/A)
Vibration (X, Y, and Z)	Shock and vibration (X, Y, and Z) @ 800 Hz	Peak axial, average axial, and peak lateral, Shock Bins (5–100,100–200 g_n), Inclination
Rotation (X, Y, and Z)	Rotation speed @ 20 Hz	Minimum rpm, maximum rpm, mean rpm, bit stick-slip severity,and bit stalling
Temperature	Temperature	N/A

3. Post-job Data Analyzer

As a part of the post-job analysis, a Python data analyzer is used to process the data and provide a detailed report for users. As Figure 3 shows, the process of the Python data analyzer is demonstrated. Firstly, the libraries of pandas, numpy, and matplotlib are imported. Pandas is the main library for data processing; Numpy is a multidimensional matrix operation library, and Matplotlib is a python plotting library. Then, the data from the recorder is read into the memory. When a data entry is empty, it will be marked as NAN in the Pandas format. The NAN values in the data are processed accordingly. As there are two sampling frequencies used, the time difference between two adjacent rows is obtained to determine the time duration in each operation mode.

Once installed to a motor in a maintenance shop, the vibration recorder starts sampling and recording, and it does not require any initialization or configuration again at rig sides. The recorder does not affect the downhole drilling operation of motors, and no extra or additional work is required from rig crews. The recorder measures, processes, and logs surface and downhole dynamics data continuously. As soon as a drilling motor is back to a maintenance shop, the recorder is removed from the tool. Once the data is downloaded by using a UI (User Interface) program, a python analyzer can be run to produce a job summary report containing the usage of drilling motors and actual downhole conditions. This report provides essential data for maintenance teams to understand the operating conditions of drilling motors quickly and make the right decision to perform appropriate levels of services and preventative actions accordingly.

Moreover, the analyzer can generate more statistics on temperatures, axial vibration, lateral vibration, rotation speed, and stick-slips. The statistics are presented in different viewing formats, such as time plots, histograms, etc.

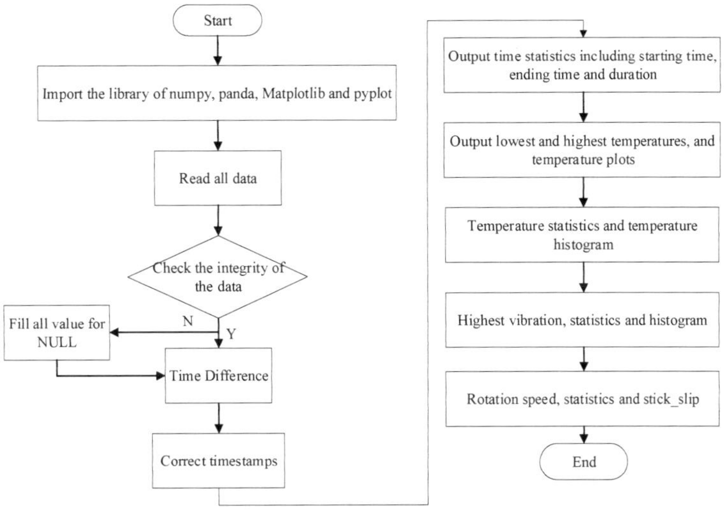

Figure 3. The data processing flowchart of Python data analyzer

4. Case Study

The field case study relates to a well located in Texas, US. In Figure 4, the recorder data captured the overall operation cycle of the drilling motor throughout 31 days from 29th September to 31st October. From temperature and vibration, the downhole drilling time can be easily differentiated from its surface activities.

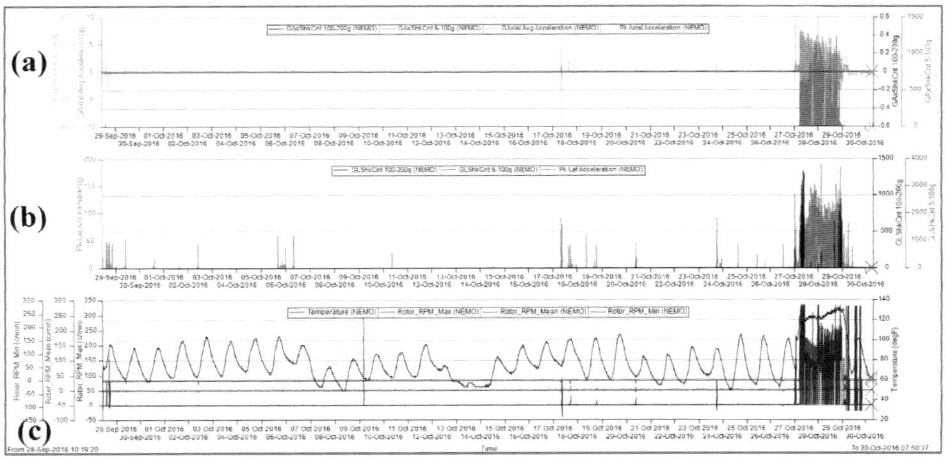

Figure 4. Data over the 31 days in the study

In Figure 4, the plot (a) shows peak axial shock, mean axial shock, and shock counts in the bins of from 5 gn to 10 gn and from 100 gn to 200 gn. During downhole operation, peak axial shock values were around 40 gn; during the time when the tool on surface, several high-amplitude axial shock events were observed at around 35 gn, and these may relate to transportation activities or rig-side tool movements. The plot (b) shows peak lateral shock, and shock counts in the bins from 5 gn to 100 gn and from 100gn to 200 gn. During downhole operation, peak lateral shock values were of 100 gn amplitude with the maximum shock of 180 gn amplitude; on the surface, several high-amplitude lateral shock events were observed at around 70 gn, and these shocks coincided with the high-amplitude axial shock events; therefore, these may also occur due to transportation activities or rig-side movements. For example, on October 17th, a peak lateral shock of 91 gn and a coincided peak axial shock of 31 gn were observed. The plot (c) shows rotation speed and temperature statistics, including maximum, mean, and minimum values. The temperature data has day-and-night patterns when the motor was on the surface. When the drilling motor was tripped into the well, the temperature gradually increased to 50°C and then rose to around 56°C when drilling to a deeper depth.

5. Analysis by Python Analyzer

By using Python Analyzer, the post-job analysis is further performed. Figure 5 shows all the data channels of the recorder. From top to bottom, the plots are temperatures, peak axial acceleration, peak axial acceleration, average axial acceleration, rotation speed, and stick-slip values.

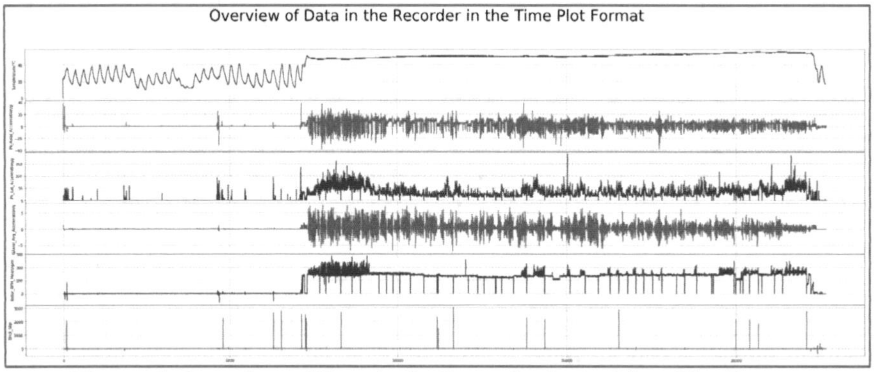

Figure 5. Overview of Data in the Recorder in the Time Plot Format

Table 2 presents the statistical information of the run generated by the Python Analyzer. Figures 6, 7, and 8, show the histogram plots of statistical data, including temperature, peak axial acceleration, and peak lateral acceleration.

Table 2. Statistical results

Parameters	Statistical results	Parameters	Statistical results
Starting time	2016-09-28 10:18:27	Pk_lateral_max	189.27g
Ending time	2016-10-30 06:49:45	Pk_axial_min	-38.7805g
Running time	31 days 20 hours 31minutes 18seconds	GAxShkCnt_5_100g	3037873
Max. temperature	56°C	GAxShkCnt_100_200g	0
Lowest temperature	0°C	GLShkCnt_5_100g	20432028
Pk_axial_max	39.4146g	GLShkCnt_100_200g	25873

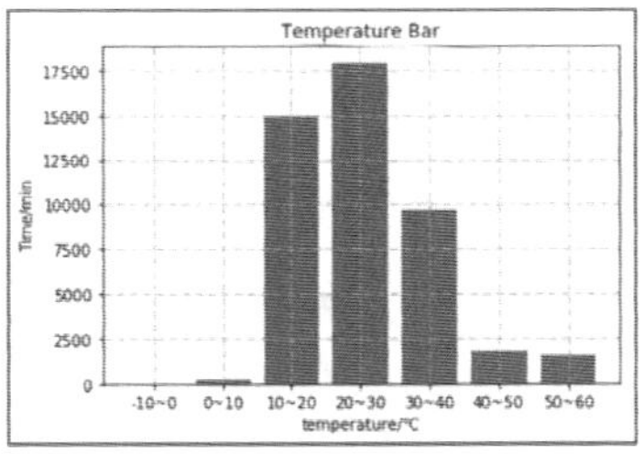

Figure 6. Temperature Histogram

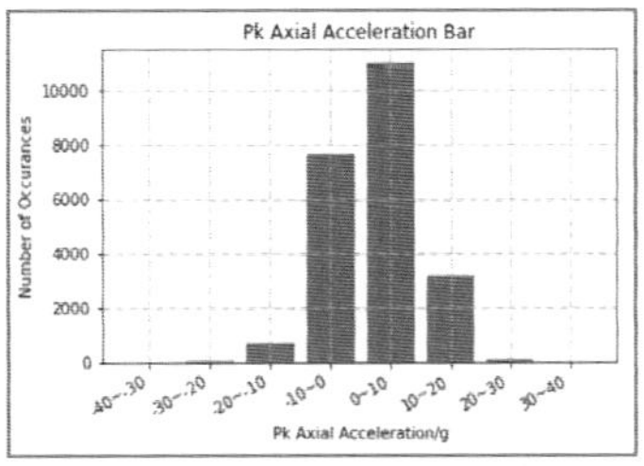

Figure 7. Pk Axial Acceleration Histogram

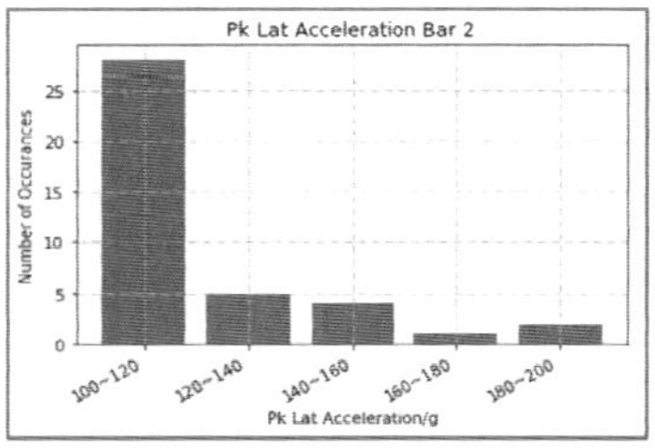

Figure 8. Pk Lateral Acceleration Histogram

6. Conclusions

The vibration recorder is a small battery-powered data-logger device, which can be installed in motor rotors at a maintenance shop. The recorder can log more than tens of days covering shipment, rig-side handling, and downhole operation. Surface data can witness any potential damages caused during surface activities. Besides, the data that the recorder collects have been shown to enable characterizing downhole vibration measurements and statistics, bit rotation speeds, stick-slip and stall occurrences, temperature readings, and timestamps.

A data analyzer has been developed to provide post-job data analysis. These measurements and statistics provide valuable insights into the vibration of drilling motors. One case study was presented, and data from the vibration recorder was presented in different plots and statistical values. These data can help asset owners better understand the working condition of the motor with the installed recorder.

References

[1] Wikipedia. (2018) Mud motor.

[2] Craig, A.D., Goodship, R.S., and Shearer, D.R. (2010) High-Frequency Downhole Dynamic Measurements Provide Greater Understanding of Drilling Vibration in Performance Motor Assemblies. In: IADC/SPE Drilling Conference and Exhibition. New Orleans.

[3] Shen, Y. *et al.* (2017) The Origin and Mechanism of Severe Stick-Slip. In: SPE Annual Technical Conference and Exhibition. San Antonio.

[4] Bowler, A., Harmer, R., Logesparan, L., Sugiura, J., Jeffryes, B. and Ignova, M. (2016) Continuous High-Frequency Measurements of the Drilling Process Provide New Insights Into Drilling-System Response and Transitions Between Vibration Modes. SPE Drill. Complet., 31: 106–118.

[5] Jones, S., Sugiura, J., Rose, K. and Schnuriger, M. (2017) Drilling Dynamics Data Recorders Now Cost-Effective for Every Operator - Compact Embedded Sensors in Bit and BHA Capture Small Data to Make the Right Decisions Fast. In: SPE/IADC Drilling Conference and Exhibition. Hague.

[6] Jones, S., Sugiura, J. and Wood, H. (2017) Recorded Dynamics Measurements While Coring Enhances Performance and Recovery. In: SPE/IADC Drilling Conference and Exhibition. Hague.

[7] Suart, J., Dave, M. and Mike, S. (1994) Putting a Damper on Drilling's Bad Vibrations. Oilfield Rev., 6.

[8] Macpherson, J.D., Paul, P., Behounek, M. and Harmer, R. (2015) A Framework for Transparency in Drilling Mechanics and Dynamics Measurements. In: SPE Annual Technical Conference and Exhibition. Houston.

[9] Sugiura, J., Samuel, R., Oppelt, J., Ostermeyer, G.P., Hedengren, J. and Pastusek, P. (2015) Drilling Modelling and Simulation: Current State and Future Goals. In: SPE/IADC Drilling Conference and Exhibition. London.

[10] Heinisch, D., Oueslati, H., Popp, T.M., Meyer-Heye, B., Schepelmann, C. and Reckmann, H. (2016) Testing and Characterization of Shock and Vibration Loads to Enhance Drilling Tool Reliability and Efficiency. In: Abu Dhabi International Petroleum Exhibition & Conference. Abu Dhabi.

IOP Conf. Series: Materials Science and Engineering **799** (2020) 012048 doi:10.1088/1757-899X/799/1/012048

[11] Wilson, J.K. and Noynaert, S.F. (2017) Inducing Axial Vibrations in Unconventional Wells: New Insights through Comprehensive Modelling. In: SPE/IADC Drilling Conference and Exhibition. Hague.
[12] Perez Garcia, D., Lockley, J.R. and Hanley, C. M. (2007) Application of Small Vibration Logging Tool Yields Improved Dynamic Drilling Performance In: SPE/IADC Drilling Conference. Amsterdam.
[13] Jacobsen Plutt, L.A., Pere, A., Pardo, N.O. and Rodriguez, E.M. (2009) Achieving Improved Performance Through Drilling Optimization and Vibration Management at a GoM Development Project. In: SPE/IADC Drilling Conference and Exhibition. Amsterdam.

IOP Conf. Series: Materials Science and Engineering **799** (2020) 012049 doi:10.1088/1757-899X/799/1/012049

Circular Economy: Integrated Design of Fish Tank- Flower Nursery

Xuexia Li, Yu Wang, Tianyong Bao, Yirui Xiang

Northwestern Polytechnical University, Mingde College, Xian, Shaanxi, 710124, China

*Corresponding author's e-mail: xsnow99@sina.com

Abstract. The internet of Everything has come into reality. Smart home has entered our life and changed our life style. Using traditional methods, consumption of time and labour needs to be considered for family breeding and conservation. Based on analysis and research on the application status of household flower nursery and aquarium equipment, an integration fish tank-flower nursery ecosystem was designed. An interaction between aquarium automatic water exchange maintenance and automatic watering system of flowers and plants can be designed via several techniques, where contains single-chip computer technology, open source Arduino technology and multi-sensor fusion technology. Furthermore, the goal of recycling of water can be fulfilled by using the rich water of fish culture to water flowers. Experiments show that the integrated ecosystem can replace the traditional manual maintenance, save resources and bring people a better home experience.

1. Introduction

With the progress of science and technology and the continuous improvement of living standards, people continue to pursue a better home environment. The cultivation of ornamental fish and various flowers and plants can not only beautify the environment, but also cultivate one's own temperamen so that people can release their pressure and feel comfortable at home after a whole day of intense work. In developed countries such as Europe and America, the popularity rate of family aquarium has reached 20-33%. In China, the annual consumption growth rate of aquarium industry has exceeded 40% [1], and the annual consumption growth rate of flowers and plants has exceeded 50%.

The improvement of people's living environment is often accompanied by the consumption of more social resources. Under a situation that global resources are increasingly exhausted and environmental pollution is more and more serious, the more green and low-energy work and life style is the inevitable requirement of this era for us. However, the traditional maintenance of family plants and fish is time-consuming and laborious. Due to the lack of maintenance knowledge and improper management of most ordinary families(i.e., family travel or business trip).it cannot be guaranteed to provide a suitable living environment for the plants and fish in the flower garden, resulting in the withering or drowning of the plants and the death of the fish due to the deterioration of the water quality environment [2].

At present, researches on fish farming in family aquarium and automatic watering and maintenance of flowers and plants has been quite mature, but both of them are based on the design and research of two independent systems, which are FishBit intelligent aquarium ecological monitoring system, various intelligent flowers and plants watering system, etc. Therefore, it is of great significance to design a system that integrates and interacts the intelligent water quality management of fish tank and the intelligent maintenance system of flower garden to realize the recycling of water. Inspired by the idea of product system design, Smart phones, water quality monitoring, soil temperature, humidity monitoring and other sensors are regarded as carriers. This paper constructs a fish tank- flower garden integrated ecosystem with easily operations and designs corresponding applications and supporting hardware products to help users to solve the problems of water quality management and soil humidity management in family fish farming and improve the user experience.

2. Overall function and structure of the system

Fish tank- flower garden integrated ecosystem consists of three sections, where the first one is fish tank intelligent water quality monitoring and management subsystem, the second one is flower garden soil automatic monitoring and watering subsystem, the last one is mobile application. The relationship between them is shown in Figure 1.

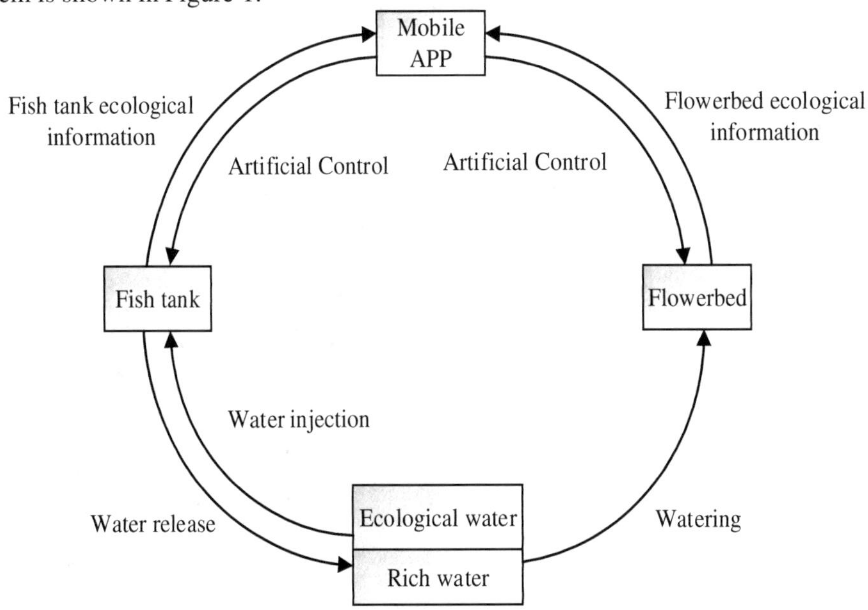

Figure 1. Diagram of aquarium-nursery ecosystem

Through monitoring, the fish tank subsystem will make the water which is not meet the requirements of fish growth flow into the water storage, meanwhile, it injects the ecological water that meets the requirements. In the flower garden subsystem, the spraying device starts as long as the soil humidity below the standard. The spraying device starts as long as the soil humidity below the standard. Fish rich water is used to irrigate the flowers and plants under appropriate conditions so as to realize the recycling of water resources.

The intelligent detection hardware of the fish tank and flower garden subsystem can be connected to the mobile app through Bluetooth. Hence, the user can set the parameter limits of each subsystem through the mobile phone, meanwhile, the water quality and soil related data can be obtained. The information of water quality of fish tank and soil state of flower bed can be presented to the user in the form of on-site display and app so that the user can know the water quality and soil state in time, and take corresponding measures to ensure the safety of fish, flowers and plants.

3. System hardware design

3.1. Garden subsystem

The flower garden subsystem is a system constructed by 8-bit single-chip microcomputer STC89C52, where its architecture is shown in Figure 2. The whole function is composed of environment parameter acquisition module and control module. The environmental parameter acquisition module detects the soil humidity, soil temperature and other parameters in the flower bed through various terminal sensors [3]. The controller analyzes and processes the collected data [4], which is displayed in real time on site and transmitted to the mobile app through the Bluetooth module. According to the data analysis and processing results, the controller intelligently decides whether to water under appropriate conditions.

Y1-69 is used to detect soil moisture in the flower garden subsystem. The principle of the sensor is humidity sensitive capacitance. The capacitance is proportional to the soil moisture. It has the characteristics of high sensitivity and fast response. As a soil temperature sensor, the resistance of the thermistor is inversely proportional to the soil temperature. The nixie tube is displayed as a parameter, the limit value of professional plant soil temperature and humidity is preset by pressing the key, and the collected soil humidity value is compared with the preset limit value of soil humidity [5]. When the soil humidity value is lower than the preset lower limit value, the system controls the water storage part to water through the control relay, and the buzzer will send out a synchronous alarm tone. When the system detects that the soil moisture value reaches the set upper limit through the soil moisture detection device, the system controls the relay again to control the watering system to stop watering. According to the parameters of soil temperature and humidity, the system water under suitable conditions to avoid the transpiration of plants [6], which is conducive to plant growth.

The analog quantity information measured by yl-6 soil humidity sensor and soil temperature sensor is converted into digital quantity by ADC0832 which has two analog quantity inputs after conditioning circuit.

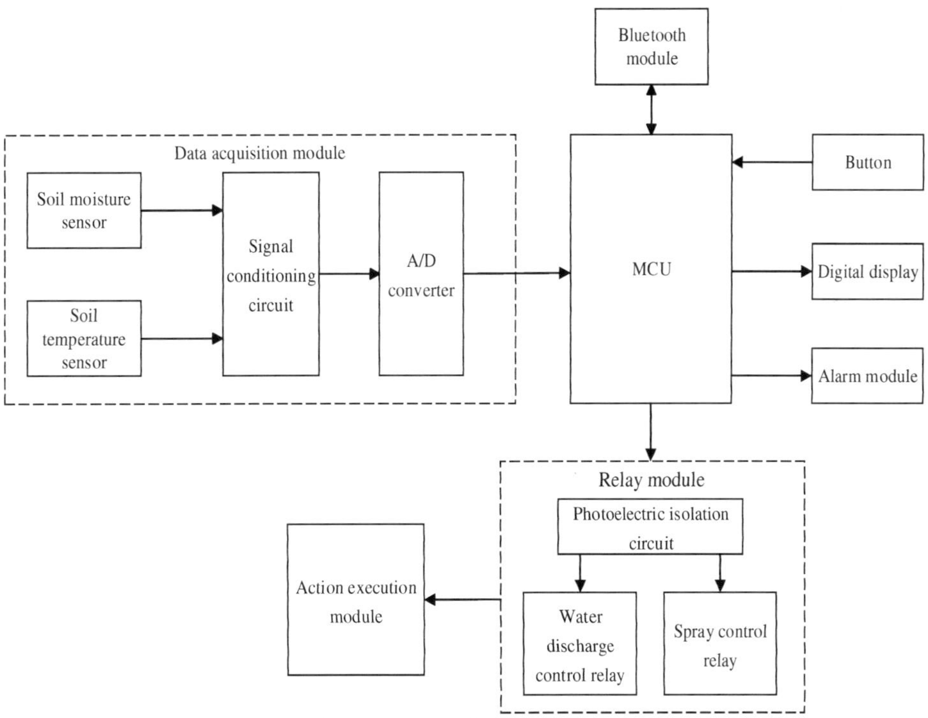

Figure 2. Flower nursery hardware block diagram

3.2. Fish tank subsystem

The aquarium subsystem is a system based on Arduino single chip microcomputer with open source code. The system architecture is shown in Figure 3, which is composed of environment parameter collection module, control module and Bluetooth communication module. Arduino single chip microcomputer has digital interface and analog interface, and it supports SPI, IIC, UART and other communication modes. It can connect various types of sensors, which has rich library functions, as well as convenient programming characteristics.

The analog port A1 of Arduino is connected with the PH value sensor produced by the DFROBOT to detect the PH value of water. When the PH value exceeds the set limit, The water change operation is started: turn on the relay 1, and the water pump will pump the rich water in the fish tank into the water storage equipment through the preset pipeline. At the same time, the water level sensor connected with the analog port A0 of Arduino monitors the current water level. When the water level of the fish tank reaches the set point, the system stops injecting water, and relay 2 opens. The water solenoid valve starts to inject the prepared ecological water for fish farming. When the water level of the fish tank reaches the set upper limit, the water stops. At this time, if the pH value test result is still not up to the standard, the system will automatically repeat the above water change operation again until the pH value is up to the standard. The analog parameters collected by the system sensors are processed by the internal integrated A/D conversion module of Arduino, without the need to expand the A/D conversion chip.

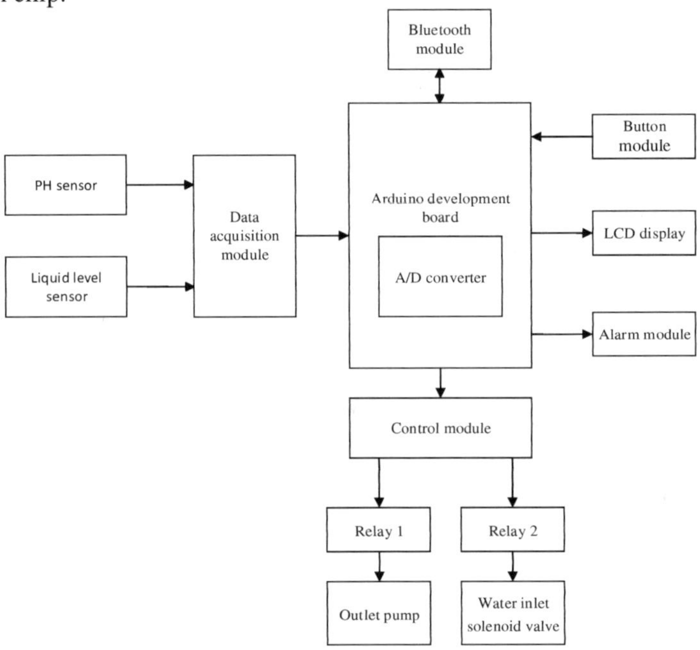

Figure 3. Hardware block diagram of fish tank

3.3. Bluetooth module

Bluetooth technology [7] establishes a relatively close wireless connection for mobile and fixed device communication environment, which can easily establish network connection. It can also carry out data communication or access the Internet through mobile terminals [8]. The system uses the British CSR company hc-05 Bluetooth module which relies on Bluetooth specification v2.0 with EDR Bluetooth protocol data transmission module. The working frequency band is 2.4ghzism, and the maximum transmitting power is 4dbm. The PCB antenna on board can realize 10 meter distance communication. The module supports UART, SPI, USB and other interfaces. Moreover, it supports spp Bluetooth serial port protocol and contains the function of integrating master and slave. It can connect 7 devices [9]. Since the system can set the mode to wireless transmission mode when initializing through UART transmission layer interface, Bluetooth can be realized The raw data received by the module is directly output through UART [10].

The hc-05 module is connected with the serial port of single chip microcomputer and arduno respectively. The connection mode of system Bluetooth module is shown in Figure 4.

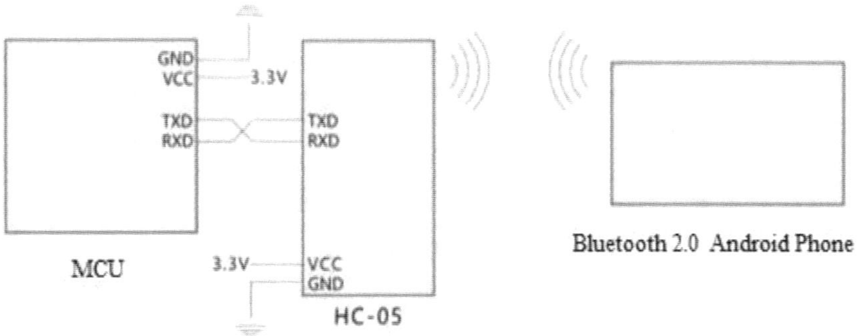

Figure 4. Bluetooth module connection diagram

4. System software design

4.1. System software structure

Two subsystems of fish tank and flower garden are programmed independently. The aquarium subsystem uses IDE, which is a professional Arduino development tool, to write and develop the system program besides ISP online programming. Arduino has rich library functions, which is very convenient for the realization of system functions. The system uses C + + programming language to obtain pH value and water level sensor value respectively by calling analog quantity collection library function. The pH value sensor is collected every 1 second, and the program is as follows:

```
void loop()
{
   Static unsigned long timepoint = millis();
 if(millis()-timepoint>1000U){
          timepoint = millis();//time interval: 1s
voltage=alogRead(PH_PIN)/1024.0*5000;  // read the voltage
phValue=ph.readPH(voltage);  //conver
voltage to pH
   Serial.print("^C   pH:");
   Serial.println(phValue,2);
     }
     ph.calibration(voltage,temperature);        // calibration process by Serail CMD
}
```

The flower garden subsystem is programmed by C language and debugged by keil uvisin4. The main flow diagram of fish tank and flower bed subsystem is shown in Figure 5.

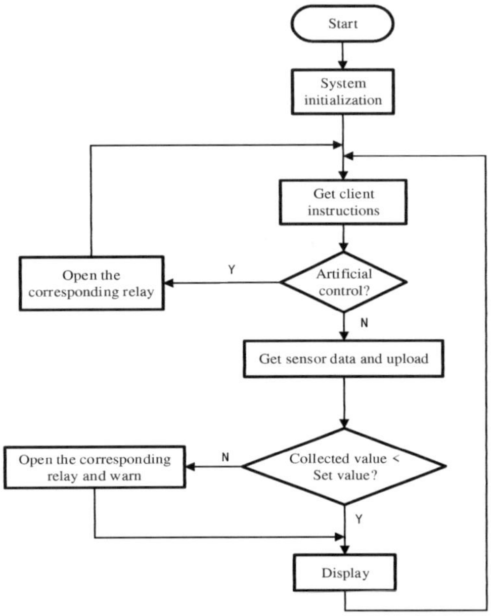

Figure 5. Main program flow of subsystem

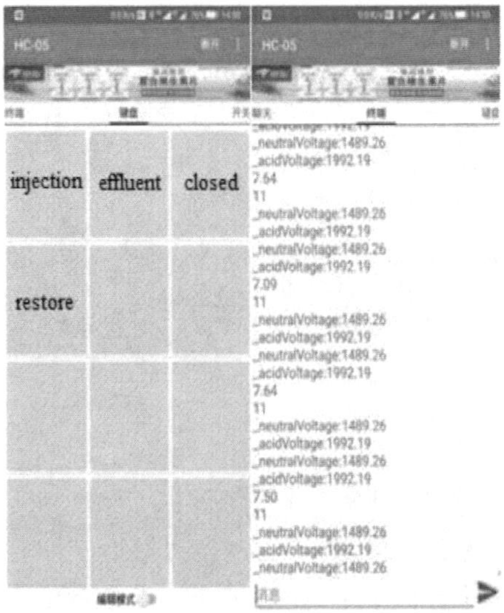

Figure 6. APP work interface

4.2. Mobile app

Users can realize the real-time monitoring and artificial control of the aquarium garden system through the mobile app.

Before Bluetooth 4.0, communication was realized through soket socket. When one-to-many master-slave communication was realized, polling mode was adopted, which greatly increased the power consumption of both sides of communication, and increased the interference factors, making the communication unreliable. The system uses the form of broadcast packet to inform the host [11]. There are two subsystems that collect the soil parameters of field flower garden and any other parameters which can be written into the broadcast packet bytes, respectively. Then, information of mobile APP can be uploaded in terms of broadcast packets within a specified time. It can also greatly reduce the consumption of the system energy since other time is at a sleep state.

The mobile app interface is shown in Figure 6.When the user connect the circuit as shown in Figure 4, turn on the Bluetooth of the mobile phone, the pairing code is the same (the default is 1234), that is, the baud rate is the same, and the connection can be automatically made after power on. After the user connects successfully, the current pH value can be obtained in the mobile phone interface, and the water can be changed manually, and the current liquid level can be obtained during the water change.

5. Conclusion

Through the interaction of human, flower and fish tank information, the fish tank- flower garden system realizes the seamless connection between the scientific maintenance of plants and the efficient utilization of water resources. A fish tank flower garden ecosystem can also be well designed. After a period of operation, the measurement data error is small, the operation is stable, and the sysytem has a very reliable utilizaition effect. Our system, instead of manual system, can be automatic operated, which integrates flower garden watering and fish tank water exchanger. It can also make fish and flower garden plants complementary symbiosis [12] and achieve the goal of saving green energy. The design of the system is based on the small family life scene, which reduces the difficulty of breeding and maintenance and improving the fun [13]. At the same time, the design has been improved to some extent, such as changing Bluetooth communication to WiFi, which brings people greater convenience. It can also be used in agricultural production to make agricultural production intelligent and realize the modern "Sangji fish pond" [14] agricultural mode.

Author
About the author: Xuexia Li, (1969 -) female, bachelor's degree, associate professor, the main research direction is embedded microcontroller and intelligent control. Yu Wang, (1996 -), male, Shangluo, Shaanxi Province, majoring in automation. Tianyong Bao, (1996 -) male, born in Baoji, Shaanxi Province, majoring in automation. Yirui Xiang, (1997 -), male, from Ankang, Shaanxi Province, majoring in automation.

Mailing address: Mingde college, Northwestern Polytechnic University zip code: 710124

Li Xuexia Tel: 13891899172

Email: xsnow99@sina.com

Acknowledgment
Supported by University Student Innovation and Entrepreneurship Training Program of Shaanxi of China (2018550002).

References
[1] Wang D.F. (2010) "Shui Nationality" into thousands of households -- Thinking about the development of China's aquatic industry. China Fisheries., 04: 5-10
[2] Li G.Y. (2014) Analysis of the ecological factors affecting the water body of cultured fish. Heilongjiang Fisheries., 03: 3-5
[3] Yang H.B., Chang R.K., Wang Y.H. (2018) Intelligent flowerpot system design based on STM32 and Arduino. Heilongjiang Agricultural Science., 10: 170-171
[4] Wang C.S., Zhang C.M., Xu J.G. (2018) Design and implementation of intelligent fish tank system based on Internet of things. Information and Computer (theoretical Edition)., 19: 141-143
[5] Zhai Y.F., Zhang X.T. (2010) Design of flower automatic watering system based on AT89S52. Microcomputer Information., 26 (20): 94-95
[6] Ciofalo M., Di L.M.,. Gurreri L., La C M., Scelsi L., Micale G. (2019) Mass transfer in ducts with transpiring walls. International Journal of Heat and Mass Transfer., 4: 1074-1086
[7] Qian Z.H., Liu D. (2012) Data transmission overview of Bluetooth technology. Journal of Communications., 22 (7): 49-51
[8] Li L., Wang A.Y., Zhang R.X. (2007) Implementation of Bluetooth technology scheme based on mobile terminal. Communication Technology., 12 (40): 35-47
[9] Guo W.Y., Li H.J., Fang J.J. (2019) Design and analysis of wireless temperature and humidity acquisition system based on Bluetooth. Journal of Inner Mongolia University, Natural Science Edition., 50 (1): 79-83
[10] Zhang C., Gu X.F., Hao W.J. (2009) Embedded system design based on Bluetooth module. Electronics and Packaging., 9 (12): 20-22
[11] Fang T.E., Qi J., Ma J.H. (2017) Research and design of intelligent Bluetooth thermometer based on Android. Application of Electronic Technology., 43 (3): 73-76
[12] Xie F.,Li Y.J., Wang M. (2015) Study on the development path of Nanchong characteristic benefit agriculture from the perspective of symbiosis and complementarity between government and market . Journal of Hubei Institute of Engineering. ,35 (5): 122-126
[13] Liu Z.P. (2018) Design and research of intelligent water quality management system for ornamental aquarium. Guangdong University of technology.
[14] Liu T., Cheng J., Su S.Q., (2017) Study on the current situation and innovative development of Sangji fishpond in the Pearl River Delta. Journal of Ecological Environment., 26 (10): 1814-1820

IOP Conf. Series: Materials Science and Engineering **799** (2020) 012050 doi:10.1088/1757-899X/799/1/012050

Research on Color Image Watermarking Algorithm Based on Quaternion Fourier Transform

Yongqiang Ma[1*], Shunli Wang[1], Wei Sun[1], Jing Bian[1], YinBo Liu[1]

School of Computer Science, Jining Normal University, Wulanchabu, Inner Mongolia, 012000, China

[*]Corresponding author's e-mail: nsd-myq@126.com

Abstract. This paper mainly proposes a color image digital watermarking algorithm based on quaternion Fourier transform (QFT). We can describe color image as a quaternion matrix, then we can use QFT to get its frequency domain information. By using Fourier transform, we can embed the watermarking information into the frequency domain information. By using inverse QFT, we can get a color image that contain watermarking information. All of the steps can be reversible, so the watermarking extraction process is opposite to the embedding process. The advantage of processing color image by quaternion theory is that three color channels of the image can be treated as a whole.

1. Introduction

With the rapid development and maturity of the Internet and multimedia technology, the efficient storage management, fast transfer sharing and real-time analysis and processing of multimedia resources have gradually become essential parts of people's work and life. In view of the advantages of digital information being open, compatible, and shared, most media resources will exist in digital form and spread in the Internet environment. The rapid development of the Internet has made it easier and more convenient for people to obtain the digital multimedia resources they need.

But everything has dual character. The Internet is also a "double-edges sword", which brings convenience to people but also a lot of negative impacts. The security of digital media resources has been greatly threatened and challenged in the open Internet environment. Many resources have been copied, tampered and disseminated without the permission of the copyright owner. Many criminals have made huge profits by this method, which not only damages the legitimate rights and interests of the copyright owner, but also seriously affected the healthy development of the digital media industry. In addition, e-commerce and telemedicine have developed rapidly in recent years, and a large number of forged transaction bills and patient medical records have appeared in the Internet, which undoubtedly caused a crisis of trust in the security of these network applications. Therefore, strengthening the security protection of digital media resources has become a research hotspot, and copyright protection and content authentication of digital multimedia have also become important research topics.

The traditional information security protection of digital multimedia generally used digital encryption and digital signature technology. Although the multimedia information security is protected to a certain extent, there are many defects in practical applications. In digital encryption technology, if the digital media works are decrypted, the problems of illegal copying, tampering and re-dissemination cannot be solved. Digital signature technology can only be used to determine the authenticity of digital media content, and it is difficult to mark the exact location of tampering.

In response to the shortcomings of traditional technology, a more effective digital media information security protection technology, which called digital watermarking, has been proposed in recent years, and it has received increasing attention from researchers all over the world.

Digital watermarking technology is an important branch of the current information security field. As an active authentication method for copyright protection and authenticity authentication, it has become an important means of protecting multimedia information security. Digital images are the most important part of multimedia resources, and digital image watermarking has also become the core research direction in digital watermarking technology. It has played an important role in digital image content authentication and copyright protection. In the open Internet environment, digital image watermarking technology hides the watermarking information into the image content which should be protected. When implementing the copyright protection process, it is necessary to ensure that the visual quality of the carrier image is not significantly reduced. After the attack, most of the hidden watermarking information can still be extracted.

2. Color image

Images are an important carrier for humans to obtain information. With the rapid development of computer hardware equipment and the rapid increase of computing speed, digital image processing technology has also developed very well. Color images carry more information than grayscale images, and color image processing has become an important research area in signal processing disciplines. Unlike grayscale images, color images have three color channels, and there is a spectral relationship between them. Many methods for processing grayscale images cannot be used to process color images directly, which also makes color image processing more difficult.

Color images have three color channels, so we need to use three-dimensional vectors to represent them. We most commonly use RGB color space to represent color images. It is based on the principle of three primary colors. In the RGB color space, light of any color can be obtained by superimposing three colors of red, green and blue. Other color spaces include HSV space and YUV space.

The application of quaternion theory and color image processing are receiving increasing attention. Based on quaternion theory, ELL and others in the UK have done a lot of work in color image processing. In 2001, S.C.Pei[1] and others used traditional two-dimensional Fourier transforms to implement QFT based on the work that ELL and Sangwine did with QFT, It was pushed and summarized in detail. In recent years, some domestic scholars have also made some achievements in quaternion color image processing. Xing Yan[2] of Hefei University of Technology has proposed a color image decomposition algorithm and a digital watermarking algorithm. This algorithm requires that the watermarked image should be a grayscale image. Gai Qi[3] of Nankai University proposed a color image information hiding technique.

The above results show that the three components of a color image should be considered as a whole, which allows us to understand problems that were difficult to solve in the past, obtain the characteristics of color images, and obtain new research methods and corresponding results. However, the research on quaternion representation and processing of color images is still in its infancy, and its theoretical system needs to be further improved.

3. Quaternion Fourier Transform

Quaternion was proposed by Irish mathematician W.R.Hamilton in 1843, who generalized complex space to four-dimensional space[4][5].

The quaternion is a number in a four-dimensional space that has a real part and three imaginary parts. Where q represents the quaternion, and we defined it as

$$q = a + bi + cj + dk \tag{1}$$

Where a, b, c, d is the real part and i, j, k are the imaginary part. The conditions it holds can be expressed as

$$i^2 = j^2 = k^2 = ijk = -1 \tag{2}$$

$$ji = -k, kj = -i, jk = -j \tag{3}$$

$$ij = k, jk = i, ki = j \tag{4}$$

When $a = 0$, we can consider q as a pure quaternary imaginary number.
The $|q|$ can be given by

$$|q| = \sqrt{a^2 + b^2 + c^2 + d^2} \tag{5}$$

There are two numbers q_1 and q_2, they can be expressed by

$$q_1 = a_1 + b_1 i + c_1 j + d_1 k \tag{6}$$

$$q_2 = a_2 + b_2 i + c_2 j + d_2 k \tag{7}$$

In the RGB color space, any color can be represented by a three-dimensional vector, so each pixel can be represented by a pure quaternary imaginary number. The three components R G and B are used as i, j, k coefficients [6][7], so the color image can be represented by a quaternion matrix. In subsequent research, we can process the color image as a whole instead of dividing the whole into three grayscale images for processing.

For a general complex-valued function, its Fourier transform is given by

$$F(u,v) = \int_{-\infty}^{+\infty}\int_{-\infty}^{+\infty} f(x,y)e^{-i(ux+vy)}dxdy \tag{8}$$

Quaternion Fourier Transform (QFT) can be expressed by

$$F_{LR}(u,v) = \int_{-\infty}^{+\infty}\int_{-\infty}^{+\infty} e^{-\mu 1ux} f(x,y)e^{-\mu 2vy}dxdy \tag{9}$$

$$F_L(u,v) = \int_{-\infty}^{+\infty}\int_{-\infty}^{+\infty} e^{-\mu 1(ux+vy)} f(x,y)dxdy \tag{10}$$

$$F_R(u,v) = \int_{-\infty}^{+\infty}\int_{-\infty}^{+\infty} f(x,y)e^{-\mu 1(ux+vy)}dxdy \tag{11}$$

$$f(x,y) = f_r(x,y) + f_i(x,y)i + f_j(x,y)j + f_k(x,y)k \tag{12}$$

In image processing, the Fourier transform we use can be expressed as a discrete form, so the Quaternion Fourier Transform (QFT) can be given by

$$F(u,v) = \frac{1}{\sqrt{MN}}\sum_{m=0}^{M-1}\sum_{n=0}^{N-1} e^{-u_1 2\pi\frac{mu}{M}} f(m,n)e^{-u_2 2\pi\frac{nv}{N}} \tag{13}$$

$$f(m,n) = \frac{1}{\sqrt{MN}}\sum_{u=0}^{M-1}\sum_{v=0}^{N-1} e^{u_1 2\pi\frac{mu}{M}} F(u,v)e^{u_2 2\pi\frac{nv}{N}} \tag{14}$$

4. Basic framework of digital image watermarking

Digital watermarking technology is essentially to rationally optimize these key steps such as watermark preprocessing, embedding location selection, embedding method design, and extraction method design, and seek to optimize design problems under constraints such as robustness, imperceptibility and security. For a complete digital watermarking system, it usually consists of two steps: digital watermarking embedding and digital watermarking extraction [8][9].

4.1. Watermarking embedding process

We should define a digital image product O, the key of the Embedding method can be expressed by K,the embedding method can be expressed by F, Then, the data information W that represented the copyright of the carrier is combined with the carrier by the F method to obtain a new carrier W′, and the process is represented by

$$W' = F(O,K,W) \tag{15}$$

The key K and watermark embedding function F are known only by the designer who designs the embedding algorithm. When the multimedia product some body used was theft, if they who want to steal the information have not the key that we defined in advance, they could not crack the embedding algorithm, it is hard to obtain the useful information of the digital image generally. The embedding process of digital watermarking technology is shown in figure 1.

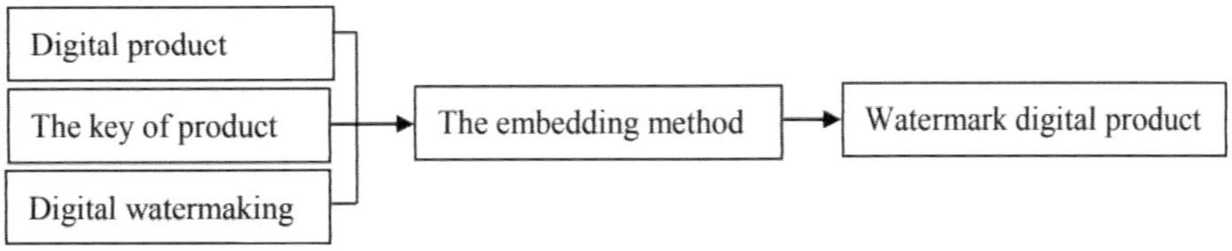

Figure 1. Watermarking embedding process.

4.2. Watermarking extracting process

Non-blind extraction and blind extraction form the extracting operation of watermarking. The non-blind extracting watermarking refers to the process of extracting watermark information with the help of the original carrier and the digital product embedded with watermark. On the contrary, blind extraction means that the original image is not needed. Both of the two methods, the key of the embedded method needs to be recorded in advance, and the embedded information is extracted by using the extraction scheme and the recorded key, the selected method is recorded as E and the extracted watermark we can call it W'', and it is given by

$$W'' = E(I, W', K) \tag{16}$$

Watermarking extraction process can be expressed in figure 2.

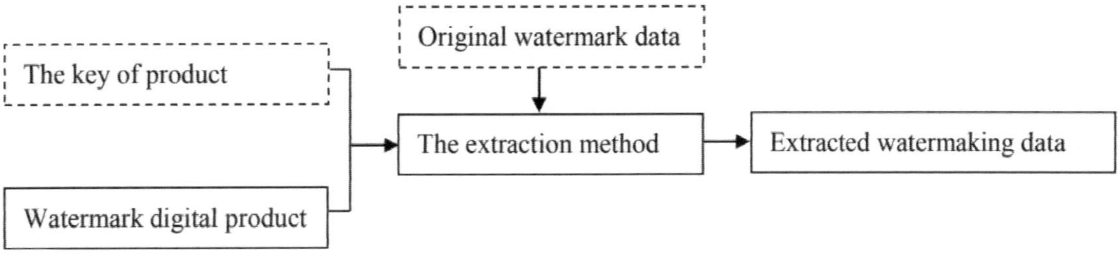

Figure 2. Watermarking extraction process.

5. Conclusion

This paper proposed a new method based on the advantages of quaternion and Fourier transform. Its aim was to improve the robustness against attacks and it was used for color image watermark embedding and extraction. The quaternion discrete Fourier transform were applied in digital watermark algorithm.

We can describe color image as a quaternion matrix, then we can use QFT to get its frequency domain information. By using Fourier transform, we can embed the watermarking information into the frequency domain information. By using inverse QFT, we can get a color image that contain watermarking information.

This study point out the advantage of processing color image by quaternion theory. Three color channels of the image can be treated as a whole.

Acknowledgments

We would like to thank professor WANG and professor SONG for stimulating discussions with respect to the topic of this paper. This work was supported by Research Program of Science and technology at Universities of Inner Mongolia Autonomous Region (NJZZ20246).

IOP Conf. Series: Materials Science and Engineering **799** (2020) 012050 doi:10.1088/1757-899X/799/1/012050

References

[1] Pei, S.C., Ding, J.J., Chang, J.H. (2001) Efficient Implementation of Quaternion Fourier Transform, convolution and correlation by 2-D FFT. IEEE Transactions on Signal Processing., 49: 2784–2798.
[2] Xing, Y. (2009) Research on Quaternion and Its Applications in Graphics and Image Processing. Hefei University of Technology Publishing, Hefei.
[3] Gai, Q. (2009) Color Image Information Hiding Technology Based on Quaternion Theory. Nankai University Publishing, Tianjin.
[4] Chen, S.X., Feng, Y.B. (2014) Double Zero-Watermarking Algorithm Based on Quaternion and for Color Image. Computer Applications and Software., 31: 263–267.
[5] Sun, J., Yang, J.Y., Fu, D.S. (2014) Color Images Watermarking Algorithm Based on Quaternion Frequency Singular Value Decomposition. Information and Control., 33: 144–148.
[6] He, B. (2016) Zero Color Image Watermarking Based on Radon Transform and Quaternion Real Matrix Representation. Optical Technique., 42: 166–171.
[7] Lukac, R., Smolka, B., Martion, K., Plataniotis, K.N. (2005) Vector Filtering for Clor Imaging. IEEE Signal Processing Magazine, 22: 74–86.
[8] Liu, F.M., Zhai, H.C., Yang, X.P. (2003) Kinoform-based iterative random phase encryption. Acta.Phys.Sin., 52: 2462–2465.
[9] Sangwine, S.J., ELL, T.A.J., Moxey, C.E. (2001) Vector phase correlation. Electronics Letters., 37: 1513–1515.

AUTHOR INDEX